Soil Movements Induced by Tunnelling and their Effects on Pipelines and Structures

Soil Movements Induced by Tunnelling and their Effects on Pipelines and Structures

P.B. ATTEWELL
Professor of Engineering
University of Durham

J. YEATES
Design Engineer
Northumbrian Water Authority

A.R. SELBY
Lecturer in Engineering
University of Durham

Blackie

Glasgow and London
Published in the USA by
Chapman and Hall
New York

Blackie and Son Ltd
Bishopbriggs, Glasgow G64 2NZ
7 Leicester Place, London WC2H 7BP

Published in the USA by
Chapman and Hall
in association with Methuen, Inc.
29 West 35th Street, New York, NY 10001

© 1986 P.B. Attewell, J. Yeates and A.R. Selby
First published 1986

All rights reserved.
No part of this publication may be reproduced,
stored in a retrieval system, or transmitted,
in any form or by any means,
electronic, mechanical, recording or otherwise,
without prior permission of the Publishers.

British Library Cataloguing in Publication Data

Attewell, P.B.
  Soil movements induced by tunnelling and
their effects on pipelines and structures.
  1. Pipe lines 2. Tunneling
  3. Structural engineering 4. Tunneling
  I. Title II. Yeates, J. III. Selby, A.R.
  621.8'672   TJ930

ISBN 0-216-91876-6

Library of Congress Cataloging in Publication Data

Attewell, P.B.
  Soil movements induced by tunnelling and their
effects on pipelines and structures.
  Bibliography: p.
  Includes index.
  1. Tunneling. 2. Earth movements and building.
3. Soil mechanics. I. Yeates, J. II. Selby, A.R.
III. Title.
TA815.A87 1986   624.1'9   85-31419
ISBN 0-412-00911-0

Photosetting by Thomson Press (India) Limited, New Delhi
Printed in Great Britain by Bell and Bain Ltd., Glasgow.

Preface

This book is written to provide an expanded treatment of a subject covered by Attewell and Yeates in Chapter 6 of *Ground Movements and their Effects on Structures* (eds. Attewell, P.B. and Taylor, R.K.), published by Surrey University Press in 1984. The present treatment of the subject serves the requirements of the design engineer more fully, and the numerous worked examples that have been included in the text should prove helpful in this respect. It is assumed, however, that the reader has a basic understanding of soft-ground tunnelling methods, since these are not covered here; also, since the book has been written for the design engineer rather than the academic, no attempt is made to explain the associated ground movements in terms of soil mechanics. The subject of site investigation for tunnels is referred to only briefly, and contractual matters not at all, but adequate references are given to these important aspects of tunnelling.

The authors thank Mrs Wendy Lister and Mrs Christine Wright of the University of Durham for typing the manuscript, and the publishers for promoting the project.

P.B.A.
J.Y.
A.R.S.

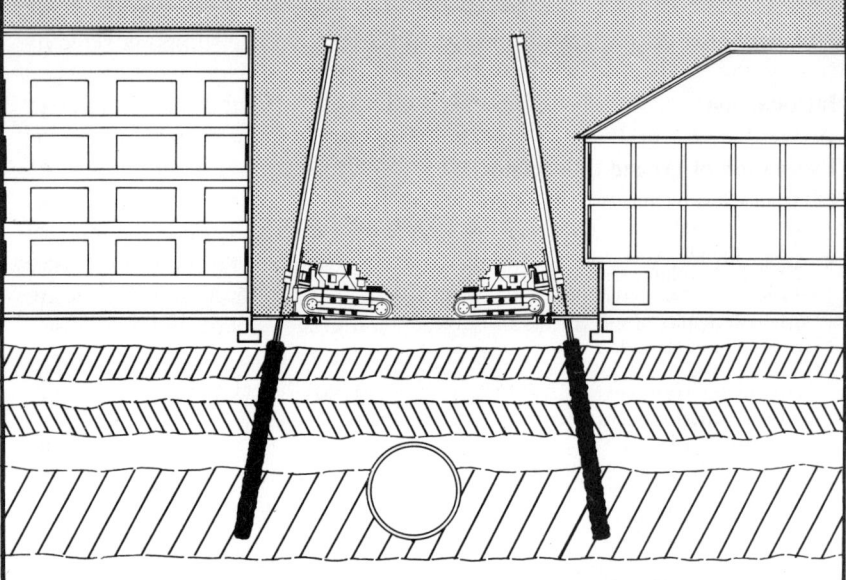

GKN KELLER JET GROUTING

Jet Grouting is a highly advanced, highly versatile process. It enables grouting in most soils – silts, clays, weak rocks – and allows the construction of blocks of grout to predetermined size and strength.

This means that Jet Grouting techniques can be applied to solve a great many below ground problems such as tunnelling. Here you see continuous rows of Jet Grout columns constructed to protect buildings with shallow foundations by restricting the ground surface movements resulting from tunnelling through soft soils.

For more examples of Jet Grouting, contact:–

GKN Keller

48/49 RUSSELL SQUARE, LONDON WC1B 4JP.
TEL: 01-580 7091. FAX: 01-637 1298. TELEX: 31302.
Also in West Germany, USA, Canada, Australia, Singapore, Saudi Arabia.

Contents

List of symbols	xi
1. Introduction	**1**
2. Estimation of ground movements	**8**
2.1 Volume loss parameter, V_t	8
2.1.1 Clay soil	23
2.1.2 Granular soil	33
2.1.3 Total volume loss	34
Example 2.1 Analysis of ground losses in a clay soil	34
2.1.4 Mechanics of soil deformation	40
2.2 Reduction of ground loss	41
2.3 Transmission of ground losses to ground surface	50
2.4 Ground deformation and strain equations	53
2.4.1 Practical estimation of volume-loss parameter, V_t, and surface settlement volume, V_s	60
2.4.2 Settlement trough dimension parameter, i	64
Example 2.2 Settlement calculation	65
2.5 Design-curve graphs	66
2.6 Ground movements and structures—use of design curves for preliminary assessment	74
2.6.1 Application of design curves—use of overlays	74
Example 2.3 Building foundation assessment	75
Example 2.4 Buried pipeline assessment	78
2.6.2 Pipes aligned obliquely to the direction of tunnel advance	80
2.7 Relations between i_y and i_x	81
2.8 Time-dependent settlement	87
2.8.1 Total maximum settlement estimation from simple overload factor	88
2.8.2 Total maximum settlement estimation using a compression index	91
2.8.3 Form of the terminal (ground loss plus consolidation) transverse settlement profile	94
2.9 Numerical methods	96
2.10 Site investigation for tunnels in soil	98
2.11 Measurement of ground movements	105
2.12 Measurements on structures	114
2.13 Shafts	117
2.14 Pretunnelling protection of structures	120
3. Ground movements and buried pipelines	**122**
3.1 Introduction	122
3.2 Ground movement transverse to a pipeline	127
3.2.1 Subgrade reaction analysis	129
3.2.2 Basic theory	131
3.2.3 General method of analysis	133
3.2.4 Solutions for a rigidly-jointed pipeline transverse to a tunnel drive	135

Example 3.1 Calculation of bending strain in large-diameter transverse pipeline	138
Example 3.2 Calculation of bending strain in small-diameter transverse pipeline	140
3.2.5 Solutions for a rigidly-jointed pipeline parallel to a tunnel drive	141
Example 3.3 Calculation of bending strain in large-diameter parallel pipeline	144
Example 3.4 Calculation of bending strain in small-diameter parallel pipeline	145
Example 3.5 Calculation of effect of ground movement on parallel offset pipeline	146
3.2.6 Solutions for pipe deflection and soil pressure	147
Example 3.6 Calculation of deflection and soil pressure for large-diameter pipeline	149
3.2.7 Effect of non-uniformity in soil-pipe system	150
Example 3.7 Calculation of allowance for bending strain in large-diameter pipeline associated with non-uniformity	152
Example 3.8 Calculation of allowance for bending strain in small-diameter pipeline associated with non-uniformity	153
3.2.8 Effect of pipe joint rotation	153
Example 3.9 Calculation of potential pipe-joint rotation	155
Example 3.10 Joint relaxation in small-diameter pipeline	159
Example 3.11 Joint relaxation in large-diameter pipeline	159
3.2.9 Ultimate soil pressure	164
3.3 Ground movement parallel to a pipeline	166
3.3.1 Elastic analysis	170
3.3.2 Method of analysis	170
3.3.3 Solutions for a rigidly-jointed pipeline transverse to a tunnel drive	172
Example 3.12 Calculation of axial strain in large-diameter transverse pipeline by elastic method	173
Example 3.13 Calculation of axial strain in large-diameter transverse pipeline by 'load-transfer' method	175
3.3.4 Solutions for a rigidly-jointed pipeline parallel to a tunnel drive	177
Example 3.14 Calculation of axial strain in parallel pipeline by elastic method	178
Example 3.15 Calculation of axial strain in parallel pipeline by 'load-transfer' method	179
3.3.5 Solutions for pipe displacement and soil-pipe shear stress	180
3.3.6 Effect of pipe-joint extension	181
Example 3.16 Calculation of potential pipe-joint pull-out	182
Example 3.17 Calculation of axial strain in transverse pipeline	184
Example 3.18 Calculation of axial strain in parallel pipeline	184
3.3.7 Soil-pipe slip	185
3.4 Application of analysis to practical problems	186
3.4.1 Soil properties	187
3.4.2 Pipeline properties	189
3.4.3 Pipe-joint properties	198
3.4.4 Transverse ground movement	204
3.4.5 Parallel ground movement	208
3.4.6 Pipeline networks	209
Example 3.19 Vertical bending of branch at connection to main pipeline	209
Example 3.20 Horizontal bending of branch at connection to main pipeline	212
Example 3.21 Axial pulling of branch off main pipeline parallel to and offset from tunnel	216
3.4.7 Performance of the soil-pipe interaction model	221
3.4.8 Other considerations	222
3.5 Design to accommodate ground movement	224
3.5.1 Consultative procedures	226
Example 3.22 Delay to tunnelling and cost associated with pipeline renewal in advance of construction	232
3.5.2 Measures to protect old pipe distribution systems	234
3.5.3 Design of new pipelines to accommodate ground movement	235

4. Structural response to tunnelling settlement — **237**

4.1 Winkler ground model—a manual technique — 238
 Example 4.1 Calculation of stresses in a long wall of uncracked reinforced concrete — 240
4.2 Structure matrix stiffness plus Winkler ground model — 242
4.3 Finite-element analysis of ground and structure — 244
4.4 A stress-based criterion for onset of damage — 248
4.5 Brick walls on clay soil — 250
 Example 4.2 1m-high brick wall on clay soil — 250
 Example 4.3 2m-high brick wall on clay soil — 251
 Example 4.4 2m-high brick wall with opening on clay soil — 251
 Example 4.5 House-front brick wall on clay soil — 251
4.6 Brick walls on sand — 255
 Example 4.6 1m-high brick wall on sand — 255
 Example 4.7 2m-high brick wall on sand — 255
 Example 4.8 2m-high brick wall with opening on sand — 256
 Example 4.9 House-front brick wall on sand — 256
4.7 Crack widths in brick walls — 258
 Example 4.10 Estimates of crack width — 259
4.8 Concrete walls on sand — 261
 Example 4.11 Reinforced concrete retaining walls on sand — 261
4.9 Modification of the results of wall examples for application to different soils, tunnels and walls — 263
 Example 4.12 Incorporation of a changed wall cross-section — 264
4.10 Building frame structures — 267
 Example 4.13 Steel frame on clay soil — 268
 Example 4.14 Steel frame on sand — 272
 Example 4.15 Heavy steel frame on clay soil — 272
 Example 4.16 Two-storey frame on clay soil — 272
4.11 Infilled frames — 274
 Example 4.17 Infilled frame—Willington Quay — 274
4.12 Piled foundations — 277
4.13 Building damage—practical appraisal — 279
4.14 Property schedules — 289
4.15 Bridge structures — 295
 Example 4.18 Responses of bridge components to tunnelling — 298

Appendix A — 304
Appendix B — 308
Appendix C — 309
References — 310
Author Index — 319
Subject Index — 323

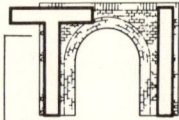

TUNNEL INVESTIGATIONS LTD.

336 CLEPINGTON ROAD
DUNDEE, DD3 8RZ
Tel. (0382) 826680
Telex 76243

Tunnel Surveys and Profiling
Geotechnical Problem Analysis
Tunnel Stability Monitoring
Tunnel Movement Monitoring
Track Movement and Foundation Stability Assessment

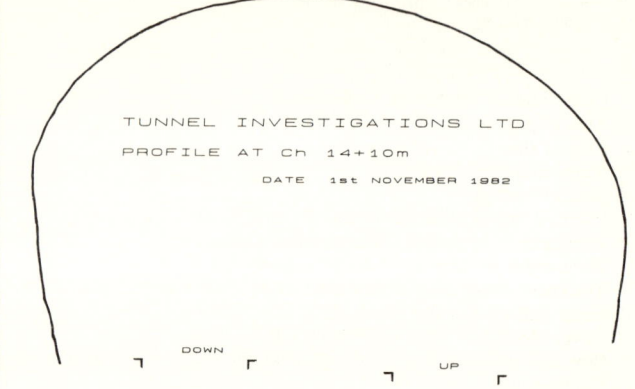

Tunnel profiling by **Tunnel Investigations Ltd.** PATENT PROFILER
Profile obtained by laser scan giving 200 shot points in 60 seconds.

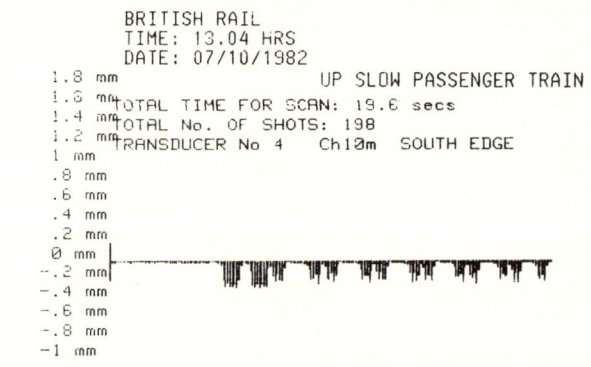

Track movement monitor scan indicating displacement with the passage of successive axles.

Tunnel Investigations Ltd. can identify deformation in tunnel linings; establish whether the deformations are 'live' and monitor movement as a protection service during subsequent remedial works

List of symbols

A	Consolidation settlement coefficient
A_p	Area of pipe wall
A_s	Area of shield and tail extrados
a	Outer radius of shield
a	Area of footing
a_1	External radius of tunnel lining
B	Parameter in consolidation settlement analysis
B	Plate width in subgrade reaction loading test
b	Thickness of any overcutting bead or equivalent device fitted to the shield
b	Beam half-width, after Biot (1937)
C	Cover (depth to tunnel soffit) $= z_0 - R$
C_a	Pipe/lead adhesive (shear) strength
C_c	Compression index
C_1	Parameter in total (ground loss plus consolidation) settlement analysis
C_2	Parameter in total (ground loss plus consolidation) settlement analysis
c	Soil cohesion
c_a	Soil–pipe adhesion
c_u	Undrained shear strength of soil
D	Diameter of extrusion hole in laboratory extrusion test
D_{10}	Size below which 10 per cent of particles in a soil are finer
d	Excavated diameter (of tunnel)
d	Pipe outside diameter
d	Width of a base
d_C	Thickness of caulked lead in pipe joint
d_L	Depth of caulked lead in pipe joint
d_S	Depth of pipe spigot in pipe joint
d_W	Pipe spigot withdrawal
d_0	Diameter of (assumed) spherical soil particles
d_p	Maximum size of particle capable of passing in suspension through the minimum pore cross-section of a soil
d_s	Displacement function of a continuous structure, or displacement of a footing
$d_{s(y=0)}$	Central deflection of a structure
$d_{s\max}$	Maximum displacement of a structure (vertical)
d'	Diameter of hole in distribution main for service tapping
$\mathbf{d}_f$	Vector of foundation displacements
$\mathbf{d}_g$	Vector of ground displacements from the datum of free-ground displacements
$\mathbf{d}_s$	Vector of absolute structural displacements (and rotations)
E	Elastic modulus
E_b	Beam elastic modulus, after Biot (1937)
E_g	Soil secant elastic modulus (Chapter 3)
E_g	Elastic modulus of the ground (Chapter 4)
E_p	Pipe secant elastic modulus
E_s	Elastic modulus of a structure
E_u	Undrained deformation (Young's) modulus of soil
E'	Drained (effective) deformation (Young's) modulus
E'_g	Drained soil secant elastic modulus
e	Void ratio
e_0	Initial void ratio

LIST OF SYMBOLS

f_{cu}	Cube strength of concrete
g	Scaling factor for plans
H	Height of water table above the base of a soil stratum to be dewatered
H	Thickness of soil offering a drainage path
H	Wall height
h	Height from the base of the stratum to be dewatered to the point of maximum drawdown at the well
I	Second moment of area
I_b	Beam second moment of area, after Biot (1937)
I_f	Influence factor
I_p	Pipe second moment of area in longitudinal bending
I_w	Plasticity index
I_1	Influence factor
i	Parameter defining the form and span of the settlement trough on the assumption that the semi-transverse (y-axis) settlement profile can be described by a normal probability equation
i_s	Scaling factor for plans
i_x	x-axis i-parameter
i_y	y-axis i-parameter
K	Modulus of subgrade reaction (FL^{-2}) for soil/foundation and soil–pipe system
K_R	Empirically-determined constant
K_{eff}	Effective modulus of subgrade reaction $= K_1 + K_2$
K_y	Modulus of axial load-transfer (FL^{-2}) for pipeline transverse to tunnel
K_1	Modulus of subgrade reaction below pipeline
K_2	Modulus of subgrade reaction above pipeline
K_∞	Modulus of subgrade reaction for infinitely long beam, after Vesic (1961)
K'_{eff}	Effective modulus of subgrade reaction for drained soil
K^*	Soil–pipe stiffness factor for axial interaction
$\bar{K}$	Modulus of subgrade reaction for square plate
K_g, K_{gd}	Stiffness matrices of the ground
$\mathbf{K}_s$	Stiffness matrix of the structure
k	Soil intrusion distance per unit length of tunnel advance
k	Permeability (hydraulic conductivity) of the ground (assumed to be isotropic)
k	Coefficient of subgrade reaction (FL^{-3}) $= K/d$ in soil–pipe interaction analysis
k_h	Horizontal permeability
k_s	Structural stiffness
k_v	Vertical permeability
k_1	Factor to take account of dome-configuration soil intrusion at the tunnel face proper
k_2	Parameter defining the degree to which the bead is closed by soil deforming radially inwards towards the shield
k_∞	Coefficient of subgrade reaction for infinitely long beam, after Vesic (1961)
$\bar{k}$	Coefficient of subgrade reaction for square plate
L	Length of pipe between two points (1,2)
L	Free length of beam beyond loaded length, after Vesic (1961)
L	Length of a structure
L_J	Spacing of pipeline joints
L_w	Liquid limit
l	Tunnel advance distance
l	Direction cosine
l_c	Critical pile length, after Randolph (1981)
l_s	Length of shield and tail
l_u	Unsupported, ungrouted tunnel length behind the shield tail
M	Bending moment
M_J	Bending moment at pipeline joint
M_{Jn}	Bending moment at pipeline joint number n
M_{max}	Maximum bending moment
M_x	Bending moment at position x

LIST OF SYMBOLS

$M_0, M_1, M_2 \ldots$	Bending moments at the centre of a beam, and at $1, 2 \ldots m$ from the centre
M_{25}	Bending moment causing an elastic stress of 25 per cent UTS in a pipe
m	Average amount of soil movement (displacement) at the tunnel
m	Direction cosine
N_c	Bearing capacity factor with respect to cohesion
N_q	Bearing capacity factor with respect to surcharge
N_γ	Bearing capacity factor which accounts for influence of soil weight
$\mathbf{N}$	Stiffness matrices of a Winkler foundation
n	Power of $(z_o - z)$ to which i_x, i_y, i are proportional (parameter for calculation of transverse displacements)
P	Axial force on a pipeline
P	Force on a structure
P	Point load
P_{FIX}	Axial force at fixed end of pipeline
P_J	Axial force at pipeline joint
P_y	Axial force at position y in pipeline analysis
$\mathbf{P}_e$	Equivalent load vector
$\mathbf{P}_f$	Force vector on foundation nodes
$\mathbf{P}_g$	Vector of interface nodal forces acting upon the ground
$\mathbf{P}_s$	Vector of interface nodal forces acting on the structure
p	Difference in elevation between the front and rear of the shield and tail minus the design grade of the tunnel
p	Dimension of a building
p	Bearing pressure $= p_1 + p_2$
p_0	Initial vertical pressure at the tunnel crown before measurable soil consolidation occurs
p_u	Ultimate bearing pressure
p_1	Bearing pressure below pipeline
p_2	Bearing pressure above pipeline
Q	Water flow rate
Q	Shear force
Q_J	Transverse shear force at pipeline joint
Q_c	Water flow rate for a confined aquifer
Q_u	Water flow rate for an unconfined aquifer
q	Distributed surface surcharge pressure
q	Loading intensity (FL^{-1}) applied to pipeline
q_u	Ultimate bearing capacity
R	Excavated radius of circular cross-section tunnel
R_A	Pipe area ratio $= A_p/(\pi d^2/4)$
r_0	Radius of drawdown depression
r_w	Radius of well
S	Plan scale factor
s_i	Winkler spring stiffness at point i
t	Pipe wall thickness
t_s	Time during which an element of soil adjacent to the shield can deform inwards through the bead gap towards the shield
u	Ground displacement in x-coordinate direction (unconstrained by any structure)
u_{max}	Maximum value of u
u_p	Horizontal displacement of pipe parallel to tunnel
V_a	Soil intrusion parameter $(\pi a^2 m)$
V_b	Volume of ground lost radially through the bead and over the shield and tail
V_c	Soil volume compression (granular soil)
V_d	Soil volume increase (dilation)
V_f	Volume of ground lost at the tunnel face proper
V_g	Volume of ground lost after contact-grouting the lining
V_p	Ground loss caused by the shield being driven with a 'look-up' attitude
V_s	Volume of ground settlement trough per unit distance of tunnel advance

LIST OF SYMBOLS

V_t	Volume of ground lost at the tunnel
V_u	Volume of ground lost behind the shield and tail, and before contact-grouting the lining rings
V_y	Ground loss caused by lateral twisting or yawing at the sides of the shield, and due to shield squat
v	Ground displacement in y-coordinate direction (unconstrained by any structure)
v_{max}	Maximum value of v
v_p	Horizontal displacement of pipe transverse to tunnel
w	Ground displacement in z-coordinate direction (unconstrained by any structure)
w_{max}	Maximum value of w
w_{maxc}	Maximum (centre line) consolidation settlement
w_{maxt}	Maximum total (ground loss plus consolidation) settlement
w_p	Vertical deflection of pipe
w_t	Total settlement (function of y)
$\mathbf{w}$	Vector of free-ground vertical displacements
x	Cartesian coordinate of any point in the ground deformation field
x_J	Position of joint in pipeline parallel to tunnel
x_f	Face or final tunnel position ($y = 0$)
x_i	Initial or tunnel start point ($y = 0$)
y	Cartesian coordinate of any point in the ground deformation field
y_J	Position of joint in pipeline transverse to tunnel
y_p	Offset distance to pipeline parallel to tunnel
z	Cartesian coordinate of any point in the ground deformation field
z_0	Depth from ground surface to tunnel axis
z_p	Pipeline depth from ground surface to axis level
z_s	Zone of 'swelling' soil, after Poulos and Davis (1980)
$\bar{z}$	Centroidal height of a section
α	Rigid body rotation (tilt)
α_1	Initial pipe-joint rotational stiffness $= M_J/\theta_J$
α_2	Secondary pipe-joint rotational stiffness
β	Soil–pipe damping factor (L^{-1}) for parallel ground movement
γ	Soil unit weight (FL^{-3})
γ_w	Unit weight of water (FL^{-3})
γ_{xy}	Ground shear strain in xy-(horizontal) plane
$\gamma_{xy\,max}$	Maximum value of γ_{xy}
γ'	Submerged unit weight (FL^{-3})
Δ	Relative deflection
ΔH	Change in thickness of soil element having initial total thickness H
ΔV	Volumetric strain (vertical strain ε_z for one-dimensional consolidation)
Δe	Change in void ratio
Δp	Change in ground pressure above the tunnel crown as a result of drawdown and consolidation
δH	Change in axial length of sample in laboratory extrusion test
δe	Extrusion displacement in laboratory extrusion test
δ_g	Maximum ground movement, after Poulos and Davis (1980)
ε	Strain = change in length per unit length
ε_p	Pipe strain
ε_{pmax}	Maximum pipe strain
ε_x	Ground strain in x-coordinate direction
ε_{xmax}	Maximum value of ε_x
ε_{xp}	Strain in pipeline parallel to tunnel
ε_y	Ground strain in y-coordinate direction
ε_{ymax}	Maximum value of ε_y
ε_{yp}	Strain in pipeline transverse to tunnel
ε_{ytmax}	Maximum tensile strain in y-direction
ε_z	Ground strain in z-coordinate direction

LIST OF SYMBOLS

$\varepsilon_{z\max}$	Maximum value of ε_z
ζ_b	Section modulus of the bottom surface of a section
ζ_t	Section modulus of the top surface of a section
θ	Angle formed by buried pipe with direction of tunnel advance
θ	Rotation
θ_J	Differential rotation at pipeline joint
$\theta_{J\max}$	Maximum differential rotation at pipeline joint
θ_p	Angular deflection of pipeline = slope in radians
λ	Soil–pipe damping factor (L^{-1}). λ^{-1} is the soil–pipe stiffness parameter for transverse soil–pipe interaction. Also damping factor in analysis of beams on elastic foundations
λ'	Soil–pipe damping factor for drained soil
ν	Poisson's ratio
ν_g	Poisson's ratio of the ground
ν_u	Undrained Poisson's ratio
ν'	Drained (effective) Poisson's ratio
σ	Normal stress = force per unit area
σ_b	Tensile or compressive stress due to bending
σ_c	Compressive stress
$\sigma_{e(f)}$	A critical axial stress level in the laboratory extrusion test
σ_f	A critical axial stress level in the laboratory extrusion test
σ_i	Internal support pressure at the tunnel
σ_t	Tensile stress
σ_{xp}	Stress in pipeline parallel to tunnel
σ_{yp}	Stress in pipeline transverse to tunnel
σ_z	Vertical stress. Also axial stress in laboratory extrusion test
τ	Shear stress
τ_a	Soil–pipe shear strength
τ_x	Soil–pipe shear stress for pipeline parallel to tunnel
τ_y	Soil–pipe shear stress for pipeline transverse to tunnel
φ	Soil angle of shearing resistance
ω	Angular distortion

COLCRETE

INTERNATIONAL GEOTECHNICAL ENGINEERS

- Dam Grouting
- Ground Anchorages
- Mini Piling
- Ground Treatment
- Diaphragm Walls
- Jet Grouting
- Structural Repair
- Fabric Formwork
- Injected Concrete
- Underpinning
- Consolidation of old Coal Workings

- Manufacture and Hire of Grouting and Guniting Equipment

Colcrete Ltd

Head Office:
Thorp Arch Trading Estate
Wetherby, West Yorks LS23 7BJ
Tel: Boston Spa (0937) 844066
Telex: 557857

Manufacturing Division
11-15 High Street
Strood, Rochester, Kent ME2 4AB
Tel: Medway (0634) 722571
Telex: 965789

1 Introduction

Excavation below ground causes local relaxation of existing stress. Dilation into the opening accompanies decompression, and is only partially restricted by the eventual insertion of ground support. The volume of ground relaxing into the excavation is usually expressed in terms of unit distance advance of the excavation that causes the relaxation (i.e. cubic metres per metre advance) and is known as 'ground loss'. Since most tunnels in soil, for transportation or sewage purposes, are of circular cross-section, the ground losses can be partitioned into those associated with radial movement and those with 'axial' movements at the tunnel face proper. The magnitude of these movements is a function of soil type, the presence of water, rate of tunnel advance, tunnel size, and form of temporary and primary support.

Ground disturbance at the tunnel triggers off a chain of movements up to ground surface. The volume losses caused by ground moving inwards at the tunnel may be transmitted quickly to the surface, but because they become spread much more widely during their passage upwards the magnitude of the ground movement in any given direction attenuates. In cohesive soils there is some evidence to suggest that virtually the whole of the soil volume lost at the tunnel appears as a surface settlement depression at ground surface. With non-cohesive soils, this may not be so. Some of the volume lost in a dense granular soil may not be transmitted to ground surface but may be absorbed as a permanent loosening of the ground close to the perimeter of the tunnel. Conversely, tunnelling disturbance in a looser granular soil could encourage a redistribution of particle contacts leading to rather denser overburden and a surface settlement depression that could be of somewhat greater volume than the ground-loss volume at the tunnel.

Ground movements at the tunnel and consequential surface settlements are much more difficult to estimate on the basis of precedent for non-cohesive soils than for cohesive soils. Non-cohesive ground, especially below the water table, is much more difficult to control at the tunnel, and so special measures such as the use of compressed air or temporary dewatering may be required if tunnelling is to progress efficiently and economically. The quality of workmanship at the tunnel face tends also to be a factor affecting surface settlement much more than in the case of cohesive ground.

Although ground losses at the tunnel are quite quickly reflected as settlement at ground surface, there may be longer-term vertical movements caused by consolidation of the ground above the tunnel. The tunnel itself, or, even if the tunnel is well sealed, the disturbed (sheared/remoulded/dilated) ground around the tunnel, provides a drainage route as a result of porewater-pressure reduction for any superadjacent soil water. Consolidation settlement will be additive to ground-loss settlement, but at the present time there are no established methods of predicting it for tunnels. Some analytic and empirical guidance on consolidation estimation is, however, given in this book.

An important requirement for the design of a tunnel in soil is that its construction should cause as little damage as possible to overlying and nearby buildings and services. Attempts should be made to estimate the possible effects of tunnelling on building foundations and buried pipelines. Movement estimates based on earlier measurement programmes usually relate to 'free ground', that is, to ground unencumbered with building foundations, basements, roads, pipelines and so on. These movement estimates must then be applied to specific cases of ground–structure interaction, where the presence of the structure must affect the local movements. This difficult problem is addressed in Chapters 3 and 4.

If the estimates of possible ground movements and structural strain suggest that problems could arise, the tunnel designer is faced with several options:

Rerouting the tunnel away from 'sensitive' buildings or buried services (not always possible in an urban area, and it may be found that a rerouting into ground unfavourable for construction could incur a higher cost penalty than maintaining an original route and bearing the added cost for structural protection).

Ground stabilization (usually grouting) combined with a carefully-specified construction programme on the original route.

Protecting the existing buildings (including protective shoring and/or perhaps some form of underpinning), and the acceptance of some inevitable compensation damage, together with the quite costly relaying of old (usually cast iron) pipes in the form of new polyethylene segments or the internal lining of the existing brittle pipes with plastic material.

Relocating buried pipelines (an expensive option but one which cannot be dismissed in the case of high-pressure gas mains).

The most common method of soft ground tunnelling in the UK is still hand excavation, using pneumatic clay spades, over twelve-hour shifts within the protection of a Greathead-style shield. Figure 1.1 is an example of such a medium-diameter shield being checked out at the works before delivery to site. The hood provides additional protection for the miners. Face jacks at the front end of the shield are used for boxing up the face during stand periods in order to resist movement of ground (ground loss) into the tunnel. Primary support for the tunnel comes in the form of lining segments erected within a 'tail' that is welded to the rear of the shield. The shield in Figure 1.1 includes a mechanical erector arm for the segments. Smaller segments for smaller-diameter tunnels are erected manually.

Figure 1.1 Medium diameter tunnelling shield for hand excavation. Photograph courtesy Stelmo Ltd.

Some open shields, if size permits, contain mechanical excavation arms (back-hoes) to cut down on the manual effort and increase advance rates. With mechanical drum diggers the face is closed and excavation is accomplished by means of a slowly-rotating cruciform head. Other excavation and advance systems—closed shields with breasting doors, closed slurry shields and earth-pressure balance shields, and pipe (tunnel)-jacking—are mentioned in section 2.2.

Linings themselves represent between about 40 per cent and 70 per cent of the cost of a tunnel. For convenience, linings can be divided into two main types—flexible and rigid—although linings are never absolutely flexible nor absolutely rigid. A 'rigid' lining relies in principle, though probably not in practice, on ring bending moments to maintain its original circular shape. A 'flexible' lining relies on the stiffness and strength of the ground to maintain its circularity, in principle resisting only hoop (tangential) forces, although bending moments can arise through non-uniformity of loading or of ground restraint. Under non-hydrostatic states of ground stress before tunnelling, a 'flexible' lining will deform in such a manner as to equalize directionally the local postconstruction ground stress at the expense of a final ovaloid tunnel cross-section.

A lining is required to support the ring load, circumferential bending (and possible longitudinal bending when passing through different strata), local buckling, shield-jacking loads, asymmetrical loading at junctions and enlargements, construction handling and erection loading, and is required to resist corrosion.

The most common forms of primary linings used in the UK are the following.

1. *Precast reinforced concrete segments.* These usually comprise five or seven segments, bolted along the longitudinal and circumferential joints, together with a key segment in the crown. They are erected within, and under the protection of, the tail of the shield and then contact grouted with a cement–water paste, injected through a hole in the centre of each segment, every shift or every ring, depending upon the contract specification.

Sometimes, in order to achieve a partial equalization of local ground stress, the bolts on the longitudinal joints are not fully tightened, so that some segment articulation in ring bending can take place about the joints. It is also often assumed that the lining can be made more rigid by 'rolling' the key segment and thereby eliminating the continuity of the longitudinal joints from ring to ring.

There are patent 'One-Pass' lining systems which comprise concrete segments for the purposes of primary support but which also eliminate the need for a secondary lining. The segments, which are smooth both internally and externally, are secured together rigidly by steel bars running through internal channels and then sealed along the joints. This type of lining, having solid segments heavier than those needed for a conventional primary lining, is used, although not frequently, for sewer tunnels. An example of concrete segmental and 'One-Pass' linings is shown in Figure 1.2.

In the case of transportation tunnels, for example, where a secondary lining is not required, decorative finishes sometimes have to be fixed. This can be done by casting mild steel plates integrally on the inside (to the tunnel) faces of the segments. After contact grouting of the lining, cover strips can be welded between the mild steel plates and over the joints to provide a sensibly

Figure 1.2 Bolted chamber with 'One-Pass' tunnel junction (main tunnel is lined with bolted precast concrete segments; branch tunnel has patent 'One-Pass' lining). Photograph courtesy Charcon Tunnels Ltd.

watertight internal surface. After grit-blasting, decorative finishes can then be applied directly to the lining.

In ground with a good 'stand-up' capacity, such as the stiff London Clay, a precast concrete boltless lining may be used. This type of lining is erected behind a tail-less shield, the segments being expanded by hydraulic pressure against the ground and then maintained in position by the insertion of a wedge block, usually in the crown. Under non-hydrostatic soil pressures the lining segments are subsequently free to articulate about the longitudinal joints, the surfaces of these joints being designed to different configurations but always being carefully reinforced to accommodate the large direct, bending and shear stresses. Compared with bolted segmental linings, this type of lining more readily deforms in compression along a diameter which is subjected to high compressive stress and attempts to dilate along an orthogonal diameter that is less highly stressed. The effective shedding of some diametral compressive stress and the concomitant increase in local compression helps to achieve more readily the external state of hydrostatic stress locally applied to the lining extrados, thereby reducing the bending stress in the lining and subjecting it mainly to hoop (tangential) compression which it is best able to support. Linings will always tend to 'squat' (decrease on the vertical diameter) even in cases, such as that of London Clay, where the coefficient of earth pressure at rest for the soil is known to exceed unity. This squat, however, rarely exceeds 0.5 per cent of the original diameter in soft clays and silts and 0.2 per cent in stiff clays.

To accommodate those instances where a site investigation (see section 2.10) has indicated ground having a good-stand-up capacity, for which a wedge-block expanded type of lining would be suitable, but which changes along the route into ground for which a bolted segmental lining is required, C. V. Buchan (Concrete) Ltd of England have introduced a 'bolted wedge-block' ring. The bolted ring is identical to the same company's standard wedge-block, and can be handled and erected without any modification being needed to the back-up equipment. Changeover from wedge-block to a bolted lining can, it is claimed, be accomplished within the time taken to erect a single ring, and so tunnelling can continue with the minimum of disruption.

2. *Cast iron segments.* These have often been used for lining tunnels beneath critical areas such as rivers (the River Tyne sewage siphon tunnel in north-east England, for example; see Attewell *et al.*, 1976), important buildings and so on. The quality of machining and the tolerances on the flanges of the iron compared with the equivalent features on concrete segments produce rather better sealing against water ingress. They are also thought to be able to resist collapse loadings rather better, although this situation is rarely likely to arise. Grey iron has been the traditional material for 100 years and, because of its high compressive strength and resistance to corrosion, it is used today for BS 1452 Grade 12 linings in bad ground. Spheroidal graphite, or nodular iron, which is cast iron to BS 2789 Grade SNG 37/2, seems to be increasingly used as a substitute for grey iron in segmental linings. With steel-like mechanical properties it has a much higher tensile strength, and can therefore resist bending moments rather better than can grey iron. It has good castability, machinability and corrosion resistance.

3. *Liner plates.* The concept of a segmental lining freely articulating on knuckle joints to give a flexible response to differential soil stresses can be carried further. Since flexural stiffness is not needed in the system it follows that the lining need only be thick enough to support hoop compression and to prevent instability developing from local negative curvature. This leads to the concept of a steel lining consisting simply of plates that have been machined or cut to size and then rolled to the required radius for the tunnel. Such plates can be transported flat and then rolled to shape on site. Differential ground deformations through stress relaxation can be accommodated by elastic and

ductile straining of the lining ring as a whole. Locating lugs on the plates aid erection, and joints are made watertight by sealing welds. A typical lining ring might comprise four plates—invert, crown and two at the springline. Each plate may have perhaps two pairs of locating lugs on two of its sides (longitudinal joints and circular joints). Lugs protruding beyond the lining extrados provide the space for contact grouting. It is argued that the ductility of the steel enables any non-uniformity of bearing to be taken up by local yielding. The steel would also be required to be resistant to long-term corrosion.

4. *Box heading.* This is strictly a *tunnelling method* for sewage and water systems, but is mentioned here because it can potentially lead to large ground movements. The soil is excavated by hand to a square or rectangular cross-section and the ground supported by timber or steel props and bars, with ribs and lagging. A pipe—usually steel, iron or polyethylene—is then laid to grade in the heading, and the space around the pipe backfilled. The primary support is either left in the heading or withdrawn progressively as backfilling takes place.

This book seeks to help the tunnel designer assess the effects of construction upon building foundations and buried pipelines. It also provides a means whereby property owners and statutory undertakers, having acquired information from the tunnel designer, can perform their own checks, ask the right questions, and ultimately ensure that their property and services continue to function satisfactorily.

2 Estimation of ground movements

Estimation of settlement caused by the withdrawal of subsurface support is a much more imprecise operation than the estimation of settlement caused by surface loading. The latter has benefited from analytical examination and field observation over a period of many years, but there is as yet no analytical solution for the former. Either some more-or-less empirical assumptions have to be made about ground behaviour, or numerical solutions must be sought. Whichever method is adopted, there is a requirement for tunnel designers to reassess their estimates carefully in the light of their own experience and that of others. Designers should ask whether their estimates seem sensible, whether the assumptions (for example, of transverse settlement symmetry) are really valid in the field circumstances that apply, and whether their specification of input conditions is overcautious. Experience suggests, in fact, that there is a tendency to err rather too much on the side of caution when specifying the input parameters for estimation. Of course, a major requirement is for good site investigation information on ground conditions; knowledge of adjacent building foundations; and knowledge of the location, construction and condition of buried pipelines that could be affected by the tunnel excavation. These matters are discussed later in the book.

To aid ground movement estimation, a body of settlement case history data is included in Table 2.1. This was originally published by Attewell (1978*a*), updated by Tsutsumi (1983) and further extended for this book. It includes information on tunnel size and depth, maximum settlement, settlement trough width, volume and slope, ground description, geotechnical properties and method of working.

2.1 Volume loss parameter, V_t

For any attempt at analysis it is convenient to distinguish between cohesive clay soils and non-cohesive granular soils, while recognizing that not all soil materials likely to be tunnelled fall comfortably into these categories. It is also necessary to recognize that the thrust of the analyses in this book relates to soils which, when relaxed, deform under a degree of self-control. Such soils usually possess some cohesion through the presence of clay minerals and/or a more amorphous cementing agent and/or moisture under suction (negative)

Table 2.1 Case history data on some tunnels and tunnelling conditions. Information is compiled, with additions, after Peck (1969), Attewell (1978a), Tsutsumi (1983), where detailed references should be sought. z_0 is usually taken as tunnel axis depth; c_u is the undrained shear strength of the ground and γ is its bulk unit weight; $\gamma z_0/c_u$ is the stability ratio; V_t is the excavated volume expressed per unit advance distance; V_s is the surface settlement volume expressed per unit tunnel advance distance; $V_s\%$ expresses V_s as a percentage of V_t; i is the distance from the tunnel centre line to the point of inflexion on an assumed normal probability form of transverse surface settlement trough; w is the 'practical' half-width of the transverse surface settlement trough; β is the angle formed to the vertical by a line tangent to the tunnel extrados and joining the point $y = 3i$ or $2.5i$; an asterisk against any value denotes an estimated quantity.

Tunnel	Tunnel data			Maximum recorded surface settlement		Ground geotechnical properties			Volumes			Settlement trough width				Tunnelling method and soil conditions
	Depth z_0 (m)	Diameter $2R$ (m)	$\dfrac{z_0}{2R}$	w (mm)		c_u (kN/m²)	$\dfrac{\gamma z_0}{c_u}$		V_t (m³/m)	V_s (m³/m)	V_s (%)	i (m)	i/R	$3i$ (m)	$w = 2.5i$ (m)	
1. London Transport Fleet Line, Green Park (Attewell and Farmer, 1974 a, b)	29.3	4.15	7.06	6.17		270	2.1		13.52	0.199	1.4	12.6	6.1	37.8	32.0 ($\beta = 46°$)	Shield construction. Cast iron lining, 7 segments per ring, erected in shield tail. Annulus behind rings contact grouted every shove. Stiff, fissured, overconsolidated London Clay. Tunnel horizon blue clay overlain by weathered brown clay.
2. N.W.A. Sewerage Scheme, Hebburn, Tyneside (Attewell et al., 1975)	7.5	2.01	3.9	7.86		75	2.02		3.17	0.077	2.42	3.9	3.71	11.7	9.75	Shield construction. 5-segment precast concrete lining per ring erected in shield tail. Annulus behind rings contact grouted every 3 rings. Laminated clay overlain by stony clay.
3. N.W.A. Sewerage Scheme, Willington Quay Syphon, Contract 32 (Attewell et al., 1978)	13.375	4.25	3.15	81.5		33	9.4		24.37	1.86	13.1	9.1	4.28	27.3	22.7 ($\beta = 57°$)	Shield construction. Compressed air pressure 90 kN/m². 7-segment precast concrete lining per ring erected in shield tail. Annulus behind rings contact grouted every 3 rings. Silty alluvial clay with sand and gravel lenses containing water at artesian pressure.

Contd.

Table 2.1 (Contd.)

Tunnel	Tunnel data			Maximum recorded surface settlement		Ground geotechnical properties		Volumes			Settlement trough width				Tunnelling method and soil conditions
	Depth z_0 (m)	Diameter $2R$ (m)	$\frac{z_0}{2R}$	w (mm)		c_u (kN/m²)	$\frac{\gamma z_0}{c_u}$	V_t (m³/m)	V_s (m³/m)	V_s (%)	i (m)	i/R	$3i$ (m)	$w = 2.5i$ (m)	
4. N.W.A. Sewerage Scheme, Howdon, Tyneside (Glossop, 1978)	14.18	3.625	3.91	11.2		100	2.98	0.37	0.194	1.9	6.9	3.81	20.7	17.25 ($\beta = 47°$)	Shieldless construction. 5-segment precast concrete lining per ring erected up to the face. Annulus behind rings contact grouted every three rings. Boulder/stony clay.
5. London Transport Fleet Line, Regent's Park Northbound Tunnel (Barratt and Tyler, 1975)	20	4.15	8.2	7		230	1.70	13.52	0.18	1.3	10.3	4.96	30.9	25.75 ($\beta = 50°$)	Shield construction. Expanded concrete segmental lining. Stiff fissured, overconsolidated London Clay.
6. London Transport Fleet Line, Regent's Park, Southbound Tunnel (Barratt and Tyler, 1976)	34	4.15	8.2	5		230	1.7	13.52	0.19	1.4	15.2	7.32	45.6	38 ($\beta = 47°$)	Shield construction. Expanded concrete segmental lining. Stiff fissured, overconsolidated London Clay.
7. London Transport Experimental Tunnel, New Cross (Boden and McCaul, 1976)	10	4.15	2.4	21.5		—	—	13.52			5	1.82	15	12.5 ($\beta = 46°$)	Slurry (bentonite) shield. Sandy gravel.

Case / Reference												Description
8. TRRL Tunnelling trials, Chinnor (Hignett and Boden, 1974; McCaul et al., 1976)	8	5	1.6	8	—	—	—	7	3	21	17.5 ($\beta = 7°$)	Full-face tunnelling machine; cruciform head with drag picks. Jacking against a two-section reaction ring. Lining comprised mining arches at 1 m spacing. Highly discontinuous Lower Chalk formation.
9. Washington D C Metro, Lafayette Square (Butler and Hampton, 1975)	11.6	6.4	2.25	112.8	—	4.60	1.21	3.8	4.5	1.4	13.5	11.25 ($\beta = 35°$)	Shield construction with bucket digger. Primary liner of steel ribs and hardwood lagging boards expanding during and after the shove. Sand–cement–bentonite grout injected originally through liner but later through shield at the front. Sand and gravel, interbedded silty sand, sand and clay.
10. Washington D C Metro, Project A-2, 1st Tunnel													Shield construction with bucket digger. Primary liner of steel ribs and timber lagging expanded during and after the shove. Partial dewatering with wells 60 m apart on centre line. Medium dense silty sand and gravel, interbedded with sandy, silty clays.
C Line	14.6	6.4	2.3	152	75	4	1.7		4.5	1.4	3.5	11.25 ($\beta = 28°$)	
B Line	14.6	6.4	2.3	139	75	4	1.5		4.2	1.3	12.6	11	
A line	14.6	6.4	2.3	76	75	4	1.0	3.5	5.4	1.7	16.2	14 ($\beta = 36°$)	
11. Washington D C Metro, Treasury Yard (Hansmire, 1975)	11.6	6.4	1.8	280	75	3	1.4	4.3	1.9	0.6	5.7	5 ($\beta = 9°$)	Shield construction with ripper bucket digger. Primary liner of steel ribs (4 section) placed on 4 ft centres with full timber lagging. Lining expansion during and after shove. Medium dense silty sand and gravel interbedded with sand, silty clays.
12. Washington DC 1st Tunnel													Articulated shield with bucket digger. Steel segments erected in tailskin and grouted before shove. Dense sand and gravel, very dense, slightly cemented sand.
Section A	20.9	6.4	3.3	6	—		0.1		5.1	1.6	15.3	12 ($\beta = 23°$)	
Section B	23	6.4	3.6	3	—		0.1	0.3					

Table 2.1 (*Contd.*)

Tunnel	Tunnel data		Maximum recorded surface settlement		Ground geotechnical properties		Volumes			Settlement trough width				Tunnelling method and soil conditions
	Depth z_0 (m)	Diameter $2R$ (m)	$\frac{z_0}{2R}$	w (mm)	c_u (kN/m²)	$\frac{\gamma z_0}{c_u}$	V_l (m³/m)	V_s (m³/m)	V_s (%)	i (m)	i/R	$3i$ (m)	$w = 2.5i$ (m)	
13. Frankfurt Shield. Fohrgasse (T-9) (Chambosse, 1972; Sauer and Lama, 1973; Breth and Chambosse, 1972)	12.4	6.5	1.9	70	—	—	2.23	0.86	2.6	4.9	1.5	14.7	12 ($\beta = 35°$)	Shield construction. Bolted concrete segments. Sand with some limestone and clay marl lenses
14. Frankfurt, Domplatz (Authors as in 13)	15	6.5	2.3	23	130–550	0.6–2.5 (1.5 av.)	0.47	0.39	1.2	6.8	2.1	20.4	17 ($\beta = 42°$)	Shield construction. Bolted concrete segments, Frankfurt clay marl with some limestone and sand lenses.
15. Frankfurt Shield, Dominikanergasse (Authors as in 13)	10.3	6.5	1.6	140	—	—	5.58	1.36	4.1	3.9	1.2	11.7	10 ($\beta = 33°$)	Shield construction. Bolted concrete segments. Sand with some limestone and sand lenses
16. Frankfurt No shield, Baulos 17 (Authors as in 13)	13.3	6.5	2.1	13	130–550	0.6–2.5 (1.5 av.)	0.16	0.23	0.7	7.1	2.2	21.3	18 ($\beta = 48°$)	Shieldless construction: heading and bench. Soil anchors; shotcrete and light steel ribs for support. Frankfurt clay marl with some limestone and sand lenses.
17. Frankfurt No shield, Baulos 18a Tunnel 13 (Authors as in 13)	13.3	6.5	2.5	10	130–550	0.6–2.5 (1.5 av.)	0.09	0.18	0.5	7.1	2.2	21.3	18 ($\beta = 43°$)	Shieldless construction: heading and bench. Soil anchors; shotcrete and light steel ribs for support. Frankfurt clay marl with some limestone and sand lenses.

ESTIMATION OF GROUND MOVEMENTS

Case														Remarks
18. Heathrow Cargo Tunnel (Wood and Gibb, 1971; Smyth-Osbourne, 1971)	13.3	10.9	1.2	12	72–295	1–4 (2.5 av.)	0.04	0.19	0.2	6.5	1.2	19.5	16 ($\beta = 38°$)	Shield construction and hand excavated. No tail to the shield-concrete segmental lining expanded behind shield. Upper section of London Clay with 3.6 m of clay cover under wet gravel.
19. São Paulo Metro, Bôa Vista, (Costa et al., 1974)	11.8	5.5	2.2	70	—	—	6.0	1.2	5	6.9	2.5	20.7	17 ($\beta = 50°$)	Shield construction with compressed air. Sand and clay lenses.
20. Brussels Metro (Vinnel and Herman, 1969)	16	10	1.6	150	—	—	5.0	2.0	2.5	5.5	1.1	16.5	13 ($\beta = 26°$)	Shield construction, hand excavated. Lining segments built in tail. Upper half of tunnel uniform cohesionless sand; lower half of tunnel clayey sand.
21. Mexico City Syphon II, Gonzalez (Tinajero and Vieitez, 1972)	11.7	2.9	4.0	105	40	5	78.9	2.1	38	7.8	5.4	23.4	20 ($\beta = 58°$)	Shield with oscillating cutters. Steel lining. Lining grouted 8 m behind shield. Cutters offer support to three-quarters of face. Ground dewatering before tunnelling. Plastic lacustrine clay.
22. Lower Market St. B.A.R.T., San Francisco (Kuesel, 1972)	19.0	5.5	3.4	36	40	14	1.73	0.64	2.7	6.9	2.5	20.7	17 ($\beta = 37°$)	Shield with a rotating cutter wheel. Compressed-air support. Grouted segmental lining. Soft plastic clay.
23. Washington DC. F2a-1 Route Tunnels (Cording et al., 1976):														Articulated (3-segment) shield construction. Excavation by large, half-moon shaped, hydraulically-operated digger spade. Tunnelling below the water table, but ground dewatered by deep well pumping in advance of tunnel construction. Segmental steel lining erected in tail of shield. Serves as both a primary and secondary or temporary support. Very variable medium stiff-to-hard clays; clayey sands–sandy clays; coarse sand and gravel.
1 IB	20.3	5.5	3.7	5	—	—	0.06	0.12	0.5					
3 IB	20.3	5.5	3.7	3	—	—	0.02	0.07	0.3					
9 IB	22.5	5.5	4.1	8	—	—	0.16	0.2	0.8	8.52	3.1	25.56	21 ($\beta = 39°$)	
10 IB	21.4	5.5	3.9	13	—	—	0.42	0.32	1.3				26	
11 IB	22.0	5.5	4.0	10	—	—	0.23	0.23	1.0	9.90	3.6	27.7	25 ($\beta = 45°$)	

Table 2.1 (Contd.)

Tunnel	Tunnel data		Maximum recorded surface settlement		Ground geotechnical properties		Volumes			Settlement trough width				Tunnelling method and soil conditions
	Depth z_0 (m)	Diameter $2R$ (m)	$\frac{z_0}{2R}$	w (mm)	c_u (kN/m²)	$\frac{\gamma z_0}{c_u}$	V_t (m³/m)	V_s (m³/m)	V_s (%)	i (m)	i/R	$3i$ (m)	$w = 2.5i$ (m)	
24. Mission Line, B.A.R.T., San Francisco (Peck, 1969)	10.97	5.33	2.06	10.5 (9–12)	—	—	22.31	0.11	0.5	4.2	1.57	12.6	10.5	Mechanical shield tunnelling with 90 kN/m² compressed air. Dense, silty fine sand (SPT·N = 30) with occasional thin lenses of peat. Dewatering by deep wells.
25. Toronto Subway (Peck, 1969)	11.89 (13.41 10.36)	30.5					21.07	0.21	1.0	2.7	1.04	7.1	6.7	Shield tunnelling, hand excavation. Medium-to-fine uniform dense sand (SPT N-value = 40 to 60) above the water table.
26. Mission Line, B.A.R.T., San Francisco (Peck, 1969)	10.97	5.33	2.06	1.5	—	—	22.31	0.03	0.13	8.0	3.0	24.0	20.0	Mechanical shield tunnelling with 62 kN/m² air pressure. Slightly cemented dense silty fine sand (SPT N-value = 40 to 60). Dewatering by deep wells.
27. Wilson Tunnel, Hawaii. (Peck, 1969)	15.24	10.06	1.51	21.3	—	—	79.48							Horseshoe-shaped small drifts, hand excavated and supported with ribs and lagging. Residual saprolitic tropically-weathered volcanic granular soil, readily cut by compressed-air spades.
28. Wilson Tunnel, Hawaii. (Peck, 1969).	30.48	10.06	3.03	61.0	—	—	79.48							As for 27.
29. Garrison Test Tunnel (Burke, 1957)	36.88	10.97	3.36	18.29 (6.1–24.4)	958	0.77	94.51							Ribs and lagging support. Full-face blasting in a clay shale having an unconfined compressive strength of 958 kN/m².

ESTIMATION OF GROUND MOVEMENTS 15

Case														Remarks
30. Subway Contract D 3, Chicago. (Peck, 1969)	23.47	7.31	3.2	36.6	38–78	8.09	41.97	0.08	0.20	0.9	0.25	2.7	2.55	Hand-excavated, horseshoe-shaped cross-section tunnel. Face benched and tunnel supported by ribs and liner plates. Compressed air pressure of 30 kN/m^2. Bottom half of tunnel in hard clay. Stiff clay (unconfined compressive strength = 96–192 kN/m^2) for 3 m above crown. Soft-to-medium clay (u.c.s. = 36–96 kN/m^2) above that.
31. G.N.R.R., Seattle (Hussey et al., 1915)	37.49	11.89	3.15	18.3	—	—	111.03	2.88	2.6	63.0	10.59	181.0	157.5	Hand-excavated using small drifts with a central core. Timbered support for hard clayey till. Ravelling at the crown; poling bars used.
32. Kyoto, Tokyo, Subway (Shiraishi—personal communication to Peck)	22.55	7.01	3.22	12.2	77	5.85*	38.59	1.66	4.3	54.4	15.52	163.2	132	Hand-excavated sectional shield. Face breasted and lining segments erected in shield. Normally consolidated sensitive clay (u.c.s. = 72 kN/m^2) requiring no compressed air support.
33. B.A.R.T., San Francisco (Peck, 1969)	17.98	5.48	3.28	46.0	77	4.67*	23.67	1.02	4.3	8.9	3.24	26.7	22.2	Shield tunnelling with breasted face. Liner segments erected in the shield. Moderately sensitive clay (u.c.s. = 77 kN/m^2) requiring no compressed air support.
34. Ottawa Sewer (Eden and Bozozuk, 1968)	18.29	3.05	6.00	6.1	354	1.03*	7.31	0.12	1.6	7.9	5.18	23.7	19.7	Mechanical shield excavation. Liner segments erected behind the shield. Sensitive Leda clay (u.c.s. = 354 kN/m^2) required 28–34 kN/m^2 compressed air support.
35. Toronto Subway (Matish and Carling, unpublished)	13.11	5.33	2.46	22.0 (13–29)	67	3.91*	22.31	0.21 (0.13–0.3)	0.95	3.8	1.42	11.4	9.5	Shield tunnelling, hand excavation. Air pressure of 69–83 kN/m^2. Silty clay (u.c.s. = 77 kN/m^2) at invert level.
36. Chicago D-5 (Peck, 1969)	11.89	6.10	1.95	39.6 (18.3–61.0)	67	3.55*	29.22	0.28 (0.23–0.3)	0.95	2.8	0.92	8.4	7.0	Hand-excavated benched heading with rib and liner plate support. Glacial lake clay (u.c.s. = 57 kN/m^2 at axis level and 33 kN/m^2 at 3 m depth). Nearer surface, ground is stronger.

Table 2.1 (Contd.)

Tunnel	Tunnel data		Maximum recorded surface settlement		Ground geotechnical properties		Volumes			Settlement trough width				Tunnelling method and soil conditions
	Depth z_0 (m)	Diameter $2R$ (m)	$\frac{z_0}{2R}$	w (mm)	c_u (kN/m²)	$\frac{\gamma z_0}{c_u}$	V_t (m³/m)	V_s (m³/m)	V_s (%)	i (m)	i/R	$3i$ (m)	$w = 2.5i$ (m)	
37. Toronto Subway (Peck, 1969)														Dense sand above ground water level.
First Tunnel	10.36	5.33	1.94	85	—	—	22.31	0.42	1.9	1.9	0.73	5.7	4.7	
Second Tunnel	13.41	5.33	2.52	140	—	—	22.31	0.85	3.8	2.4	0.92	7.2	6.0	
38. São Paulo (Terzaghi, 1950)	30.48	2.74	11.12	204	—	—	5.9	2.97	50.39	5.8	4.2	17.4	14.5	Tunnelling in stiff clay with many construction difficulties.
39. Ayrshire Joint Drainage Scheme Tunnel (Eadie, 1976)														Shield tunnelling, hand excavation through water-bearing raised beach sands of Clyde Estuary, Scotland. Timber breasting with face jacks. Internal pressure 1.4 to 1.6 atmospheres absolute. Non-expanding concrete lining segments with cement–bentonite grout injected into void at end of each 12h shift.
First Tunnel	6.3	2.59	2.43	13.5	—	—	6.89	0.06	0.87	1.41	1.09	1.23	3.52	
Second Tunnel	6.2	2.59	2.39	16.0	—	—	6.89	0.06	0.87	1.60	1.23	4.80	4.00	
40. Acton Grange Sewer, Warrington														Slurry (bentonite) shield, machine excavated through a mixed face comprising mainly sand with some boulders but with a small proportion of Bunter Sandstone in the invert. Water-table level was partway up face. Bolted precast concrete segmental lining
Section C-C'	5.75	2.44	2.36	19.9	—	—	6.16	0.086	1.37	1.73	1.42	5.19	4.37	
Section D-D' (O'Reilly et al., 1980)	5.75	2.44	2.36	14.2	—	—	6.16	0.071	1.10	2.00	1.64	6.00	5.00	

ESTIMATION OF GROUND MOVEMENTS

Case														Remarks
41. White Mud Creek Tunnel, Edmonton, Alberta (Thomson and El-Mahhas, 1980)	15.25	6.05	2.52	—	—	—	—	—	—	—	—	—	—	Two moles without shields (each 6.05 m diam.). Poorly indurated clay shale interbedded with thin sandstone strata. Bolted steel segmental ribs in temporary lining and replaced by plain concrete lining.
42. 170 Street Tunnel, Edmonton, Alberta (authors as in 41)	21.5	2.56	8.40	12	—	—	—	—	—	—	—	—	—	Mole with shield. Temporary lining consisted of segmental steel ribs and later replaced by plain concrete lining. Major portion of the tunnel excavated through till.
43. Nagoya Subway (Kawamoto and Okuzono, 1977) Section A Section B Section C	17.4 19.2 16.5	6.4 6.4 6.4	2.72 3.0 2.58	48 45 46										Shield tunnelling. Twin circular tunnels of 6.4 m diameter placed side by side. Tunnel constructed through the alluvium deposit.
44. Stockton-on Tees Stage I Interceptor sewer, Measurement Section D (McCaul, 1978)	6.28	1.26	4.98	43.7	30.5	2.7	1.25	0.38	30.4	3.48	3.47	5.51	8.70	Mini-Tunnel system. Hand excavation from shield. 3-segment, smooth, precast concrete lining. Soft, silty, sandy clay.
45. Stockton-on. Tees Stage IV Interceptor sewer, Measurement Section D (McCaul, 1978)	5.86	1.26	4.65	56.3	41.7	2.2	1.25	0.52	41.5	41.7	3.68	5.84	9.22	Mini-Tunnel system. Hand excavation from shield. 3-segment, smooth, precast concrete lining. Soft, silty, sandy clay.
46. New Cross L.T.E. Experimental Tunnel (Boden and McCaul, 1974)	10	4.15	2.4	21.5	—	—	13.52	0.27	2.0	5	1.82	15	12.5	Slurry (bentonite) shield. Sandy gravel.

Table 2.1 (Contd.)

Tunnel	Tunnel data			Maximum recorded surface settlement		Ground geotechnical properties		Volumes			Settlement trough width			Tunnelling method and soil conditions
	Depth z_0 (m)	Diameter $2R$ (m)	$\dfrac{z_0}{2R}$	w (mm)		c_u (kN/m²)	$\dfrac{\gamma z_0}{c_u}$	V_1 (m³/m)	V_s (m³/m)	V_s (%)	i (m)	i/R	$3i$ (m)	$w = 2.5i$ (m)
47. Hamburg-Wilhelmburg collector (Jacob, 1978)	19.24	4.48	4.3	0–10		—	—								Well-bedded, sharp sand and gravel (0.2–100 mm), boulders to 80 cm overlain by clay, peat and fill. Water table 16 m above invert. Hydro-shield. Reinforced concrete lining. Air pressure 1.6 atmospheres.
48. Antwerp Metro (Jacob, 1978)	24	6.56	3.7	6–7		—	—								Hydro-shield. Reinforced concrete lining. Fine alluvial sand; interlayers of clay, overlying overconsolidated clay. Water table 12 m above invert, lowered to 10 m before tunnelling.
49. Agasegawa Sewer Main No 2, Katsushikaku, Tokyo (Eng. News Record, 1974).	av. 10	5.05	2.0	25–90		—	—								Slurry mole, concrete segmental primary lining. Mixed face of fine 'quick' sand and silt and clay, SPT N-value 20. Water table 7 m above crown.
50. Takudo Water main, Suguinami-ku, Tokyo (Miki et al., 1977)	27.8	3.55	7.8	21.9 max. 1.4 av.		—	—								Slurry mole; concrete segmental lining. Cemented dense sandy gravel (2–150 mm), SPT N-value 50, overlain by clay, sandy gravel and silt. Water table 11.5 m above crown.
51. Yotsugui Sewer Branch, Katsushikaku Tokyo (Miki et al., 1977)	7.4	2.40	3.1	20 max. 15 av.		—	—								Slurry mole; steel segmental lining. Loose alluvial sand with silt; SPT N-value = 5 – 20. Water table 5.4 m above crown.

Case														Remarks
52. Southern Line Proposal, Amsterdam Metro. Section Churchilham-Singelgracht (Publ. Works Dept., 1975)	19.40–9.85, 18.50 av.	6.2	3.1–1.6, 3.0 av.	27.55	—	—	—	0.3–0.6	1–2	4.38	1.41	13.13	10.95	Fully mechanized shield with full face support. Medium to very dense sand and silt, overlain by clay, peat and fill. Water table 2.0 m above invert.
53. Chicago S-6 (Peck, 1969)	10.97	6.10	1.8	25.6 (15.0–36.6)	5–7	3.85*	29.22	0.15	0.5	2.3	0.75	6.9	5.7	Hand-excavated benched heading with rib and liner plate support. 83 kN/m² compressed air support. Glacial lake clay (u.c.s. = 57 kN/m² at axis level and 33 kN/m² at 3 m depth). Nearer surface, ground was stronger.
54. Liner Plate Tunnel, Sabe SP, Brazil (Negro and Eisenstein, 1981)	8.0 appr.	3.6	2.22	25.0 av.	250	4.03	10.5	0.213	2.03	3.0	1.67	9.0	7.5	Full-face hand excavation with circular steel segmental lining plates erected immediately behind the face. Tertiary soft porous clay and clayey dense sand.
55. Horseshoe Tunnel, Sabe SP, Brazil (Authors as in 54)	9.00 appr.	3.6	2.5	15.5	250	4.03	12.5	0.171	1.37	2.5	1.38	7.5	6.25	Same as 54 above.
56. NATM Tunnel Sabe SP, Brazil (Authors as in 54)	8.5 appr.	3.96	2.14	5.0	250	4.28	12.8	0.048	0.37	3.0	1.52	9.0	7.5	Hand-excavated in three stages: heading, bench and invert. Shotcrete 10–13 cm thick, with 10 × 10 cm steel wire mesh. Soil conditions as 54 above.
57. Anglian Water Authority Maycroft Relief Sewer (O'Reilly and New, 1982) (a) (b) (c) (d)	8.0 / 5.5 / 5.5 / 6.5	2.7 / 2.7 / 2.7 / 2.7	2.96 / 2.04 / 2.04 / 2.41	95 / 60 / 58 / 97	12 / 12 / 12 / 12	13.4 / 9.2 / 9.2 / 10.8	5.73 / 5.73 / 5.73 / 5.73	0.905 / 0.481 / 0.407 / 1.046	15.8 / 8.4 / 7.1 / 18.2	3.8 / 3.2 / 2.8 / 4.3	2.81 / 2.37 / 2.07 / 3.19	11.4 / 9.6 / 8.4 / 12.9	9.5 / 8.0 / 7.0 / 10.75	Hand-excavated in shield; lined with concrete segments; compressed air applied about 20 days after excavation; lower 60% of face stiff stony clay (Grimsby Marine Warp) overlain with 2.5 m of stiff clay.

Table 2.1 (Contd.)

Tunnel	Tunnel data		Maximum recorded surface settlement		Ground geotechnical properties			Volumes			Settlement trough width				Tunnelling method and soil conditions
	Depth z_0 (m)	Diameter $2R$ (m)	$\frac{z_0}{2R}$	w (mm)	c_u (kN/m²)	$\frac{\gamma z_0}{c_u}$		V_t (m³/m)	V_s (m³/m)	V_s (%)	i (m)	i/R	$3i$ (m)	$w = 2.5i$ (m)	
58. Thames Water Authority, Sutton Sewer (O'Reilly and New, 1982)															Hand-excavated; stiff fissured London Clay. Hand-excavated; firm to stiff weathered London Clay. Full-face machine (mini-tunnel) excavated; firm to stiff weathered London Clay.
(a)	17.1	1.78	9.61	3.8	180	1.89*		2.49	0.096	3.86	10.0	1.12	30.0	25.5	
(b)	3.4	1.78	1.91	3.7	90	0.76*		2.49	0.019	0.75	2.0	2.25	6.0	5.0	
(c)	4.9	1.52	3.22	7.1	9	1.09*		1.81	0.054	2.98	3.0	3.95	9.0	7.5	
59. Bristol City Engineers Dept. Avonmouth 2 Sewerage Scheme (Toombs, 1980)	6.0	3.4	1.76	20.0	18	6.67*		9.08	0.251	2.8	5.0	2.94	15.0	12.5	Hand-excavated within shield with compressed air; soft to very soft alluvium overlain with fill for motorway embankment.
60. London Transport Interchange Subway at Kings Cross, London (West et al., 1981)	14.06	4.13	3.4	4.0	230	1.22*		13.46	0.078	0.6	7.8	3.78	23.4	19.5	Hand-excavated (no shield); cast iron lining; London Clay.
61. Thames Water Authority Oxford Trunk Outfall Sewer (O'Reilly and New, 1982)	11.7	2.82	4.15	2.2	200–400	0.78*		6.24	0.028	0.44	5.0	3.55	15.0	12.5	Full-face machine in shield; stiff heavily overconsolidated fissured Oxford Clay.

Case											Description
62. WNTDC Lumb Brook Sewer (O'Reilly and New, 1982)											Hand-excavated within shield; loose to medium sand with some gravel. Hand-excavated in medium to dense sand with some clay; cover of very stiff sandy clay. Partially stabilized medium dense sand and gravel with a little clay. Fully stabilized sand and gravel.
(a)	4.7	3.6	1.31	78.0	—	—	10.18	0.47	4.6	2.4	1.33	7.2	6.0
(b)	9.0	3.6	2.5	19.0	—	—	10.18	0.12	1.2	2.52	1.40	7.56	6.3
(c)	6.5	3.6	1.81	19.0	—	—	10.18	0.06	0.6	1.59	0.88	4.77	3.98
(d)	6.5	3.6	1.81	20.0	—	—	10.18	0.09	0.9	1.79	0.99	5.37	4.48
(e)	6.5	3.6	1.81	7.0	—	—	10.1	0.04	0.4	2.28	1.27	6.84	5.70
63. North West Water Authority, Mersey Street to Howley Sewer (O'Reilly and New, 1982)	8.4	2.0	4.2	28.0	—	—	3.14	0.23	7.1	3.2	3.2	9.6	8.0	Hand-excavated within shield using compressed air; variable loose silty sand with some clay; tunnelling about 4 m below water table.
64. Northumbrian Water Authority Sewerage Scheme, Ouseburn Valley (Spencer, 1978; Dobson et al., 1979)	13.0	3.47	3.75	81.0	—	—	9.64	1.48	15.6	7.29	4.20	21.87	18.23	Hand-excavated within shield; recent fill materials, rubble, timber, household waste and ash in soft clay matrix.
65. Budapest Metro (Ulrich, 1974)														Shield tunnelling, hand-excavated in Oligocene clay overlain by sandy silt. Bolted concrete segmental lining.
(a) N-S Line Rubbing Tunnel	30	5.5	5.45	26	—	—	23.76	0.83	0.29	9.23	3.35	27.66	23.05
(b) E-W Line Running Tunnel	30	5.5	5.45	37	—	—	23.76	2.2	0.11	30.25	11.00	90.75	75.62
66. Tyne and Wear Passenger Transport Executive, Running Tunnel, Eldon Square, Newcastle (O'Reilly and New, 1982)	14.2	5.21	2.73	7.5	200	1.4*	21.32	0.132	0.6	7.0	2.69	21.0	17.5	Partial-face machine excavated in shield with compressed air. Glacial till, firm/stiff clay with some sand and gravel lenses.

Table 2.1 (Contd.)

Tunnel	Tunnel data			Maximum recorded surface settlement		Ground geotechnical properties		Volumes			Settlement trough width				Tunnelling method and soil conditions
	Depth z_0 (m)	Diameter $2R$ (m)	$\dfrac{z_0}{2R}$	w (mm)		c_u (kN/m²)	$\dfrac{\gamma z_0}{c_u}$	V_t (m³/m)	V_s (m³/m)	V_s (%)	i (m)	i/R	$3i$ (m)	$w=2.5i$ (m)	
67. Sewage Pipeline No. 352, Uchiku-cho Ibaraki (Miki et al., 1977)	8.6	2.55	3.4	12.9 max. 5 av.		—	—								Slurry mole, segmental lining. Cemented fine (0.4 mm) clayey sand, overlain by sand, clay and silt. Water table 5.4 m above crown.
68. Belfast Sewerage Scheme, Sydeham, Belfast. (Glossop and Farmer, 1977)	5	2.74	1.82	37.5		2.1	8.3	5.9	0.12	2.0	2.1	2.75	2.00	8.00	2.74 m diameter shield, 2 m long + 1 m tailskin. Compressed air spade excavation by hand. 41 kN/m² compressed air pressure for ground lining segments, 0.6 m long. Each ring grouted individually immediately after shield shove; 3–4 rings erected per shift. Belfast 'sleech'—soft organic silty clay with a high moisture content.

ESTIMATION OF GROUND MOVEMENTS 23

pressure, noting that cohesion is enhanced by negative pore pressures. Granular soils without cohesion are more prone to collapse at the tunnel face, creating large vertical deformations which are less easily analysed and estimated for the purposes of support design at the tunnel (lining) and the protection of near-surface structures (buried pipelines and structural foundations). On this latter problem, reference may be made to Deere *et al.* (1969) and for a 'behavioristic' (*sic*) classification of various soils—ranging from cohesive to granular—see Figure 11.1, Appendix page 11–7 in that report.

2.1.1 *Clay soil*

It may be assumed that, during tunnelling in clay soil, the ground moves into the unsupported parts of the excavation at a constant rate. This particular problem was modelled by Attewell and Boden (1971) in the laboratory and is further described in Attewell (1978*a*). For shallow tunnels the overburden stress (on a depth times unit weight basis) can be regarded as supplying the driving pressure, and the rate of inward movement axially at the face proper is considered to be the same as that radially around the cut boundary. A laboratory-derived deformation rate of 0.0055 mm/min for London Clay was the same as that actually measured, using deep settlement probes, above the crown of a London Transport tunnel under construction (Attewell and Farmer, 1974*a*).

If clay soils tend to intrude into a tunnel at a constant rate for a given driving pressure, then it is obvious that the *rate of tunnel advance* can be a critical factor controlling ground loss. This rate of advance determines the time during which an element of soil at or near the opening is able to relax, and so the slower the rate of tunnel advance the greater is the total volume of soil intrusion at the tunnel for a material of particular rheological properties surrounding a tunnel of specified depth.

Referring to Figure 2.1, it is useful to consider the total ground loss V_t associated with open (hand) shield tunnelling, a bolted segmental lining constructed inside a tailskin welded to the rear of the shield body, and a cementitious contact grout injected behind newly-erected lining segments. These are the conditions that most frequently apply at present to British soft-ground tunnelling. The source contributions comprising the ground loss are as follows.

Face loss, V_f. This is the axial loss at the tunnel face proper. It contributes both to the transverse spread and to the extent of the forward longitudinal span of the surface settlement trough (see Figure 2.2).

Shield loss, V_b. This is the radial ground loss around the perimeter of a shield and its tail due to the presence of an overcutting bead or equivalent overcutting device installed to reduce ground friction with the extrados of the shield and to facilitate shield steering. There are also other losses V_p and V_y, discussed subsequently.

24   SOIL MOVEMENTS

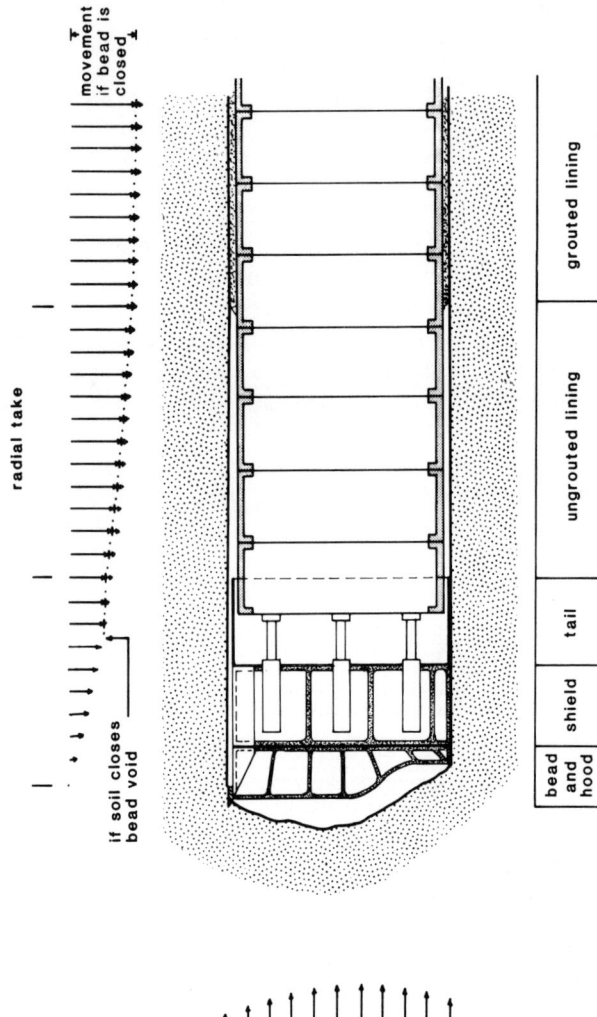

Figure 2.1 Simplified vertical cross-section along the centre line of a hand-driven shield-supported tunnel showing the likely zones of ground loss. For the purposes of clarity the possible deformations of the exposed soil surface are not shown within the cross-section but are indicated in exaggerated form outside the cross-section. Below the soffit, the radial movements will generally be somewhat smaller than those developing at the soffit.

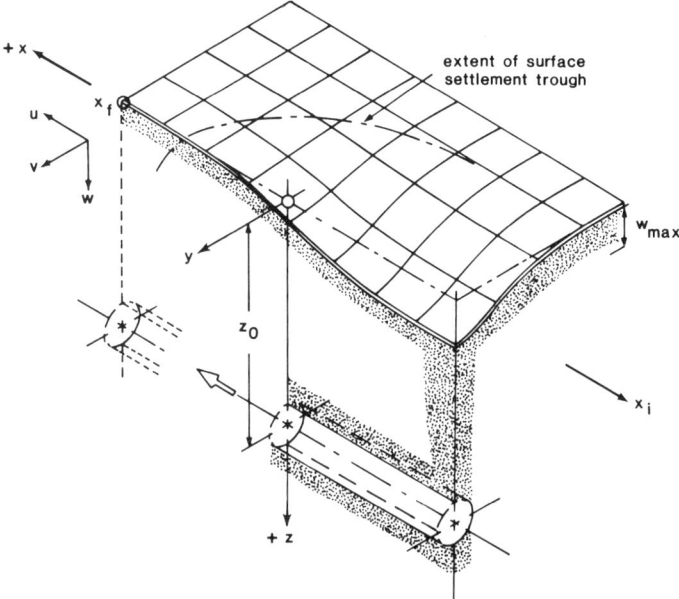

Figure 2.2 Tunnel face advancing in a $+x$ direction, creating a settlement (w) trough having a long axis of assumed cumulative probability form in the xz plane and a transverse axis of normal probability form in the yz plane. Axes x, y, z are orthogonal; $x = y = z = 0$ at ground surface vertically above the centre of the tunnel face. The tunnel start and final face positions are denoted by x_i and x_f, respectively. Ground displacements u, v, w, induced by tunnel excavation, occur in directions x, y, z, respectively.

Postshield/pregrout loss, V_u. This radial ground loss occurs behind the tailskin. It may be caused by uninterrupted continuous movement which began over the shield as a result of bead presence or, if the clay soil had encountered the shield, a reactivation of movement as the shield moves forward.

Postgrout loss, V_g. This radial loss occurs behind the tail of the shield until the set grout resists further inward movement of the ground.

The following parameters are used to analyse the problem.

m is the average amount of soil movement (displacement) at the tunnel and is more usefully expressed as dm/dt, the average rate at which the material moves unrestrained into the excavation. It is convenient to take, as an approximation, the same rate inwards axially at the face and radially over and behind the shield. Experimental work (Figure 2.3), specific to the problem, aimed at evaluating deformation rates in clay soils was performed by Attewell and Boden (1971) and others (see Attewell, 1978a, p. 871).

l is the tunnel advance distance, but again is more usefully expressed as dl/dt, an *average rate* of tunnel (and shield) advance.

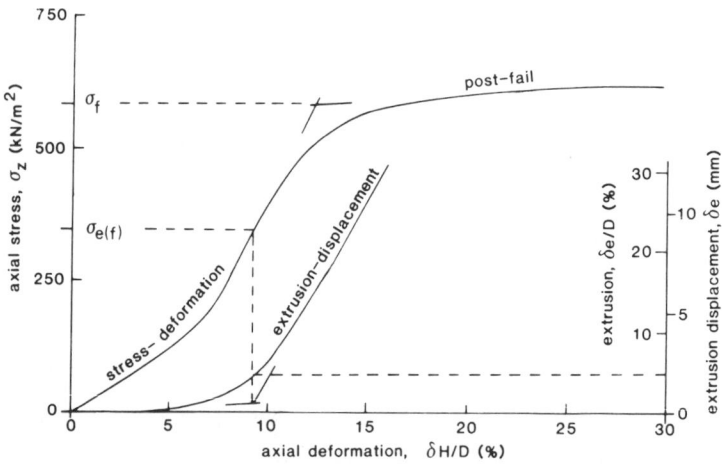

Figure 2.3(a) Equipment for estimating deformation rates in 100 mm undisturbed clay soil samples. The sample is extruded directly from a U100 tube into a brass cylinder and then compressed axially at a constant rate. Extrusion laterally through a hole of definable diameter is analogous to intrusion at a tunnel face, and is monitored by a linear variable differential transformer transducer.

ESTIMATION OF GROUND MOVEMENTS

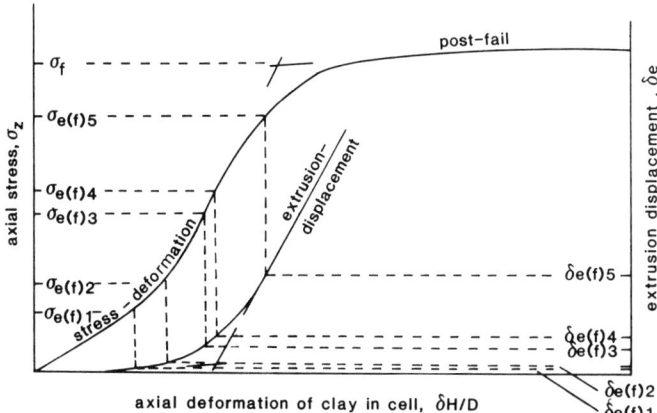

Figure 2.3(b) Result of a typical constant axial deformation rate test on a laminated clay (constant axial deformation rate of 0.593 mm/min). δH is the change in axial length of the sample in the cell under axial stress $\sigma_z(=\gamma z_0)$. Lateral extrusion δe occurs through hole of diameter D. Knowing the unit weight γ of the soil and the axis depth z_0 of the tunnel, rates of soil extrusion (tunnel face intrusion) can be assessed for different overburden pressures.

Surface situation	Critical overburden pressure	Definition of critical overburden pressure
Super-critical (e.g. important surface structures very sensitive to differential settlement)	$\sigma_z = \sigma_{e(f)1}$	Onset of extrusion acceleration—end of pre-acceleration 'linearity'
Critical or near-critical (e.g. less important surface structures, much less sensitive to differential settlement)	$\sigma_z = \sigma_{e(f)2,3,4}$	Range of $\sigma_{e(f)}$ values defined by the following lines constructed from point of intersection of pre- and post-acceleration curve:
	$\sigma_z = \sigma_{e(f)2}$	Horizontal back-projection (sub-maximum d^2e/dt^2)
	$\sigma_z = \sigma_{e(f)3}$	Angular bisector (approx. $d^3e/dt^3 = 0$)
	$\sigma_z = \sigma_{e(f)4}$	Vertical projection (Post maximum d^2e/dt^2)
Sub-critical (e.g. in an urban environment, under housing estates, etc.)	$\sigma_z = \sigma_{e(f)5}$	End of extrusion acceleration
Non-critical (e.g. under open land)	$\sigma_z = \sigma_f$	Ultimate stress-deformation yield—this could be apportioned in the same way as above, but there would be little merit in doing so

l_s is the length of the shield and its tail.
a is the outer radius of the shield.
b is the thickness of any overcutting bead or equivalent device fitted to the shield.

Face loss, V_f. Ground losses at the face proper are a multiple of the area of the face πa^2 (ignoring the small contribution of any bead) and the soil intrusion distance per unit length of advance k if soil intrusion is assumed to take place uniformly over the whole face area. In practice, because of drag at the cutting end of the shield and particularly at the hood, a clay soil will tend to intrude more readily at the centre than at the periphery. Field measurements (section 2.11) using inclinometer access tubes intercepted at the tunnel face and also measurements taken in a shaft towards an approaching tunnel face confirm such a doming effect (Attewell, 1978a; and see Figures 2.40b, 2.41b). A factor of 50 per cent could reasonably be applied to take this doming into account but, to generalize, another factor k_1, where $0 < k_1 < 1$, could be used so that

$$V_f = \pi a^2 k k_1. \tag{2.1}$$

The factor k may be expressed in terms of the average rate of soil movement at the face, $dm/dt = m'$, and the average rate of tunnel advance $dl/dt = l'$, so that

$$k = \frac{m'}{l'}. \tag{2.2}$$

Thus,

$$V_f = \pi a^2 k_1 \frac{m'}{l'} \tag{2.3}$$

or, as a percentage of the tunnel carcass volume per unit distance of advance (face area),

$$\% V_f = 100 k_1. \tag{2.4}$$

Equation (2.3) does not allow for the fact that when tunnel advance ceases the basic rate of clay inward movement should strictly remain the same. Cording *et al.* (1976) suggested, from the numerical modelling of Ghaboussi and Ranken (1975), that V_f may be approximately related to a parameter V_a, where V_a is the multiple of πa^2 and the soil intrusion m, independent of length of tunnel advance:

$$V_f = \frac{2V_a}{a} \quad \text{for } a \leq 2\text{m} \tag{2.5}$$

and

$$\% V_f = \frac{200m}{a}. \tag{2.6}$$

ESTIMATION OF GROUND MOVEMENTS

As before, allowing for any intrusive doming of the clay soil,

$$\%V_f = \frac{100\pi a^2 m}{\pi a^2} \times \frac{2}{a} = \frac{200 k_1 m}{a}. \qquad (2.7)$$

Cording et al. (1976) claim that equation (2.7) is applicable to most soils, including granular soils and stiff-to-hard clays.

The example, quoted in Attewell (1978b), of a 4.25 m diameter tunnel at an axis depth of 13.375 m in a silty alluvial clay, may be used to test these equations. Inclinometer measurements showed that face take began 3 days (6.71 m) before the tunnel face intersected a vertical inclinometer tube and increased to 3.17 mm at 0.7 m below tunnel axis level when the face was 1 day (1.83 m) away. Obviously the soil maximum inward displacement at axis level when exposed by excavation would exceed this figure of 3.17 mm. The $\%V_f$ value calculated from equation (2.4) was $0.06 k_1$. From equation (2.7), and extrapolating the measured movement to give a maximum displacement m of 4.6 mm, the estimated $\%V_f$ is $0.43 k_1$, or about 0.2 if k_1 is approximately 0.5.

Radial loss, V_b, over a shield. The construction detail of the shield tends to control the total volume of ground lost. A particular source of loss comprises a bead, or similar device, welded on to the leading edge of the hood and shield in order to overcut the ground, reduce friction during shield advance, and allow the shield to be steered more easily.

There are three usual cases: no bead ($V_b = 0$); the bead spans the upper 180° of hood and shield (Attewell and Farmer, 1974a, b); and the bead covers the full circumference (360°) of hood and shield (Attewell and Farmer, 1973; Attewell et al., 1975).

For purposes of analysis assume a 360° bead. The area of shield and tail extrados over which radial deformation may take place is

$$A_s = 2\pi l_s(a+b) \simeq 2\pi l_s a, \qquad \text{since } b \ll a. \qquad (2.8)$$

In terms of shield advance, the soil adjacent to the shield has the following time t_s in which to deform inwards:

$$t_s = \frac{l_s}{l'}. \qquad (2.9)$$

During this time the soil movement m is

$$m = t_s m'. \qquad (2.10)$$

Thus, the potential volume of soil moving radially into the excavation over the length of the shield is

$$V_b = A_s m = 2\pi l_s^2 a \frac{m'}{l'}. \qquad (2.11)$$

Expressed per unit advance,

$$V_b = A_s m/l_s = 2\pi l_s a \frac{m'}{l'}. \tag{2.12}$$

Since the value of $m (= l_s m'/l')$ must be less than or equal to b, the latter applying when the bead has been completely closed by intrusion of the relaxed soil, another parameter, k_2, is used:

$$V_b = 2\pi l_s a k_2 \frac{m'}{l'}, \tag{2.13}$$

where $0 < k_2 \leqslant 1$.

Expressed as a percentage of the tunnel face area:

$$\% V_b = 200 l_s \frac{k_2 m'}{a\ l'}. \tag{2.14}$$

This last equation has been examined by Attewell (1978a) in the light of measurements taken during construction of a London Transport tunnel (Attewell and Farmer, 1974b). The following parameters applied: l_s (shield and tailskin) = 3.348 m, a = 2.073 m, l' = 0.134 m/h, m' = 0.0055 mm/min. As noted earlier, this quoted value of m' was measured both *in situ* by means of a ring magnet system installed just above the tunnel crown in a borehole, whereby the magnet was able to move with the deforming London Clay and its position was located continuously by an audible probe activated by the magnetic field, and also by the laboratory soil extrusion method described in Attewell and Boden (1971). For this particular shield the bead covered only the upper 180° of the shield hood and so, from equation (2.14),

$$\% V_b = 100 l_s \frac{k_2 m'}{a\ l'} = 0.40 \quad \text{for} \quad k_2 = 1.$$

However, in Attewell and Farmer (1974b) it was shown that the bead of thickness 6.5 mm was closed by the clay because the potential intrusive deformation m of the clay over the shield and tail length was 8.244 mm ($= l_s m'/l'$). Thus, by proportion,

$$k_2 = \frac{6.5}{8.24} = 0.79 \quad \text{and} \quad \% V_b = 0.40 \times 0.79 = 0.31.$$

For this example, expressed as a direct volume per unit advance,

$$V_b \simeq \frac{0.31 \times \pi a^2}{100} = 0.042 \text{ m}^3/\text{m}.$$

Other losses over the shield, V_p and V_y. There are usually other, generally unquantifiable, losses at the shield that are attributable to the mechanics of shield driving.

First, a shield may tend to dive off-grade at the front end (promoted by the rotational influence of a protruding hood), or it may tend to settle as a whole in a very soft clay (see, for example, Shiraishi, 1968; Muir Wood, 1970). In order to counteract this tendency the shield must be driven with a 'look-up' attitude (an upward pitch). This resulting ploughing action creates a hole of elliptical cross-section, with the major axis vertical, and so results in a further and possibly substantial volume loss V_p.

Second, it must be expected that the shield, unstiffened by any diametral bracing, will squat towards the tail as a result of a relaxed overburden pressure exceeding a relaxed horizontal pressure if the inward deformation of the soil takes it on to the shield tail. This process will be enhanced if there is soft or remoulded clay in the tunnel sidewalls, or if the soil is granular and contains many voids, because then the ground is insufficiently stiff to mobilize passive resistance to horizontal outward deflection of the tail. On the other hand, a shield that has to manoeuvre round a tight curve may become ovaloid at the tail, with its major axis vertical. However, during curve negotiation, or even along straight lengths of tunnel, there may be a lateral twisting, or yawing, at the sides of the shield, with the elliptical cut surface this time having a major axis horizontal. The net result of these ellipticities is another source of loss V_y.

Hansmire and Cording (1972), measuring on the Washington DC Metro, found that V_y was much less than V_p. On the same tunnel Butler and Hampton (1975) noted that grade changes of 1 per cent could generate as much as 0.34 m³/m of lost ground (V_p). This latter figure was a substantial 28 per cent of the observed total surface settlement volume, but of course it is usually impractical to quantify V_y and V_p. For V_p, Cording et al. (1976) suggest

$$V_p = \frac{\pi a l_s}{2} \times \text{(excess pitch)}. \qquad (2.15)$$

'Excess pitch' is defined as p/l_s, where p is the difference in elevation between the front and rear of the shield and tail *minus* the design grade of the tunnel. Thus,

$$\%V_p = 50\frac{p}{a} \qquad (2.16)$$

expressed in terms of the tunnel volume per unit advance.

Ground loss after lining erection, V_u. If a bolted segmental lining is to be used, it will be erected within the protection of a tailskin, a circular extension welded to the rear of the shield and having the same external diameter. The length of the tail will be such as to accommodate rather more than one ring of lining segments (one ring in the UK occupying 0.61 m of tunnel lining length), the leading edge of the penultimately erected ring resting on the rear of the tail and so providing some measure of stability against the shield diving at the front end. When sufficient forward excavation of the soil has been accomplished ahead of the shield under the protection of a hood, the rams push the shield forward off the newly erected lining ring leaving a length of ground unsupported behind the lining until such time as the annulus is filled by grout, pea gravel (or an equivalent substitute[†]), or a mixture of the two. If the ground is strong enough to stand up unsupported without collapsing, the tunnelling contractor will usually delay contact grouting operations until the end of a shift or the beginning of the next in order to increase his productivity. Three or four erected lining rings may thus remain ungrouted, so allowing more radial ground loss—and surface settlement—to occur than if a single ring had been erected and grouted. When tunnelling beneath buildings it may be necessary for the contract documents to specify a 'build-one-ring, grout-one-ring' tunnelling regime in order to limit the ground movements, and hence eliminate or greatly reduce building and/or adjacent pipe damage. In writing such a specification the project designer will have balanced the extra cost of the works against the potential savings from damage and consequential claims.

The simplest method of estimating these losses is to assume that the soil has contacted the extrados of the tail before the shield is shoved forward. Then, in the manner of equation (2.13) with $k_2 = 1$ and l_u being the unsupported ungrouted tunnel length behind the tail,

$$V_u = 2\pi l_u a \frac{m'}{l'} \qquad (2.17)$$

or

$$\%V_u = 200 \frac{l_u}{a} \frac{m'}{l'}. \qquad (2.18)$$

Taking the example quoted earlier of the London Transport tunnel (Attewell and Farmer, 1974a,b), l_u was approximately 1.2 m (two precast concrete segmental rings ungrouted) and so, inserting values into equation (2.18) above,

$$\%V_u = 200 \times \frac{1200}{2073} \times \frac{0.0055 \times 60}{134} = 0.28.$$

Obviously this type of estimation applies only to a reasonably firm clay soil (say $c_u \gtrsim 50 \text{kN/m}^2$).

Ground loss after grouting, V_g. Radial losses continue after grouting while the grout bleeds and sets, and until such time as the stiffness of the lining and grout match that of the ground. A bolted segmental lining undergoes some ring bending as the segments articulate at the longitudinal joints, and this can be reduced somewhat by 'rolling' the key segment at the crown (eliminating the

---

[†]In the UK, this is typically 'Lytag', which comprises spheres of sintered pulverized fuel ash. It is manufactured by Pozzolanic Lytag Ltd of Hemel Hempstead, Hertfordshire, and also serves as a concrete aggregate.

continuity of the longitudinal joints) to give a compositely stiffer ring. Even in soils such as the London Clay for which K_0 has been shown to exceed unity at tunnel depths the lining invariably 'squats';[†] outward horizontal deflection of the lining compresses the ground at springline, so helping to equalize the radial pressure on, and reduce the bending moments in, the lining.

Other than by measurements *in situ*, using magnetic settlement monitors installed above the crown before tunnelling and both settlement and horizontal movement monitors at tunnel axis level to determine volumetric strain, there is no satisfactory method of estimating V_g as m' decreases. Measurements by Attewell and Farmer (1974a) in the stiff London Clay indicated that approximately 30 per cent of the total settlement occurred after the lining was grouted, but some of this percentage must be attributed to consolidation effects, which are considered in section 2.7 below. This problem of estimating V_g is further addressed via the Example 2.1 calculation in section 2.1.3.

2.1.2 Granular soil

These soils usually possess some cohesion perhaps through the presence of a little clay, some suction pressure dampness, and sometimes from iron oxide cementation. Damp sands are usually excavated with ease and maintain upright faces. Without any cohesion they would slump into the face of the shield at a slope angle controlled by the interparticle friction, thereby generating very large $\%V_f$ values, and the slumping would persist with shield advance (see, for example, the analysis of Széchy, 1970). For a slump-slope angle of α degrees to the (horizontal) invert of the shield a simple analysis would suggest $V_f \simeq \pi a^2 (2a \cos \alpha)/2 = \pi a^3 \cos \alpha$ and $\%V_f \simeq 100\, a \cos \alpha$. Before relaxation, an average medium-to-dense dry soil in its confined condition would probably have an effective friction angle of 35°, and when relaxed and saturated its slope angle would be about 25°. Particular friction angles would be assessed from the ground investigation standard penetration test results (Peck *et al.*, 1953), and from Hough (1957)—see also Lambe and Whitman (1969, Table 11.2, p. 149).

If there is a bead on the shield there will be continuous slumping of the soil on to the shield and tail extrados. V_b will then be equal to $\pi(2ab + b^2) \simeq 2\pi ab$ for a 360° bead and notionally half that figure for a 180° bead. The respective values of $\%V_b$ are approximately $200b/a$ and $100b/a$.

In a similar manner the postshield loss V_u would be approximately $\pi(a^2 - a_1^2)$, where a_1 is the external radius of the lining, and any bead thickness is ignored. $\%V_u$ is approximately $100(1 - (a_1/a)^2)$. During tunnelling for the

---

[†] In the London Transport tunnel described by Attewell and Farmer (1974a) measurements of *soil* movement suggested a slight inward lateral movement and an accompanying upward vertical movement above the crown after contact-grout set.

Washington DC Metro in the USA an expansion apparatus was designed partially to enlarge the lining during the shove in order to minimize these particular losses. Deep settlement monitors registered slight upward displacements of the ground, but it is thought unlikely that this particular facility will mitigate ground settlements and settlement distributions to any real degree. An alternative system of contact grouting over the lining when it was in the tail of the shield was developed, also for the Washington DC Metro (F2a section). After erection of a ring of steel liner plates inside the tail of a modified Robbins shield, the void between the lining and inside surface of the tailskin was packed with 100 mm diameter tubes of dense foam plastic that were held in place by the action of a thrust ring. A sand–flyash–cement grout was then pumped into the void, and it set to the consistency of a loose silty-clayey sand. As the shield shoved forward the volume of lost ground per unit distance advance was, ideally, restricted to the cross-sectional area of the tailskin.

As in the case of cohesive soils, there is currently no satisfactory method of predicting any further losses that occur over the grouted lengths of tunnel.

2.1.3 Total volume loss

Clearly, the total volume loss V_t at the tunnel will be compounded from the individual losses in the case of bolted and grouted segmental linings:

$$V_t = V_f + \underbrace{V_b + V_p + V_y}_{\text{over the shield}} + \underbrace{V_u + V_g}_{\text{behind the shield}} \quad (2.19)$$

In metric units, the volume loss values are expressed as cubic metres per metre tunnel advance. The following example shows rather more clearly how losses at the tunnel may be estimated by calculation, given information on soil relaxation rates.

Example 2.1 Analysis of ground losses in a clay soil

The following example analysis relates to a tunnel in north-east England where the face passed through both laminated clay and stony clay. Graphical information on intrusion is given in Figure 2.4a.
Tunnel depth to axis level, $z_0 = 7.5$ m.
Tunnel external diameter, $2R = 2.024$ m (hand excavation within a shield; free air).
Shield outside diameter, $2a = 2001.4$ mm.
Bead on shield (360°), thickness $(b) = 10$ mm; length $= 230$ mm.
Length of shield $= 1926$ mm. Length of tail $= 930$ mm.
Length of shield and tail $= 2856$ mm.
Length of shield and tail minus bead length, $l_s = 2626$ mm.
Rate of tunnel advance, $l' = 0.182$ m/h (on working days)
$\qquad\qquad l' = 0.113$ m/h (overall).

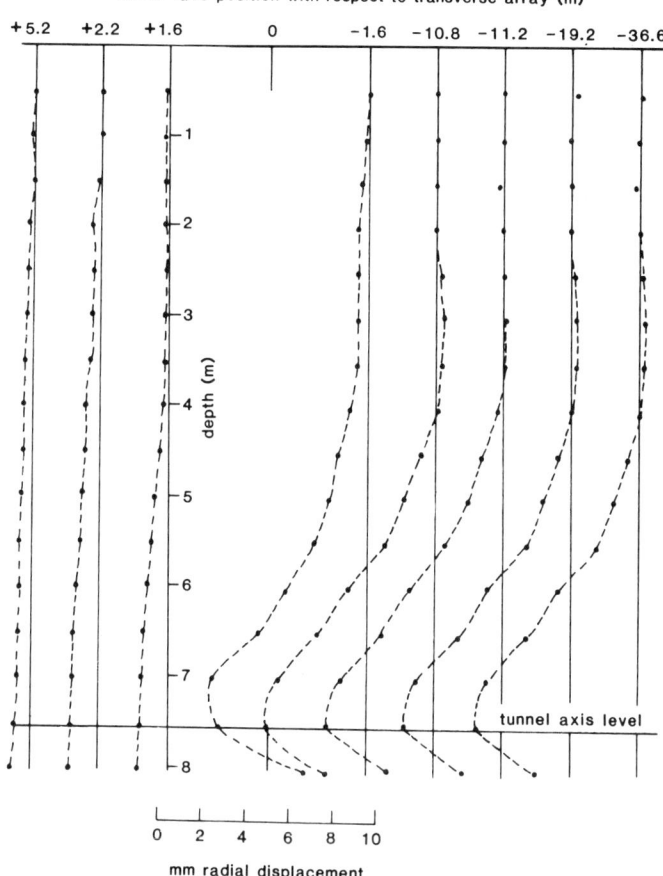

Figure 2.4(a) Radial ground deflection towards an approaching and passing tunnel face as monitored by inclinometer in borehole 4 of Figure 2.39. This borehole is slightly offset from the tunnel sidewall and so does not experience the maximum radial inward movement of the ground.

Rate of soil intrusion, $m' = 0.221$ mm/h (laminated clay).
$m' = 0.0134$ mm/h (stony clay).

1. *Using the overall rate of tunnel advance (laminated clay)*

Face take

$$V_f = \frac{\pi}{4}(2001.4)^2 k_1 \frac{0.221}{113} \text{ for laminated clay}$$

(see equation (2.3))

$= 0.00308 \text{ m}^3/\text{m}$ for k_1 assumed to be 50 per cent.

Shield/tail radial take

$$V_b = \pi \times 2626\,(2001.4 + 20)\,k_2 \frac{0.221}{113}$$

(see equation (2.13))

$$= 0.03261\,k_2 \mathrm{m^3/m}.$$

(Check for k_2. $m' = 0.221$ mm/h. Therefore, bead thickness, $b\,(= 10\,\mathrm{mm})$, will be covered by soil intrusion, m, in $10/0.221 = 45.2489$ h. Since tunnel advance rate $l' = 0.113$ m/h, length before bead closure $= 45.2489 \times 0.113\,\mathrm{m} = 5.1131\,\mathrm{m}$. This length exceeds l_s, and so $k_2 = 1$.)

Loss after lining erection

$$V_u = \pi \times 1800\,(2001.4 + 20)\frac{0.221}{113}$$

(see equation (2.17))

$$= 0.02236\,\mathrm{m^3/m}.$$

(This allows for three ungrouted concrete segmental rings, each ring width 0.61 m, but with the forward ring just resting over the end of the shield tail in order to assist shield stability.)

The measured maximum settlement was 7.86 mm. After plotting the transverse distribution of settlement as measured at the several stations (see Figure 2.39) the area under the settlement curve was calculated by adding 1 mm squares on a graph paper underlay. This area was $0.0774\,\mathrm{m^2}$. An error curve was then fitted to the measured curve (Figure 2.4b) using the fact that $y = i = 3.9$ m, where $w = 0.606 w_{\max}$. Thus, given that $V_s = \sqrt{2\pi} \times i \times w_{\max}$, then $V_s = 0.0768\,\mathrm{m^3/m}$. The difference between the two areas is less than 1 per cent.

Ignoring V_p and V_y, since in the former case there was no evidence of the shield having to be driven with a 'look-up' attitude in the stiff clay soil to counteract any tendency to dive, and in the latter case there was no evidence of any cutting or shield ellipticity, then rewriting equation (2.19),

$$V_t \simeq V_s = V_f + V_b + V_u + V_g,$$

it is possible to determine the relative contributions of the individual losses to the total loss:

Face take,

$$V_f = 4\% V_s,$$

Shield take,

$$V_b = 42.44\% V_s,$$

Postshield pregrout take,

$$V_u = 29.10\% V_s,$$

Postgrout take,

$$V_g = 24.45\% V_s \ (= 0.0188 \text{ m}^3/\text{m}) \text{ by difference.}$$

Inward ground movement radially over shield and tail

$$= l_s \frac{m'}{l'} = 5.136 \text{ mm}.$$

Inward ground movement radially over ungrouted rings

$$= l_u \frac{m'}{l'} = 3.250 \text{ mm}.$$

Inward ground movement at the face

$$= \frac{V_f}{2\pi a} = 0.490 \text{ mm (using equation (2.5))}.$$

Radial movements of the clay soil around the tunnel were measured by deep settlement ring probes (crown) and inclinometers (axis) (for the latter, see Figure 2.4a) and produced an average value for radial displacement (m) of 12.3 mm. Thus, summing the two radial movements above (5.136 mm + 3.520 mm) and subtracting from 12.3 mm gives the radial inward movement that can be presumed to have taken place over the grouted length of tunnel. Thus, ground movement on to the grout is 3.64 mm (approximately). This represents 29.6 per cent of the total average movement and, for an average 44 mm thickness of contact grout, represents a grout compression of about 8.27 per cent over the setting period. The inward ground movement at the face proper, estimated above at about 0.5 mm, is very small, but is confirmed by inclinometer tube deflections at face level when intercepted by the face (Attewell and Farmer, 1973).

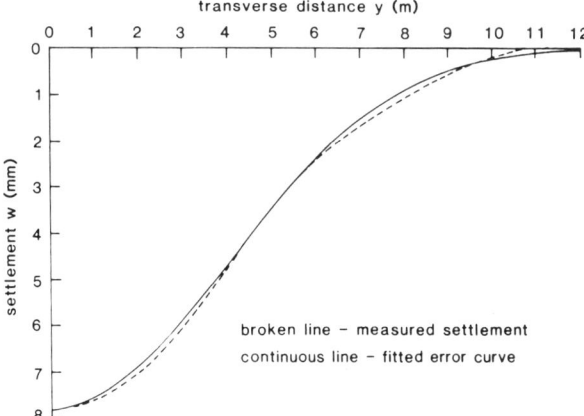

Figure 2.4(b) Settlement distribution in Example 2.1.

2. *Using the weekday rate of tunnel advance (laminated clay)*

Face take

$$V_f = 0.00191 \text{ m}^3/\text{m} = 2.49\% V_t.$$

Shield plus tail take

$$V_s = 0.02025 \text{ m}^3/\text{m} = 26.35\% V_t.$$

Postshield pregrout take

$$V_u = 0.01388 \text{ m}^3/\text{m} = 18.06\% V_t.$$

Ground movement on to the grout

$\simeq 12.3 - (3.189 + 2.186) \simeq 6.92$ mm = 56 per cent of the total average movement and represents a grout compression of about 16 per cent.

3. *Assessment for weekends (laminated clay)*

During stoppage periods only face take and shield/tail take continue, since all the rings are grouted up and it is convenient to assume that grout movement has ceased.

At the start of the weekend the 10 mm thick bead will be fully open 230 mm from the front of the shield but will have closed to a gap of only $(b - l_s(m'/l')) = (10 - 2626 \times (0.221/182)) = 6.811$ mm at the end of the shield tail. With the measured clay soil intrusion rate of 0.221 mm/h, the 10 mm annular gap will close completely over a period of $10/0.221 = 45.25$ h, which covers the weekend break.

Shield/tail take over the weekend is therefore

$$V_{se} = \pi \times 2626 \times 2021.4 \times \left(\frac{0.010 + 0.006811}{2}\right) = 0.14008 \text{ m}^3.$$

Face take over the weekend is

$$V_{fe} = \frac{\pi}{4}(2001.4)^2 k_1 \times 0.221 \times 48 = 0.01669 \text{ m}^3 \quad \text{for } k_1 = 0.5.$$

Thus, *total take over a 48-hour weekend* is $V_{se} + V_{fe} = 0.15677 \text{ m}^3$.

4. A more realistic model uses the real (working days) rate of advance plus the calculated weekend losses, although the actual weekend losses may be slightly lower than those calculated because the face would be partly boxed up during the stand period. The boxing would comprise waling boards and battens thrust against the face through the action of the shield jacks (see the four face jacks on the shield in Figure 1.1).

Face take over week

$$= V_f + V_{fe} = 0.00191 + \frac{0.01669}{7 \times 24 \times 0.113}$$

$$= 0.00279 \text{ m}^3/\text{m} = 3.63\% V_t.$$

(Note that the weekend loss must be expressed per unit advance distance. This distance is the overall advance rate (0.113 m/hour) multiplied by the number of hours (7 × 24) in a week.)

Shield/tail take over week

$$= V_s + V_{se} = 0.02025 + \frac{0.01669}{7 \times 24 \times 0.113}$$

$$= 0.02112 \text{ m}^3/\text{m} = 27.50\% V_t.$$

Postshield pregrout take over week

$$= V_u = 0.01388 \text{ m}^3/\text{m} \text{ (as in 2 above)}$$

$$= 18.06\% V_t.$$

Postgrout take

$$V_g = 0.03904 \text{ m}^3/\text{m} \text{ (by difference)}$$

$$= 50.81\% V_t.$$

Assessing the ground movement radially on to the contact grout (m_g) by proportion:

$$m_g = 12.3 \left(\frac{V_g}{V_g + V_s + V_{se} + V_u} \right) = 12.3 \left(\frac{0.03904}{0.03904 + 0.02113 + 0.01388} \right)$$

$$= 6.48 \text{ mm}.$$

Thus, by this more realistic calculation the grout is compressed 6.48 mm radially and by 14.73 per cent of its (assumed) original thickness of 44 mm.

Similar calculations could be performed for the other (stony) clay through which the tunnel has passed.

Some rather obvious comments on design stem from these analyses.

Use a shield having as low a length-to-diameter ratio as possible.

If a long shield must be used, then try to design it for articulation so that steering is eased and ovality problems arising from driving round curves are reduced.

Design for a tight diametral build of lining within a tailskin of just sufficient length. Consider using an improved tail seal and place the contact grout as the shield is shoved forward. There must be suitable provision for preventing any build-up of hardened grout on the shield/tail extrados.

Plan the deployment of labour, the soils excavation and support erection as a systems problem so that the only constraints on a high rate of advance are unexpected ground conditions, which themselves can be reduced by good prior ground investigation.

2.1.4 Mechanics of soil deformation

The application of soil mechanics principles to the deformation of soil around an advancing tunnel is a topic too lengthy to be covered in this book and, at present, insufficiently developed for all but the most conceptually simple geometries and homogeneous soils having established predictable behaviours. Atkinson and Mair (1981) have examined the problem in terms of critical-state soil mechanics, but primarily from the standpoint of the magnitude of internal support needed in the tunnel to achieve stability and avoid collapse, and the use of dimensionless stability numbers (three parameters T_c, T_γ and T_s analogous to bearing capacity factors N_c, N_γ, N_q).

Atkinson and Mair consider two general soil types in terms of their past loading history. Overconsolidated soils that are not in a collapse state at the tunnel deform, under excavation relaxation conditions, with increasing shear stress towards but inside the ultimate failure-state boundary surface. They can be analysed as for elastic deformation, Hooke's law applying to the stress–strain relations. Effective moduli E', ν' define any drained behaviour and 'total' moduli E_u, ν_u define undrained behaviour. The ultimate deformation state of a normally consolidated soil maps out the critical-state boundary surface for which the stress–strain relations need to be formulated via some elasto-plastic law.

The (directional) permeability of the soil and the rate of tunnel advance (more specifically, the rate at which a sealed lining is installed) together determine the degree of allowable soil drainage. The rate of change of total stress due to excavation could be fast compared to the rate of excess pore-pressure reduction via a drainage boundary or boundaries. At the limit, the rate of the former relative to the latter could be such as to virtually inhibit drainage, in which case there would be no volumetric strain, and so no consolidation during actual excavation and construction, but there would be undrained shear strain. The state of radial and tangential total stress for elastic conditions and a circular excavation can be estimated from the equations originally formulated by Kirsch (1898)—see also Savin (1961).

Those elements of soil which are relaxed by excavation, but are not excavated, and which consequently dilate, undergo changes in pore pressure which are usually conducive to consolidation. This is the situation that applies around the tunnel as the shield advances. Even if the total stresses, comprising radial, tangential and centre line axis stresses, remain constant behind the face of the shield and before contact grout offers resistance, changes in pore pressure cause changes in effective stress which in turn lead to strains, which

ESTIMATION OF GROUND MOVEMENTS

are due to consolidation and occur without changes in external loading. Consolidation is related to radial drainage, the seepage itself creating additional stress that must be supported.

2.2 Reduction of ground loss

In addition to the build-one-ring, grout-one-ring construction method, the ground losses at the tunnel can be reduced by adopting one or more of the following practices.

Smooth shield and tail extrados. Increasing, unquantifiable losses may occur if grout seeps behind the tail and shield, sticks to the rough metal surface, and sets. A build-up of set grout in this manner may cause weak soil around the shield to be disturbed and 'ploughed' as the shield is advanced. This problem occurred on a Tyneside (UK) sewerage scheme contract when tunnelling under compressed air in a soft, silty alluvial clay (details of the work are described in Attewell (1978*b*) and Attewell *et al.* (1978)). As a result of this experience a specially-prepared Teflon-coated shield and tail extrados was

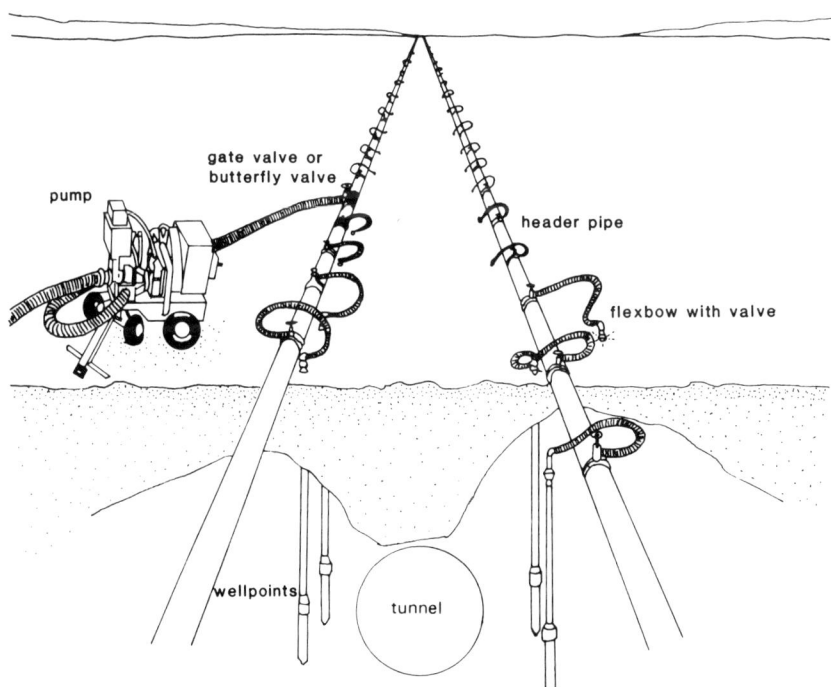

Figure 2.5 A wellpoint dewatering system.

specified by the same water authority for a later urban tunnelling contract (see Attewell, 1981, for details of this contract), the shield being manufactured by Wade and Hobson of South Yorkshire, England.

Wellpoint dewatering. For shallow tunnels in dominantly granular soils, individual wells, generally 40 mm or 50 mm diameter, are sunk to tunnel invert level, secured to the end of riser pipes, and washed into the soil at close centres using high-pressure jetted water (see Figure 2.5). These wells are set to one side of, or both sides of, the proposed tunnel and connected via a suction header pipe to a centrifugal pump fitted with an air exhauster system capable of handling large quantities of air. Pumping depresses the groundwater table via a series of overlapping cones of depression and allows excavation to take place in sensibly dry working conditions. If the groundwater table is required to be lowered by more than about 7 m, a tube well system can be used. This consists of a perforated well liner surrounded by a suitable sand filter, with an electro-submersible pump at the bottom of the hole. Suitable filter design to reduce loss of fines from the surrounding soil is important, otherwise attributable settlement might be induced.

Drawdown itself could lead to differential settlements at the foundations of adjacent property. In the ancient centres of some cities (Amsterdam, for example, where the groundwater table may be only one or two metres below ground surface), the possibility of negative skin friction effects and pile drawdown, together with the rotting of timber piles when exposed to the atmosphere, must be considered. One method of partially overcoming the problem is to drill recharge holes between the tunnel and the property and to circulate the abstracted water through them. The foundation support capacity of the water to the sides of the tunnel is thereby maintained by means of a steepened drawdown surface.

If the tunnel is too deep for complete dewatering to be achieved economically, one option is to lower the water table partially and to tunnel under the reduced head with a tolerable internal compressed-air pressure or with a slurry/earth balance shield.

Neglecting the effects of well overlaps, assuming soil homogeneity and permeability isotropy, the standard Theim equation can be used to establish water quantities requiring to be pumped in order to achieve the requisite drawdown. Letting H be the height of the water table above the base of the soil stratum to be dewatered, h the height from the base of the stratum to the point of maximum drawdown at the well, b the thickness of the aquifer, r_0 the radius of the drawdown depression, r_w the radius of the well, k the ground permeability (assumed to be isotropic), and Q the flowrate in units of volume per unit time from the well, then the radial water flow in a confined aquifer must be

$$Q_c = 2\pi k b(H - h)/2.30 \log(r_0/r_w) \qquad (2.20)$$

and, for an unconfined aquifer,

$$Q_u = \pi k(H^2 - h^2)/2.30 \log(r_0 - r_w) \qquad (2.21)$$

in order to achieve the required drawdown $(H - h)$ at the well. Reference should be made to standard textbooks for further reading on the theory of this subject, but it is preferable to seek practical advice from specialist groundwater-lowering contractors since the ideal ground conditions required to satisfy the theory will rarely pertain in practice.

Grout injection. This again is a subject on which a specialist contractor's advice must be sought, and reference may also be made to the numerous papers, technical articles and books on the topic. Some grouts act dominantly as 'water stops' to reduce the porosity (and permeability) of the soil before the use of, or as a substitute for, compressed air, the stabilized ground being excavated without too much difficulty. Other grouts greatly enhance the compressive strength of the ground, causing it to require more excavation effort. Some grouts may be used to densify loose non-cohesive soils and others may be used to fill known voids—for example, around building foundations and pipes that may have settled. Sometimes, as an alternative to the use of slurry walls for isolating tunnels from adjacent properties, it may be necessary to grout soil beneath a building in order to provide added, 'underpinned' support before tunnelling. Grouting can be from the ground surface ahead of the tunnel, or, more expensively, from a pilot tunnel or from (and ahead of) the main tunnel. Compaction grouting has been used with some success in tunnelling where large (say, greater than 12 mm) structural settlements have been anticipated in built-up areas. The aim is, after each shove of the shield, to place a low-slump cement through holes drilled from ground surface and thereby replace the volume of ground lost around the tunnel during its advance. At suitable pressures the grout tends to densify the soils surrounding the grout bulb, heaving the overlying soils and pushing down on to the tunnel lining. The cost of this operation and the possible difficulties of access for drilling have to be balanced against the projected savings in building reinstatement costs.

The types of injection grouts range from particulate (usually) cementitious grouts through silicate grouts to organic polymer grouts, the choice being determined by cost and by the particle size of the soil into which the injection is to be made. In the case of the cheaper particulate grouts, the maximum size d_p of particle capable of passing in suspension through the minimum-pore cross-section of a soil will have a diameter in the region of $0.15d_0$, where d_0 is the diameter of the (assumed) spherical soil particles. In practice, the maximum diameter d_p of injection particle may be taken as $0.1d_0$, where d_0 is equated to the D_{10} soil grain size in a mass of irregular-sized particles. D_{10} is the size below which 10 per cent of the particles are finer. Ordinary portland cements normally contain particles up to $100\,\mu$m in size, and this restricts the use of these cement grouts to medium sands and coarser. High-early-strength cements having a maximum grain size of 20 μm could be used to grout all sands with a permeability as low as 10^{-4} m/s. Prediction of groutability from *in-situ* permeability tests or vice versa may be made on the basis of the Hazen formula

$$k(\text{cm/s}) = D_{10}{}^2(\text{mm}^2).$$

Freezing. This temporary ground-improvement option is suitable for a wide range of water-bearing soil types, including mixed ground where grouting may be inapplicable or where compressed-air pressures would be too high. Freezing is normally done from the ground surface ahead of the tunnel face. Basically the choice lies between relatively slow freezing, commonly using an indirect circulatory brine system, and a direct-injection fast freeze down to about −25°C using liquid nitrogen. The former is usually cheaper, but requires adequate site space on which to house the bulkier freeze plant. Such space is not always available in city centres. Liquid-nitrogen running costs are usually higher, but the system demands less space. All freezing systems disrupt traffic flows because when (as is often the case) a tunnel follows the line of a road, that road must be closed during the freezing and tunnelling operations. During tunnelling under property it may not be economically possible to freeze at all. Liquid nitrogen is of greatest use as a standby facility when local water ingresses might occur unexpectedly. The ground can then be quickly frozen, the tunnel can proceed rapidly through the bad area, and the ground then left to thaw out. A major disadvantage of freezing is that it is very difficult to control and estimate the position of the ice front, with the possibility of affecting buried pipes and creating ground heave beneath building foundations. Services can, however, be isolated from the effects of freezing by excavating the surrounding soil. Furthermore, the frozen ground creates unpleasant environmental conditions in which the tunnellers must work, and there can be a danger if the freeze tubes are broken inadvertently when intercepted at the tunnel face. Excavation will usually be by hand-held pneumatic spades and jigger picks, but explosives may sometimes be needed. The overall effectiveness of a freezing operation will be reduced if water is actually flowing in more permeable soils that are to be frozen, since some heat is generated by the motion.

Compressed air. The introduction of an air pressure above atmospheric pressure into a tunnel behind an airlock, while technically attractive and much used, is physiologically undesirable. There are stringent controls on the use of compressed air (in the UK via the use of the 'Blackpool Tables'[†]) and

[†]Construction Industry Research and Information Association (1973). *A Medical Code of Practice for Work in Compressed Air.* CIRIA Report 44, London.

prescribed requirements for medical supervision when gauge pressures exceed one atmosphere.

Compressed air acts in three ways:

By balancing the external head of water (the compressed air going into solution with the groundwater over a diffuse zone) and resisting its intrusion into the tunnel.

By providing a direct reaction to the field forces attempting to propel the soil particles themselves into the tunnel.

By drying out a 'skin' of soil at the tunnel face and, by so increasing its effective strength, enabling it more easily to resist the passive pressures in the ground.

Compressed air is used only when the effects of groundwater would otherwise prevent the tunnel being advanced. Because of its low soil permeability, a high head of water in a clay is unlikely to create tunnelling problems. Within a glacial till any granular inclusions may contain water at artesian or sub-artesian pressures and could severely affect tunnel progress. If the site investigation evidence is not reasonably conclusive about the need or otherwise for compressed air in a tunnel, then it is prudent to make provision for it in a bill of quantities at a (1983) installation cost of circa £25 000 for a facility up to 1 bar gauge pressure and a 20 per cent increase in cost above that. Installation of a compressed-air facility at a later stage after tunnelling has begun would cause progress to be disrupted and would be more expensive.

Tunnellers have used compressed air for more than a century in order to control water intrusions from silty/sandy/gravelly soils. Air and water permeability in such soils depends very much upon the volume of fine material present in the deposit. If there is a very low silt content the water may be controlled only at the expense of unacceptably high air losses. Furthermore, compressed air depends for its effectiveness upon continuity of the deposit to be locally dewatered. If water is contained in isolated pockets or lenses of silt or sand within a dominantly clay soil stratum, for example in tills, at artesian or sub-artesian pressures, then the pocket becomes uniformly pressurized because the water cannot be driven back, and simply seeps into the tunnel. On the other hand, there may be continuity of such lenses beyond the local tunnelling area. Compressed air may then drive back the water over quite large distances, seeking out loss paths through, for example, foundations and any voids between building services and the soil in urban areas.

Experience is needed to judge the size of plant necessary to accommodate air losses. The adopted pressure (in kN/m^2, equal to 9.81 times the water head in metres) will usually be chosen to balance at tunnel axis level. However, the most sensible criterion for specifying compressed air pressure is that the pressure should be low enough to permit a water seepage of insufficient magnitude at the face to delay progress. These seepages may, however, cause some groundwater lowering and induce early consolidation settlement (see section 2.8) to be added to ground-loss settlement. Thus, decisions as to the compressed-air balance pressure should also be based on a review of possible consequential settlements and the effect of those settlements upon in-ground (Chapter 3) and surface (Chapter 4) structures.

In all cases the thickness of cover above the tunnel crown should be checked to make sure that there is no possibility of excessive ground heave from tunnel air pressure. Cover stability obviously demands special attention during subaqueous tunnelling.

Enclosed shields. Of the two principal types of enclosed shield adopted for weak water-bearing ground, one is termed 'active' and the other 'passive'. In soft clay soil or non-cohesive running ground tunnelled below the water table without the use of compressed air for temporary support, or some other form of ground improvement, a passive shield may act as a bulkhead to seal off the face but with provision for retractable hydraulic breasting doors the opening of which will allow a controlled inward movement of the ground as the shield is thrust forward. A small-diameter shield of this enclosed type is shown in Figure 2.6.

Although the rate of advance depends on the total area of opening, the thrust for advance must match the intrusion rate, otherwise grossly excessive ground losses may occur. In very mobile ground problems could arise at the

Figure 2.6 Small-diameter, fully welded shield with closed front for working in loose sand conditions. Photograph courtesy Stelmo Ltd.

seal between the rear of the shield and the lining, but the system has a major advantage in that the tunnelling crew is not exposed to the disadvantages of compressed-air working and decompressions.

Large shields comprise several compartments, each having its own door through which a miner can dig out the soil. Such shields may also incorporate vacuum ejectors for dewatering immediately ahead of the shield, and this system can be used in conjunction with deep-well dewatering from ground surface.

An enclosed shield of this type is heavy, and the additional steel at the front end pulls the centre of gravity of the shield forward. In very weak ground the shield may then tend to dive more readily off-grade, particularly if tunnelling down an incline. An upward pitch will then be necessary to counteract the dive, so producing overcut volume losses V_p. On the other hand, the additional forward weight of the shield could be offset somewhat because a shield hood might not then be needed.

Breasting doors may be designed to cover only the upper half of a tunnel face, the shield not being compartmentalized. Cording et al. (1976) have described Washington DC Metro F2a-L Route shields having six doors across the upper half of the shield (see Figure 2.7). The doors were hinged to the shield circumference and were used at the discretion of the shield driver during shield shoves in order to admit soil into the heading. However, they were not particularly effective in controlling the ravelling-to-running sand and gravel encountered in the upper half of the heading throughout most of the L Route.

This passive system has been largely superseded by the more recent bentonite-shield, slurry-shield and earth-pressure balance tunnelling systems in which the face is again sealed but in which ground intrusion is resisted on the

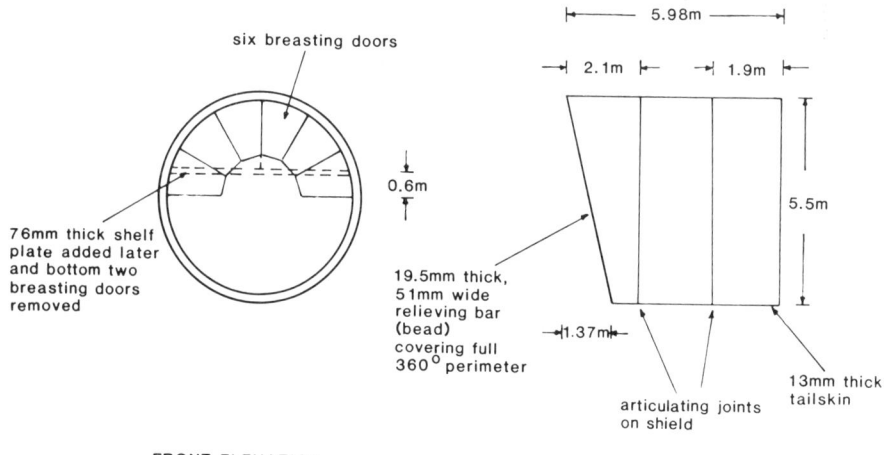

Figure 2.7 Washington DC Metro L Route shield (after Cording et al., 1976).

face side of the shield by the injection of a pressurized bentonite–water slurry or by a pressurized slurry of the soil itself. The bentonite-shield system was invented in Britain but has been exploited in Germany (Wayss and Freytag Hydroshield system) and, more particularly, as slurry-shield tunnelling in Japan and by the Japanese in other countries. Particular features of the system include the design of the rotating cruciform arms to mobilize the soil and slurry it up at the face, the design of the sealing at the tail of the shield, and the mechanical design for spoil removal at the face of the shield. The handling of boulders created severe problems for the British experimental bentonite shields. Recent Japanese shields, examples of which were used on 1984 sewer tunnel contracts on the south bank of the River Tyne in northern England and in Oldham, Lancashire (Iseki Poly-Tech Inc. 'Crunchingmole') are claimed to have a facility for crushing boulders of size up to 250 mm diameter. They are operated remotely from ground surface. An example of one such shield is shown in Figure 2.8.

There is currently little measurement evidence to confirm the likely effect that these latter shields permit very low ground losses. They will be increasingly adopted for large-diameter, relatively shallow tunnels in weak water-bearing ground where compressed-air tunnelling (with accompanying physiological and pressure-balancing problems) would otherwise be specified. They will also offer particular attractions for urban tunnelling where small-diameter man-entry- and non-man-entry- size sewers must be built.

Figure 2.8 Slurry shield. Photograph courtesy Iseki Poly-Tech Inc.

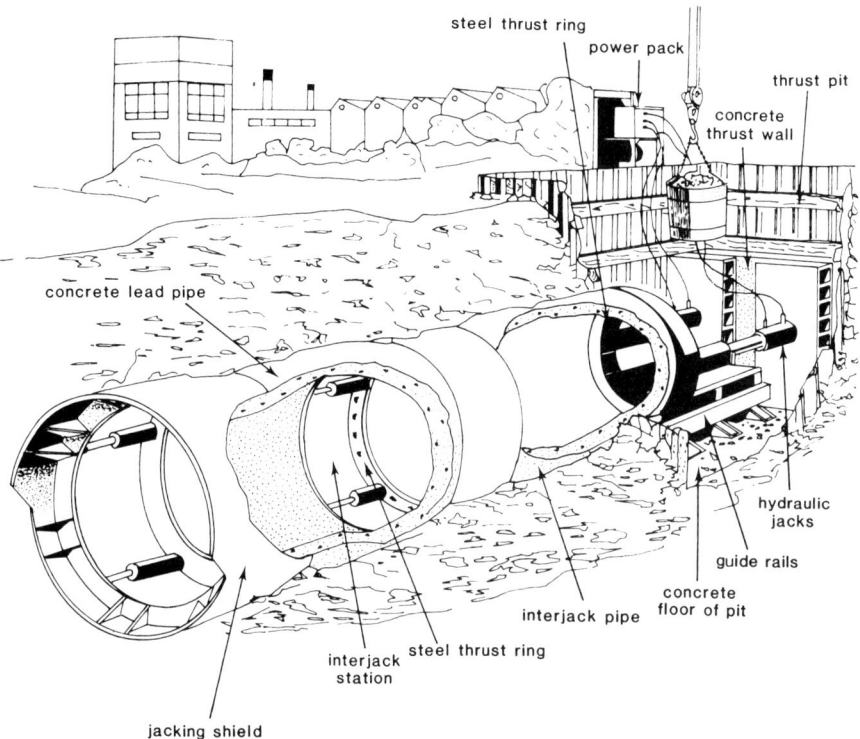

Figure 2.9 Pipe jack system for concrete pipes.

Pipe-jacking. This is a method of installing (usually circular) pipes of about 900 mm diameter and above by driving a line of them through the ground by hydraulic rams from a prepared jacking pit (Craig, 1983). The minimum practicable size for man-entry during construction is 900 mm. Excavation is carried out, usually by hand, at the forward shield end of the pipe as the pipeline is thrust from the jacking pit. After a full pipelength is pushed into newly excavated ground a further pipe is put in the pit and the process repeated.

A lubricant, such as bentonite, is sometimes used on the outside of the pipe to reduce ground/pipe friction, which absorbs a large proportion of the jacking force. For greater pipe-jack distances, intermediate jacking stations can be used to reduce the jacking force at the pit (see Figure 2.9), which usually varies from about 1500 kN to 4000 kN. Problems related to maintaining line and level have been much reduced in recent years, and pipe-jacking may be performed around curves. If there is careful control on excavation and minimal overbreak at the face, then the radial ground losses should be less with this form of construction than with 'Mini Tunnelling'[†] where three-segment

[†]Mini Tunnel is the trade mark of William F. Rees Ltd.

lining rings are erected conventionally within the protection of a shield tail, and injected pea gravel or Lytag behind the lining is not usually grouted up before the tunnel is completed.

Pipe-jacking was introduced into the UK in the late 1950s and is gathering momentum as a tunnelling technique. Slurry systems are now also being used with jacked pipes which can have diameters below 900 mm. However, the system currently most common involves hand excavation under the protection of a shield. The advantages of pipe-jacking, which provides a reinforced precast lining from 900 mm diameter incorporating flexible joints, are claimed to be strength, watertightness, fewer joints, reduced ground settlements, and elimination of secondary lining. Jacking lengths are usually restricted to about 400 m, unless the ground is homogeneous and has good self-support characteristics. *Tunnel jacking* is a similar construction technique, but related to larger-diameter excavations. Smaller-diameter pipes (non-man-entry) are usually installed by thrust bore and auger methods.

In the recently introduced Uni-Tunnel system (Richardson and Scruby, 1981) rubber bladders are inserted between the pipes. Pipe progress is achieved by inflating the bladders sequentially so that every third pipe is driven forward simultaneously. The frictional resistance of the two following pipes acts against the ground in reaction to the thrust, and so contrasts with the conventional pipe-jacking system where frictional resistance increases with drive length. With this system, in order for the full frictional resistance to be mobilized by the two thrust pipes of each set of three, it is specially important that the excavated diameter be the same as the pipe external diameter. Radial ground losses can thus be very much reduced.

Some difficult decisions often have to be made as to whether the cost of protective measures for adjacent proporties should be borne by the project promoter, in the expectation that problems will arise. If the perception of the likelihood of problems developing is uncertain, then the promoter may decide instead that it is more cost-effective to plan for the temporary relocation of occupants, the reinstatement of buildings, compensation for loss of business, and so on. Costs in this area, as indeed in engineering and life in general, depend on a personal, experienced view of the degree of risk involved in the operation.

2.3 Transmission of ground losses to ground surface

This subject is discussed in Attewell (1978*a*, pp. 855–867), reference to which is recommended. Details of measurement case histories are given for single tunnels, for adjacent parallel tunnels, and for superadjacent parallel tunnels. The following conclusions may be drawn:

(i) For most single tunnels in *firm-to-stiff clay soils* the volume V_s of the settlement depression created at ground surface is approximately equal to the volume V_t of ground lost at the tunnel. In practice, the time lapse between creation of the loss source and its contribution to ground-surface settlement may be small.

(ii) For single tunnels in *granular non-cohesive soil*, dilation V_d may occur through arching above the tunnel crown if $z_0 \geqslant 3R$ (Hansmire, 1975). This arching absorbs some of the ground loss and reduces the surface settlement volume below that predicted from any ground-loss estimates. Soil above the arch and up to ground surface must then be supported as a surcharge to the sides of the tunnel. In a loose compressible soil the surface settlement trough may thereby be widened as a result of such compression V_c. If the tunnel is of large diameter and/or near ground surface the arch may break through to the surface, subsequent losses at the tunnel becoming volumetric settlement losses. Although the general equation is $V_s = V_t - V_d + V_c$, V_d and V_c could be ignored for any estimate of V_s, remembering that in dense sands and gravels, the dilation V_d could reduce V_s and in loose disturbed sands some compression could develop above and to the sides of the tunnel to increase V_s. If large displacements are permitted to develop at the tunnel face and crown, then in granular soils a funnel-type upward transmission movement, creating a narrow but deep surface settlement trough, can occur.

(iii) The greatest concentration of vertical movement occurs at the tunnel crown, and there have been measurements, usually by adopting deeply-emplaced magnetic settlement rings, to quantify it. Clearly, the intensity of vertical movement should decrease upwards from the crown to ground surface simply as a function of the increasing transverse spread of the disturbance. By adjusting the crown movement for this spread and then comparing the adjusted movement with the settlement actually measured, conclusions can be drawn as to whether the soil between tunnel crown and ground surface has compressed in volume or has dilated. This matter is discussed in Attewell (1978a, pp. 894–900, with particular reference to Table 6, on pp. 898–9 of that paper). It would be unwise in practice to offer any specific recommendations, since the behaviour of the ground depends also on the tunnelling and lining method.

(iv) It must usually be assumed that, as the ground movement is transferred from the tunnel source to the ground surface, the transverse settlement distribution as a function of depth $(z_0 - z)$ takes normal probability form. The transverse trough width increases and settlement w decreases as $\overline{z_0 - z}$ tends to z_0 (at ground surface). For buried pipelines and building foundations, the settlement trough parameters are those appropriate to depth z, where z is finite below ground level and is zero at ground surface, with the standard probability equations describing the form of ground movement distribution (equations (2.22) to (2.27)) and with an i-parameter being entered for depth z.

(v) For the style of vertical movement suggested in (iv) above to occur, the ground movement vectors would be directed towards the tunnel, although some temporary ground-surface heave may develop ahead of a conventional hand shield when it is jacked forward. Observations from compressed-air shield-driven tunnels in weakly cohesive soils confirm the compressions to the sides of the tunnel at axis level as noted above. Slurry/earth-pressure balance shields may also push the ground temporarily and locally away from the shield. Upward movements above the crown have also been reported by Attewell and Farmer (1974b) for free-air tunnelling in stiff overconsolidated London Clay and for compressed-air tunnelling in a soft silty clay at Stockton in north-east England (Hurrell, 1985).

(vi) Settlements caused by adjacent tunnels are discussed in Attewell (1978a, pp. 863–865). The most detailed measurement evidence is from Cording and Hansmire (1975) on the driving of Washington DC Metro twin tunnels in *medium dense silty sand and gravel, interbedded with sandy silty clays.* Settlement due to a second, later, but close tunnel produces a composite settlement trough which is asymmetrical with respect to shape and location above the two tunnels. The Washington measurements showed the additional asymmetrical settlement over the second tunnel to have been caused by additional deflection of the lining of the first tunnel, by compression of the abutment between the tunnels, and by compression of the previously-dilated region over the crown of the first tunnel. Thus, the volume component of the surface settlement trough attributable to the second tunnel exceeded the volume lost at that tunnel, the difference being termed the interference volume. The sequence of recompression of dilated non-cohesive soil in the arch over a first tunnel, further vertical squat of the primary support system, lateral outward deflection of the support system and further compression of intervening pillar soil depends on the stiffness of both the ground and the tunnel support system. If the first tunnel support system is very stiff, then the interference volume will be small. The final composite settlement trough is created by the superimposition of the two tunnel individual troughs and the summation of the settlement vectors. In many cases the final settlement trough width outside the centre lines of the two tunnels will be the same as a symmetrical half-width of trough above a single tunnel.

In the case of twin tunnels driven at the same level in clay soil there will be little or no dilation of the ground above the crown of the first tunnel and thus no recompression when the second tunnel is driven. The second tunnel may, in fact, dilate ground that had previously compressed above the first tunnel. There will be vertical compression at the sides of both tunnels and some lateral interference compression at both sides of the abutment between the tunnels. With clay soils, therefore, any final settlement trough asymmetry should be less than that associated with granular soils. An example of settlement trough superposition in clay soil is given in Attewell (1978a pp. 866, 867).

In assessing the effects of interaction between tunnels and, for example, at tunnel junctions, it will be realized that subsequent construction takes place in soil that may well have deformed down to its residual strength as a result of earlier disturbance. Later soil deformations and contributions to total settlement may therefore be greater.

2.4 Ground deformation and strain equations

The primary notation to be used in the ground deformation equations is listed below for convenience, but a comprehensive symbol list is included at the beginning of the book.

x,y,z are the cartesian coordinates of any point in the ground deformation field. Theoretically, $x = y = 0$ vertically above the tunnel face on the tunnel centre line. Positive $(+)x$ is ahead of the tunnel on the centre line, and horizontal. $\pm y$ is horizontal at right angles to x. Positive $(+)$ z is vertically downwards. This coordinate system is shown in Figure 2.2.

u,v,w are the ground displacements in the x,y,z directions, respectively. u and v are always towards the origin of the cartesian coordinate system. w (settlement) is always positive downwards.

$\varepsilon_x, \varepsilon_y, \varepsilon_z$ are the ground strains in the x,y,z directions, respectively. These strains can change from tensile (positive) to compressive (negative) depending upon position in the deformation field. Tensile ground strains are more likely to have a serious effect upon the brittle foundation of a building or upon a brittle pipe than are compressive ground strains.

γ_{xy} is the ground shear strain in a horizontal plane.

z_0 is the depth of the effective source of ground loss (taken as approximating to the tunnel axis).

R is the excavated radius of a tunnel with circular cross-section.

K_R is an empirically-determined constant.

n is the power of $z_0 - z$ to which i_x, i_y, i are proportional.

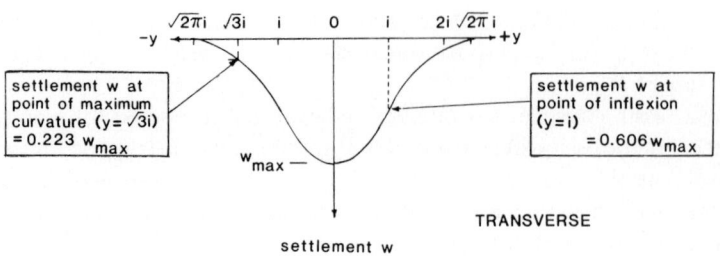

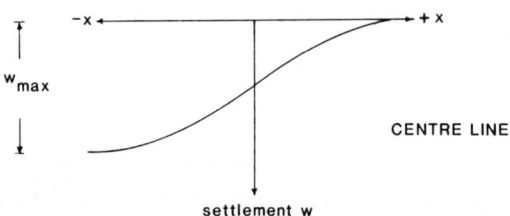

Figure 2.10 (a) Normal probability transverse and cumulative probability centre line settlement profiles (see Figure 2.2 for full impression of settlement trough).

V_s is the volume of the settlement trough per unit distance of tunnel advance, the settlement being attributable to ground losses and not incorporating any longer-term consolidation movement.

i is a parameter defining the form and span of the settlement trough, on the assumption that the semi-transverse (y-axis) settlement profile can be described by a normal probability equation (Schmidt, 1969; Peck, 1969; Attewell, 1978a). On the transverse settlement profile (Figure 2.10), i_y is half the distance between the points of inflexion (greatest slope) either side the tunnel centre line. On the centre line profile (Figure 2.10), i_x is the distance from the point of inflexion to the 15.9 per cent (of maximum) settlement point, and twice the distance from the point of inflexion to the 30.9 per cent settlement point. For the numerical examples discussed in the book, i_x is taken to be equal to i_y and written simply as i. Case history evidence on i_x and i_y is discussed in section 2.7.

x_i is the initial or tunnel start point ($y = 0$).

x_f is the face or final tunnel position ($y = 0$).

$$G(\alpha) = \frac{1}{\sqrt{2\pi}} \int_{-\infty}^{\infty} \exp\left[\frac{-\beta^2}{2}\right] d\beta$$

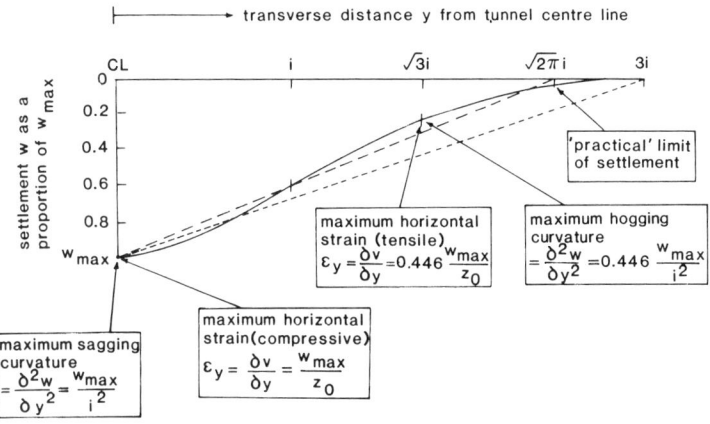

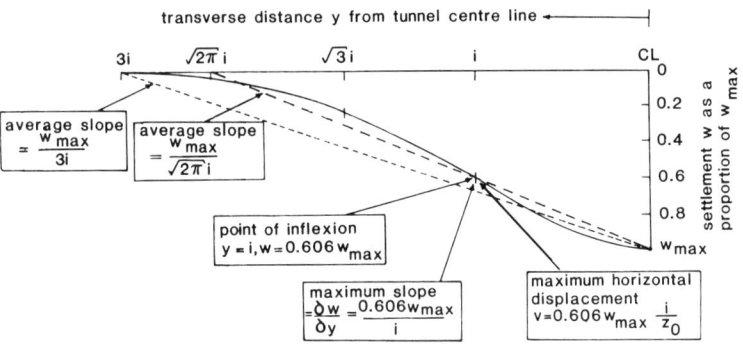

Figure 2.10(b) Transverse surface settlement trough, detail based on a normal probability curve fit to the trough profile (refer to Attewell and Woodman, 1982, for equation derivation).

and may be determined from, for example, a standard probability table such as Table 2.2. Note particularly that $G(0)$ (e.g. $x - x_f = 0$ directly above the tunnel face) gives a value of $1/2$ and $G(\infty)$ (e.g. $x - x_i \to \infty$ for substantial distances of tunnel advance from the face start position) gives a value of 1.

The equations are based on the assumption that the transverse ground settlement profile (yz-plane) is of normal probability, or Gaussian, form. Empirical evidence suggests that this is generally valid for many soils and substantially insensitive to the method of tunnelling, but not to the quality of workmanship in dominantly granular soils. Experienced judgement, however, must be applied to the application of the subsequent ground-movement equations to particular cases. Attempts have been made to match measured profiles with alternative mathematical expressions, but there seems now to be

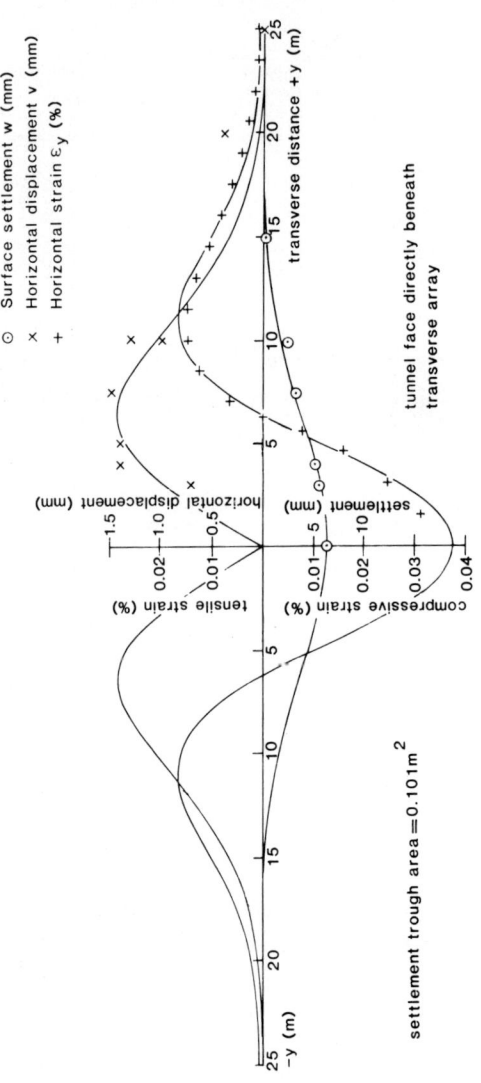

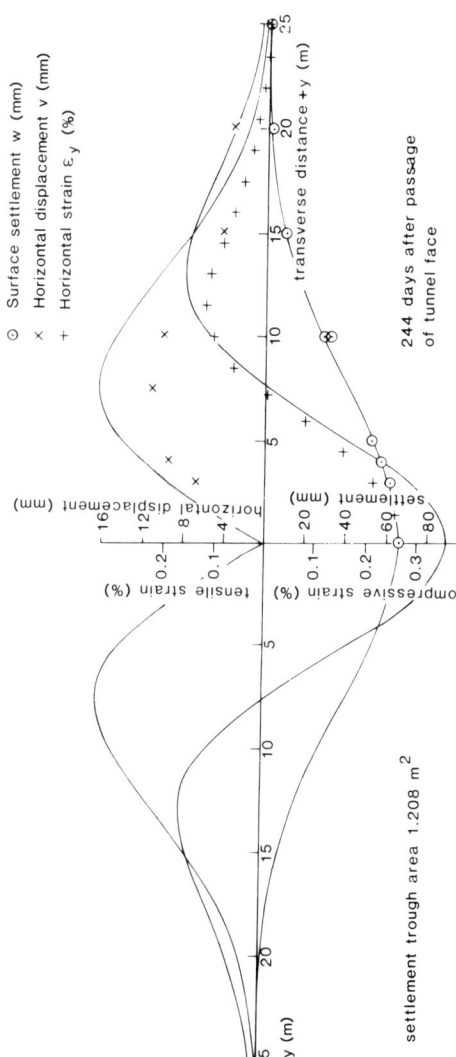

Figure 2.10(c) Illustrating the essential elements of transverse (y-axis) settlement (w), horizontal displacement (v) and horizontal strain (ε_y) distribution. These curves should be read in association with the information given in Figures 2.10(a) and 2.10(b). The *settlement curves* represent best-fit normal probability functions to actual measurement data. The displacement and strain curves are derived from the best-fit normal probability settlement curve and are not best fits to the measurement data. Information was derived from a tunnel of 2.45 m external diameter driven under compressed-air pressure at an axis depth of 13.375 m in silty alluvial clay (Sizer, 1976; Attewell *et al.*, 1978). These two graphs are the first and last members of a suite which demonstrates the effects of time and the removal of compressed air upon the ground movements.

Table 2.2 Numerical integration of the normal probability curve. Table of $G[(x - x_f)/i]$

$(x - x_f)/i$	0	1	2	3	4	5	6	7	8	9
0.0	.500	.504	.508	.512	.516	.520	.524	.528	.532	.536
0.1	.540	.544	.548	.552	.556	.560	.564	.567	.571	.575
0.2	.579	.583	.587	.591	.595	.599	.603	.606	.610	.614
0.3	.618	.622	.626	.629	.633	.637	.641	.644	.648	.652
0.4	.655	.659	.663	.666	.670	.674	.677	.681	.684	.688
0.5	.691	.695	.698	.702	.705	.709	.712	.716	.719	.722
0.6	.726	.729	.732	.736	.739	.742	.745	.749	.752	.755
0.7	.758	.761	.764	.767	.770	.773	.776	.779	.782	.785
0.8	.788	.791	.794	.797	.800	.802	.805	.808	.811	.813
0.9	.816	.819	.821	.824	.826	.829	.831	.834	.836	.839
1.0	.841	.844	.846	.848	.851	.853	.855	.858	.860	.862
1.1	.864	.867	.869	.871	.873	.875	.877	.879	.881	.883
1.2	.885	.887	.889	.891	.893	.894	.896	.898	.900	.901
1.3	.903	.905	.907	.908	.910	.911	.913	.915	.916	.918
1.4	.919	.921	.922	.924	.925	.926	.928	.929	.931	.932
1.5	.933	.934	.936	.937	.938	.939	.941	.942	.943	.944
1.6	.945	.946	.947	.948	.949	.951	.952	.953	.954	.954
1.7	.955	.956	.957	.958	.959	.960	.961	.962	.962	.963
1.8	.964	.965	.966	.966	.967	.968	.969	.969	.970	.971
1.9	.971	.972	.973	.973	.974	.974	.975	.976	.976	.977
2.0	.977	.978	.978	.979	.979	.980	.980	.981	.981	.982
2.1	.982	.983	.983	.983	.984	.984	.985	.985	.985	.986
2.2	.986	.986	.987	.987	.987	.988	.988	.988	.989	.989
2.3	.989	.990	.990	.990	.990	.991	.991	.991	.991	.992
2.4	.992	.992	.992	.992	.993	.993	.993	.993	.993	.994
2.5	.994	.994	.994	.994	.994	.995	.995	.995	.995	.995
2.6	.995	.995	.996	.996	.996	.996	.996	.996	.996	.996
2.7	.997	.997	.997	.997	.997	.997	.997	.997	.997	.997
2.8	.997	.998	.998	.998	.998	.998	.998	.998	.998	.998
2.9	.998	.998	.998	.998	.998	.998	.998	.999	.999	.999
3.0	.999	.999	.999	.999	.999	.999	.999	.999	.999	.999

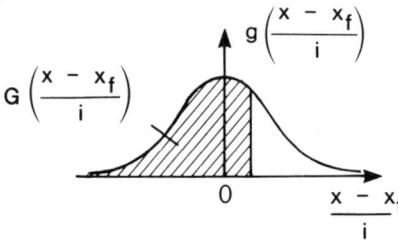

a general consensus of opinion that a normal probability curve match is more appropriate and convenient for estimation purposes. Its application to tunnelling-induced ground settlement is discussed in Schmidt (1969), Attewell (1978a), and Attewell and Woodman (1982). Some of the more important elements of this curve (which has been reproduced in several papers by different authors) are shown in Figure 2.10. These features relate to the surface of a settlement trough but, of course, the real problems, considered in Chapters 3 and 4, concern the response of buried pipelines and building foundations within the trough. It is indicated subsequently that the response of in-ground structures to ground movement may not be especially sensitive to the form of ground settlement profile.

The transverse settlement trough represents a terminal state of ground deformation induced by ground losses alone or by ground losses plus longer-term consolidation type movements (see section 2.8). However, ground movements—settlements and their derivatives—are also projected ahead of the tunnel face. These forward movements, although a temporary wave of disturbance as the tunnel face advances, also have an effect on buildings and buried pipes, which are twisted and subjected to torsion in three dimensions.

If the state of ground-surface deformation ahead of and to the sides of a tunnel is to be fully estimated, then the form of the settlement profile in an xz-plane, parallel to the tunnel centre line, must also be specified. A logical extension of the earlier assumption of normal probability form for the transverse (yz) settlement is that the tunnel centre line (xz) profile should be of cumulative probability form (see Figure 2.10). This assumption has been reasonably validated by examination of several field study reports (see Attewell and Woodman, 1982). Preliminary analysis assumes that the origin of the cartesian coordinate system, vertically above the tunnel face, coincides with the point of 50 per cent of maximum settlement. Although case history evidence from tunnels in cohesive soils suggests that in many cases only 30 per cent to 40 per cent w_{max} will have developed vertically above the tunnel face, this is really only academic, since all points at ground surface ahead of and within the boundaries of the transverse settlement trough will at some time realize their ultimate maximum movements for the translating tunnel source of ground loss.

If the above assumptions as to the form of the *ground-loss* settlement trough apply, if it is assumed that deformations occur without volume change, and if the deformation at the tunnel is approximated as a linearly translating point source of loss then the displacements and strains for any field point with coordinates $x, y,$ and z are

$$w = \frac{V_s}{\sqrt{2\pi}i} \exp\left[\frac{-y^2}{2i^2}\right] \left\{ G\left(\frac{x-x_i}{i}\right) - G\left(\frac{x-x_f}{i}\right) \right\}, \qquad (2.22)$$

$$v = \frac{-n}{z_0 - z} yw, \qquad (2.23)$$

$$u = \frac{nV_s}{2\pi(z_0 - z)} \exp\left[\frac{-y^2}{2i^2}\right] \left\{ \exp\left[\frac{-(x-x_i)^2}{2i^2}\right] - \exp\left[\frac{-(x-x_f)^2}{2i^2}\right] \right\}, \qquad (2.24)$$

$$\varepsilon_z = \frac{-nV_s}{\sqrt{2\pi}\, i(z_0 - z)} \exp\left[\frac{-y^2}{2i^2}\right] \left\{\left(\frac{-1}{\sqrt{2\pi}}\right)\left(\frac{x-x_i}{i}\right)\right.$$

$$\times \exp\left[\frac{-(x-x_i)^2}{2i^2}\right] - \left(\frac{x-x_f}{i}\right)\exp\left[\frac{-(x-x_f)^2}{2i^2}\right]\right\}$$

$$+ \left(\frac{y^2}{i^2} - 1\right)\left[G\left(\frac{x-x_i}{i}\right) - G\left(\frac{x-x_f}{i}\right)\right]\bigg\}, \qquad (2.25)$$

$$\varepsilon_y = \frac{n}{z_0 - z} w\left(\frac{y^2}{i^2} - 1\right), \qquad (2.26)$$

$$\varepsilon_x = \frac{-nV_s}{2\pi i(z_0 - z)} \exp\left[\frac{-y^2}{2i^2}\right] \left\{\left(\frac{x-x_i}{i}\right)\exp\left[\frac{-(x-x_i)^2}{2i^2}\right]\right.$$

$$\left. - \left(\frac{x-x_f}{i}\right)\exp\left[\frac{-(x-x_f)^2}{2i^2}\right]\right\}. \qquad (2.27)$$

As noted earlier, reference to Table 2.2 is required for the resolution of the functions $G(x - x_i/i)$ and $G(x - x_f/i)$. For a tunnel face that has advanced sufficiently (say, two to three tunnel depths, $3z_0$) to allow the transverse (yz-plane) ground-loss settlement profile to develop fully, the function $G(x - x_i/i)$ can be re-expressed in the above equations as unity. It remains to determine possible input values for the other non-geometric parameters.

2.4.1 *Practical estimation of volume-loss parameter, V_t, and the surface settlement volume, V_s*

The contributory sources of ground loss have been identified in section 2.1. For ground-movement estimation purposes expert advice should be sought, but the following ranges of values could be appropriate for guidance in different soils.

In *cohesive soils* it is usual to expect a ground loss of between about 0.5 per cent and 2.5 per cent of the tunnel face excavated area, depending upon the stiffness of the soil and the speed (Attewell 1978a, p. 849, and section 2.1) at which the initial support is installed. Estimation may, however, be more satisfactorily attempted on the basis of an overload factor where the driving pressure for ground loss is the multiple of tunnel depth-to-axis (z_0) and soil unit weight (γ), and this is resisted by the undrained shear strength (c_u) of the soil at the tunnel face. If a building imposes a distributed surface surcharge pressure q, then this pressure should be added to the γz_0 term. Internal support (σ_i), usually in the form of compressed air but increasingly, as noted in section 2.2, with the use of newer technology, closed shields in the form of pressurized slurry, obviously resists ground loss at the tunnel.

Figure 2.11a was originally presented in Attewell and Yeates (1984) and

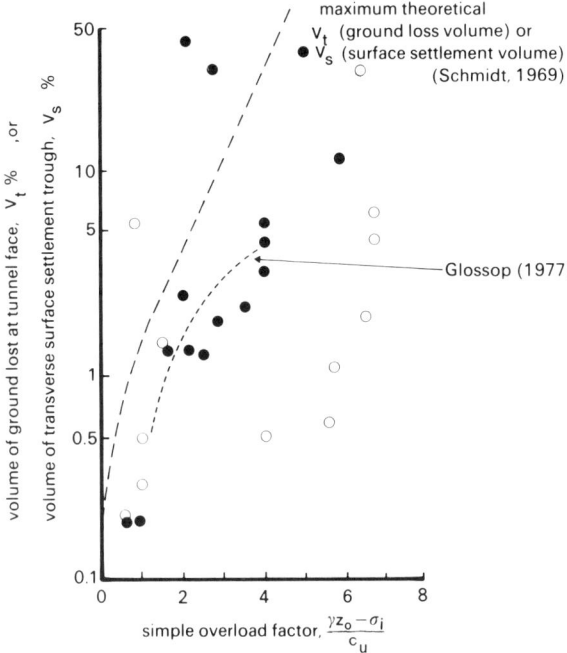

Figure 2.11 (a) Estimation of ground losses and surface settlement volumes from the overload factor for tunnels in cohesive soils. Open circles are Schmidt's (1969) data from shield-driven tunnels. Solid circles are as presented in Attewell and Yeates (1984).

indicates quite a considerable spread of the volume loss data as a function of the overload factor. Glossop has suggested the expression

$$V_s\% = 1.33 \times (\text{simple overload factor}) - 1.4 \qquad (2.28)$$

for $1.5 \leqslant \text{OFS} \leqslant 4$, which is reasonably representative for tunnelling in most clay soils. More recently the curves in Figure 2.11b have been used for design purposes. If the plotted data are thought to be sound, the unbroken line serves as an upper bound for ground loss in 75 per cent of the cases. It seems reasonable to adopt the level of 75 per cent probability for design where there is such an obvious measure of uncertainty, and it is thought unreasonable always to assume a 'worst case' for design purposes. The positions of pipe joints transverse to a tunnel are assessed, for example, on the basis of 75 per cent probability. It should also be noted in Figure 2.11b that the broken lines denoting twice and half the chosen 75 per cent design curve contain 50 per cent of the data points.

Some concern might be expressed at the high ground losses (30 per cent and 40 per cent) associated with mini-tunnelling. These two results are from soft alluvial soil and can be explained by the delay in contact grouting and high consequential radial intrusions. (Mini-tunnel three-segment rings have their

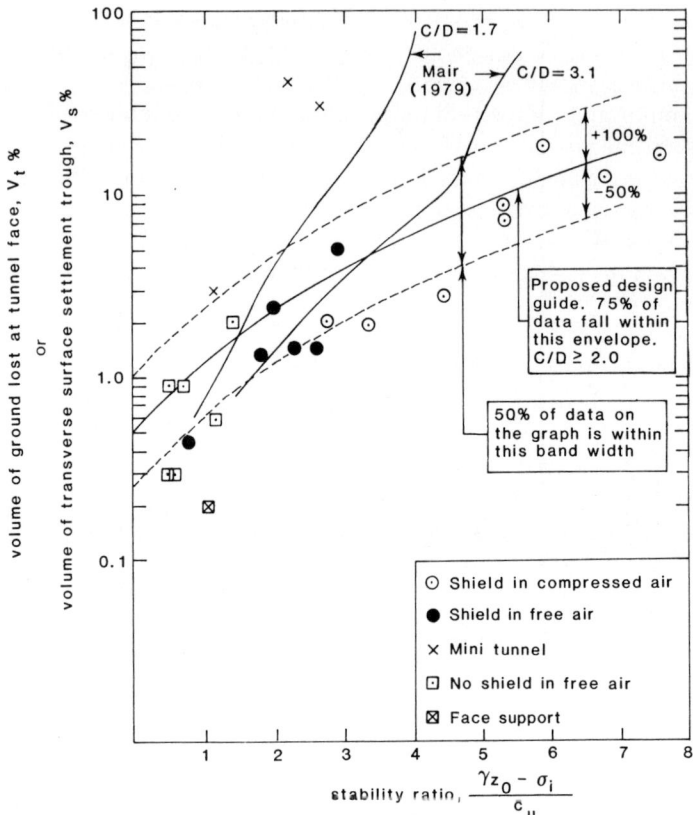

Figure 2.11 (*b*) Ground losses related to stability ratio (partly after Leach, 1985, and with additions). The curves after Mair (1979) relate to two centrifuge model tests. Symbol *C* is the cover (depth to tunnel soffit).

annular voids filled with pea gravel or Lytag on erection, but the infill is usually grouted up only when the tunnel is completed.) Thus although the excavation may be generally stable, the method is somewhat unsatisfactory for controlling ground movements. Fortunately, because of the relatively small face area of such mini-tunnels, the absolute volume loss may be tolerable in all but the most critical of urban conditions.

Attewell (1978*a*), pp. 868–878 contains quite a detailed discussion on, and further literature reference to, the subject of stability ratio or simple overload factor. For values of $(\gamma z_0 + q - \sigma_i)/c_u$ less than unity there should be little or no ground movement at the face, and so negligible attributable surface settlement. For stability ratios between 1 and 2, the deformations are regarded as being elastic and cause only very small surface movements. Ratios from 2 to about 4 cause elasto-plastic movements, and between 4 and 6 the movements are deemed to be plastic. Above a ratio of about 6 (6.28 according to Broms and

Bennermark, 1967) shear failure will be mobilized in a clay soil from the tunnel up to ground surface, and so this should be regarded as the limiting overload factor for face stability.

The importance of speed in the erection of initial support has been noted, and in critical urban areas the adoption of a build-one-ring, grout-one-ring construction regime is often advocated. However, the benefits to be derived from this procedure may in part be lost by the inevitable slowing of advance and the greater time offered for movement inwards at the face proper.

Based on the case studies summarized by Peck (1969) and Schmidt (1974), Mitchell (1983, p. 99) defines the volume loss as

$$V_t \simeq \frac{c_u}{E_u} \exp\left[\frac{\gamma z_0 - \sigma_i}{2c_u}\right] \quad (2.29)$$

where E_u, the undrained deformation modulus, is generally about 200 to 700 times c_u for soft soils. Mitchell suggests that for very sensitive soils or for poor construction control (for example, excavating ahead of the shield, and poor jacking techniques at the shield) the ground loss estimate should be increased by a factor of about 3.

There is now a considerable corpus of data available on ground losses and surface settlements in *stiff fissured clays* tunnelled with or without a shield. V_t (and V_s) for these materials is normally between 1 and 2 per cent. *Glacial tills* often contain silt/sand/gravel lenses with water at artesian or sub-artesian pressures, and so although the clay soil itself may be stiff the presence of the lenses may dictate the adoption of compressed-air temporary support measures. During shield-tunnelling of such deposits in free air, the volume losses would probably be 2–2.5 per cent, but with compressed-air support they could be reduced to 1–1.5 per cent. Recent *silty clays*, having undrained shear strengths probably in the range 10–40 kN/m², and shield-tunnelled under compressed-air pressure (an example being Attewell et al., 1978) would tend to incur ground losses in the range 2–10 per cent.

When tunnelling in *granular soils* above a water table a V_t range of 2–5 per cent is appropriate, but of course the losses in these soils, compared with the losses in cohesive soils, are more dependent on operator experience and skill. If compressed air is used to control the stability of a granular soil tunnelled below the water table, then a V_t range of between 2 and 10 per cent will apply. In both cases, a 5 per cent trial value could be adopted for preliminary calculations, but it is preferable to base any selection of trial values on site investigation standard penetration test (SPT) evidence. In structurally sensitive urban areas it is particularly important that the potential ground loss, even in loose non-cohesive soil, can be reduced by adopting a build-one-ring, grout-one-ring construction regime in a segmentally-lined tunnel.

Man-filled ground must be tunnelled in certain areas, and estimation of ground loss in such cases is made difficult by the usually quite variable composition and compaction of the fill. Ground losses of about 17 per cent have been estimated for a recent household/industrial waste fill in the north of England (Dobson et al., 1979). For guidance purposes, a lower value of about 8 per cent should probably be used for an old fill comprising natural ground material, a value of 10–12 per cent for an old, established industrial fill, and a figure of 15 per cent for a recent loose industrial or household waste fill. Good ground investigation is obviously required in order to establish a working value.

More complex expressions for the vertical pressure which drives the ground loss at the tunnel may be found in Széchy (1973). These expressions, attributable to different authors, accommodate c-ϕ soils, surface surcharge pressures, and arching relevant particularly to frictional soils.

These practical values of volume loss are usually adopted directly as surface settlement volumes and used in equation (2.39a) in order to estimate maximum surface settlement. The values are related to medium-sized (say 3 m to 3.5 m diameter) tunnels and should be increased somewhat for smaller-diameter tunnels (less than 1.5 m). It should also be noted that the movement data used to estimate future settlements and ground losses are biased, since those tunnels chosen for monitoring are expected to generate significantly large ground movements and are often in areas that are not built up (otherwise measurements would be hampered). Small or immeasurable movements never appear in the general data set. Ground movements in urban areas may thus be subject to a degree of overestimation.

2.4.2 Settlement trough dimension parameter, i

Attempts have been made to estimate this parameter empirically from case history evidence of settlement trough measurements in different types of ground. Whereas the ground loss depends upon the shear strength of the soil at tunnel level, the settlement trough dimensions will be related more closely to the soil strength nearer to ground surface. The equation that has been used, $i = RK_R \overline{(z_0 - z/2R)^n}$, has been discussed by Attewell and Woodman (1982) in the context of a stochastic model and has been shown to be unsatisfactory. It is to be expected that i should depend to some degree on the excavated diameter d, and indeed for ground-surface settlement Peck (1969) proposes the approximation

$$i = 0.2(d + z_0). \qquad (2.30)$$

On the other hand, O'Reilly and New (1982) have been unable to detect any such dependence in their survey of UK tunnel settlements. It is probably advisable (certainly in the case of cohesive clay soils) to assume that both K_R and n tend to unity, which suggests that i approximates to half the tunnel axis depth (in practice a range of between 0.4 times and 0.7 times the tunnel depth as observed for stiff clays and for soft silty clays, respectively). Alternatively,

O'Reilly and New have suggested, from case history reviews, that the following relations apply for UK tunnels for which adequate ground settlement records are available:

$$i = 0.43(z_0 - z) + 1.1 \text{ m for cohesive soils } (3 \leqslant z_0 \leqslant 34), \quad (2.31a)$$

$$i = 0.28(z_0 - z) - 0.1 \text{ m for granular soils } (6 \leqslant z_0 \leqslant 10). \quad (2.31b)$$

More recently Leach (1985) has analysed data from 23 tunnels constructed by different methods (no-shield, shield in free air, mini-tunnel, shield in compressed air) and has produced the following relations:

$$i = 0.57 + 0.45(z_0 - z) \pm 1.01 \text{ m} \quad (2.32a)$$

for those sites where consolidation effects are deemed to have been insignificant;

$$i = 0.64 + 0.48(z_0 - z) \pm 0.91 \text{ m} \quad (2.32b)$$

for those sites where consolidation effects are deemed to have been significant.

These relations, and particularly equation (2.31b) for granular soils, are not dissimilar to those derived by Atkinson and Potts (1976) from their laboratory model experiments. Reference may also be made to the graphs in Attewell (1978a) pp. 884, 885, which confirm that narrower (lower i-value) surface settlement troughs tend to be associated with granular soils, where running and ravelling may be concentrated at the tunnel crown and over the tail void of the shield.

Rather than accepting reported values of i (almost invariably i_y) for the development of the above empirical relations, it is preferable to plot the logarithm of the original recorded settlements ($\log w$) against the square (y^2) of the transverse distance and draw a regression line through the points. Maximum settlement (w_{max}) can then be defined by the intercept of the regression line with the axis $y^2 = 0$, and the value of i by the fact that i^2 is the value of y^2 where $w/w_{max} = 0.606$.

It is noted that under the normal probability profile assumptions, the width of the fully-developed transverse settlement trough is infinite, but in practice can be taken as $2\sqrt{2\pi}\, i$ (i.e. $5i$). The forward trough extends a practical distance of $\sqrt{2\pi}\, i$ (i.e. $2.5i$) ahead of the tunnel face if the latter is located directly beneath the 50 per cent maximum surface-settlement point (which will not usually be the case). The full form of the ground-loss settlement distribution is shown graphically in Figure 2.14.

Example 2.2 Settlement calculation

Consider a case history example (Attewell and Farmer, 1973) used in Attewell and Woodman (1982).

Tunnel axis depth, $z_0 - z = 7.5$ m
Settlement trough width parameter, $i = 3.9$ m.
Maximum surface settlement, w_{max} (at $y = 0$) = 7.86 mm.
Ground-loss volume, $V_t = 0.077$ m^3/m (see equation (2.22),
  with $w = w_{max}$ and $y = 0$).
Let the tunnel face be well-advanced from the start point.
Then $(x - x_i) \to \infty$ and $G(x - x_i/i) = 1$.
Take a point on ground surface having coordinates $(x - x_f) = 4$ m, $y = 1.5$ m.
Then $G(x - x_f/i) = G(1.020) = 0.846$ (from Table 2.2).

It follows from equation (2.22) that the ground settlement at the specified point is 1.13 mm.

Calculation on equations (2.23) and (2.24) produces values of -2.23 mm for v and -0.90 mm for u. (The negative signs denote displacements horizontally inwards towards the centre of the trough.) A check summation of the three displacements having regard to sign confirms the no-volume-change assumption underlying the analysis.

Further calculation produces ground strain values of $-110\,\mu\varepsilon$ (compressive), $-130\,\mu\varepsilon$ (compressive) and $+240\,\mu\varepsilon$ (tensile) for $\varepsilon_z, \varepsilon_y$ and ε_x, respectively. These values sum to zero and so satisfy the no-volume-change assumption.

2.5 Design-curve graphs

With respect to buildings and in-ground structures, information on ground-surface settlement and strain is only part of the requirement. The horizontal u, v ground displacements impose direct strains on a foundation or buried pipe, the upper bound structural strains being equated to $\varepsilon_x, \varepsilon_y$ by neglecting ground-structure differential stiffness and assuming no ground material shearing. Building foundations, however, settle differentially and incur angular distortions, the magnitudes of which depend on the position of the foundation in the settlement trough (see Attewell, 1978a). Strains associated with angular distortions are additive to the direct lateral distortional strains. It should also be realized that a foundation is subjected to a wave of torsion before the advance of the tunnel face removes the effects of the forward settlement trough. Similarly, buried pipes within a ground settlement trough experience direct strain, bending strain (related to ground curvature) and torsion. These matters are discussed in Attewell and Woodman (1982), but they are probably more easily accommodated by use of design-curve graphs. There is also more detailed discussion later in the book.

The use of equations implies a degree of precision in estimation that in fact does not exist. The approximate character of the assessment is probably better reflected by expressing the equations graphically in such a way that a foundation or buried pipe analysis can be conducted more easily and perhaps more rapidly.

Figures 2.12–2.17 were originally constructed by desk computer and graph plotter from adaptations of equations (2.22) to (2.27). The coordinates for the

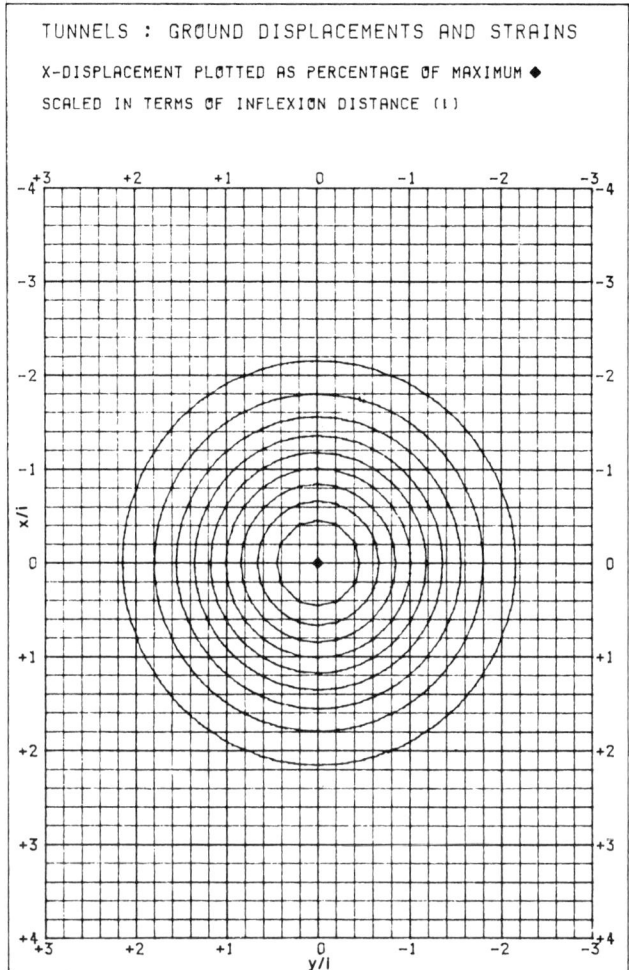

Figure 2.12 10% incremental ground displacement (u) curves, x-axis.

curves are scaled as x/i (ordinate: down the graph) and y/i (abscissa: across the graph), and the positive z-axis is directed downwards into the plane of the paper. On the graphs, the tunnel face is theoretically positioned at the origin. The curves assume that the tunnel is of infinite length (in practice, the distance of advance must be large compared to $(z_0 - z)$) and that the increments of horizontal ground displacement are radial (always towards the current tunnel face origin). Each curve on a particular graph represents 10 per cent of the parameter absolute maximum, each maximum being marked by a small infilled parallelogram.

The curves show a number of interesting features. For example, the settlement, or z-displacement (w), curves (Figure 2.14) profile the settlement

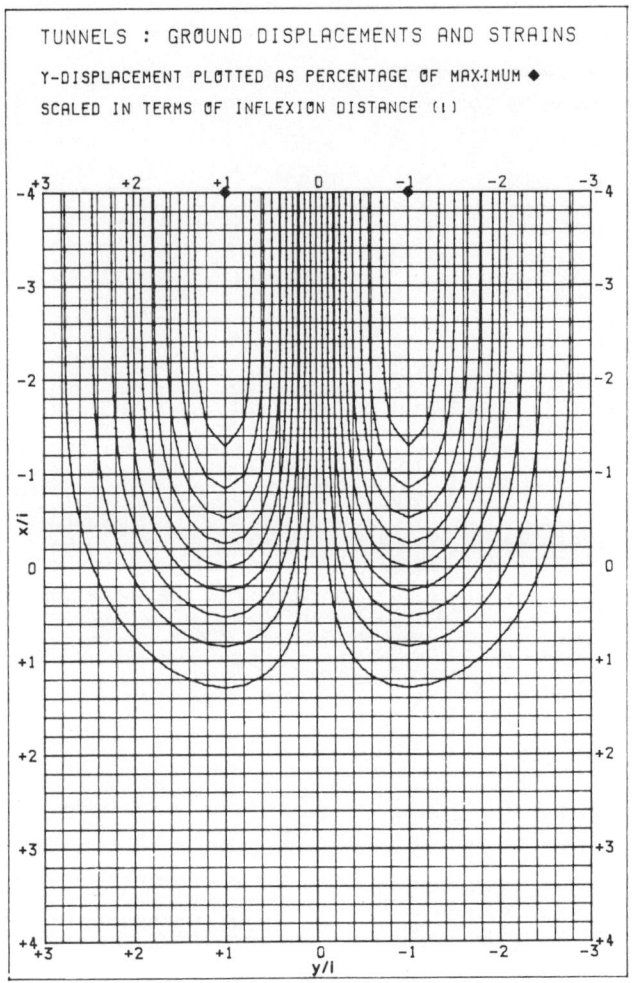

Figure 2.13 10% incremental ground displacement (v) curves, y-axis.

distribution about the source ($x/i = y/i = 0$) and show that the transverse ground-loss settlement profile becomes almost fully developed when the tunnel face has advanced about $2i$ beyond the profile section. Figure 2.15 shows that the tensile (positive) ε_x regime ahead of the tunnel face is a mirror image of the compressive (negative) ε_x behind the face. Figure 2.16 shows that the transverse ε_y has a compressive maximum above the tunnel centre line which decreases to zero with increasing y to a tensile ε_y maximum that is between 40 per cent and 50 per cent (theoretically 44.6 per cent) of the absolute maximum. The relatively complex pattern of ε_z (Figure 2.17) can be interpreted by remembering that $\varepsilon_z = -(\varepsilon_x + \varepsilon_y)$ according to the underlying

ESTIMATION OF GROUND MOVEMENTS

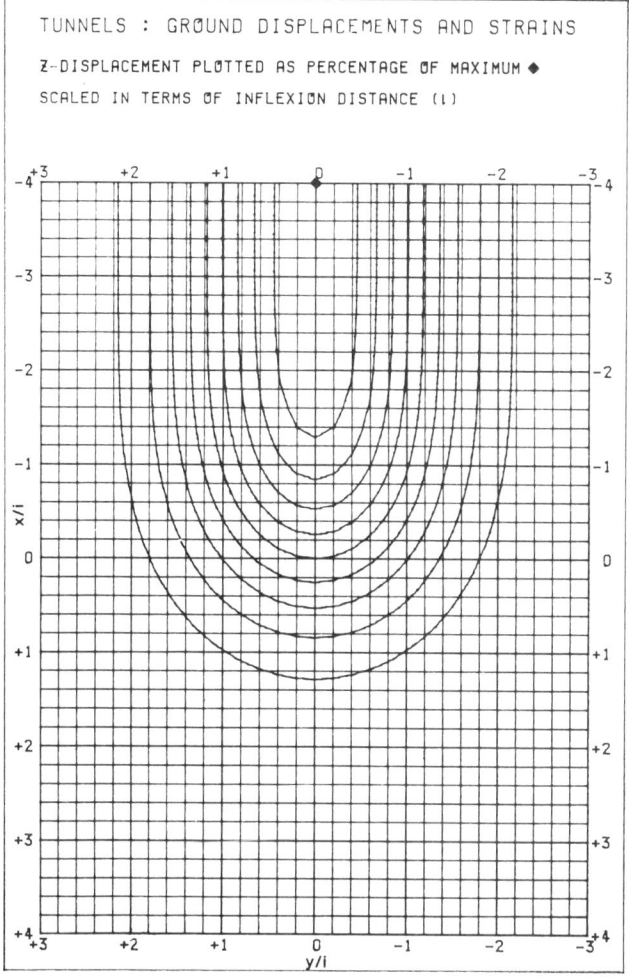

Figure 2.14 10% incremental ground displacement (w) curves, z-axis.

assumption of no volume change in the soil. There is a tensile maximum on the tunnel centre line and behind the face; it decays with radial distance to zero strain and thence to compression. However, vertical strain distribution is of marginal practical importance in the present work, and is not considered further.

Only six graphs are necessary. Although this number may be inadequate for some practical purposes, there are simple relations and equivalences between parameters which allow the use of these graphs to be extended. For greater completeness, the i subscript (x or y) is given, although in most cases an unsubscripted i-parameter ($i_x = i_y = i$) must be used. Reference should be made

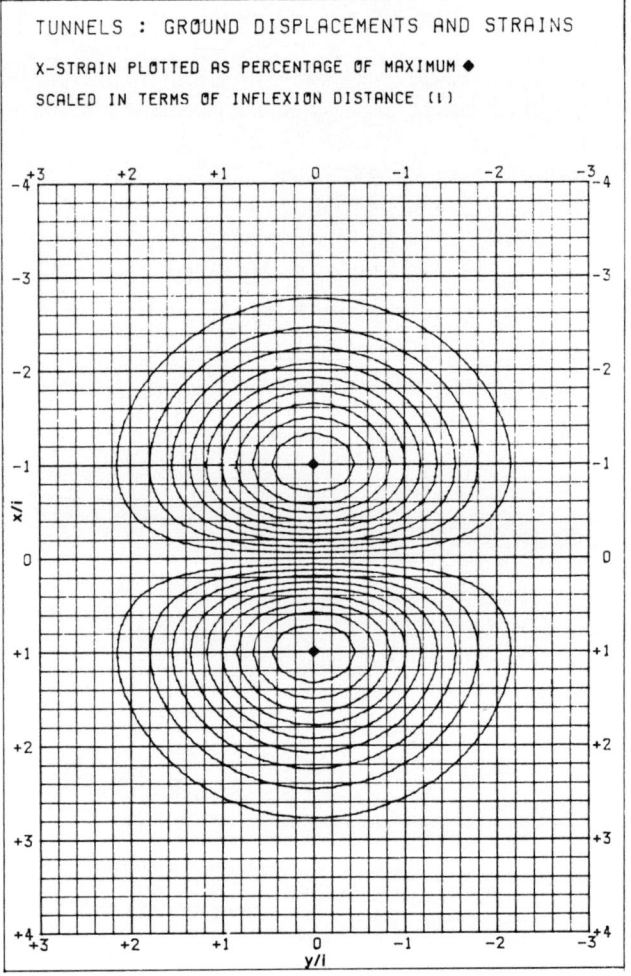

Figure 2.15 10% incremental ground strain (ε_x) curves, x-axis.

to section 2.7 for further discussion on this point.

$$\gamma_{xy} \text{ graph} \equiv \varepsilon_y \text{ graph rotated about the } z\text{-axis through } \pi/2. \tag{2.33}$$

$$xz\text{-plane gradient } \frac{\partial w}{\partial x} = \frac{z_0 - z}{ni_x^2} u \sim u. \tag{2.34}$$

$$yz\text{-plane gradient } \frac{\partial w}{\partial y} = \frac{z_0 - z}{ni_y^2} v \sim v. \tag{2.35}$$

$$\frac{\partial^2 w}{\partial x^2} = \frac{z_0 - z}{ni_x^2} \varepsilon_x \sim \varepsilon_x, \qquad \frac{\partial^2 w}{\partial y^2} = \frac{z_0 - z}{ni_y^2} \varepsilon_y \sim \varepsilon_y,$$

$$\frac{\partial^2 w}{\partial x \partial y} = \frac{z_0 - z}{n} \cdot \frac{\gamma_{xy}}{i_x^2 + i_y^2} \sim \gamma_{xy}. \qquad (2.36)$$

(These second derivatives of w are approximations to the components of curvature of the settled surface, given small displacement and gradient, and to the components of bending strain referred to the initial pre-settlement horizontal plane, under conditions of small strain and small gradient.)

Sufficient parameters have probably now been covered for building foundations, but not for buried pipes. In particular, horizontal pipes also need

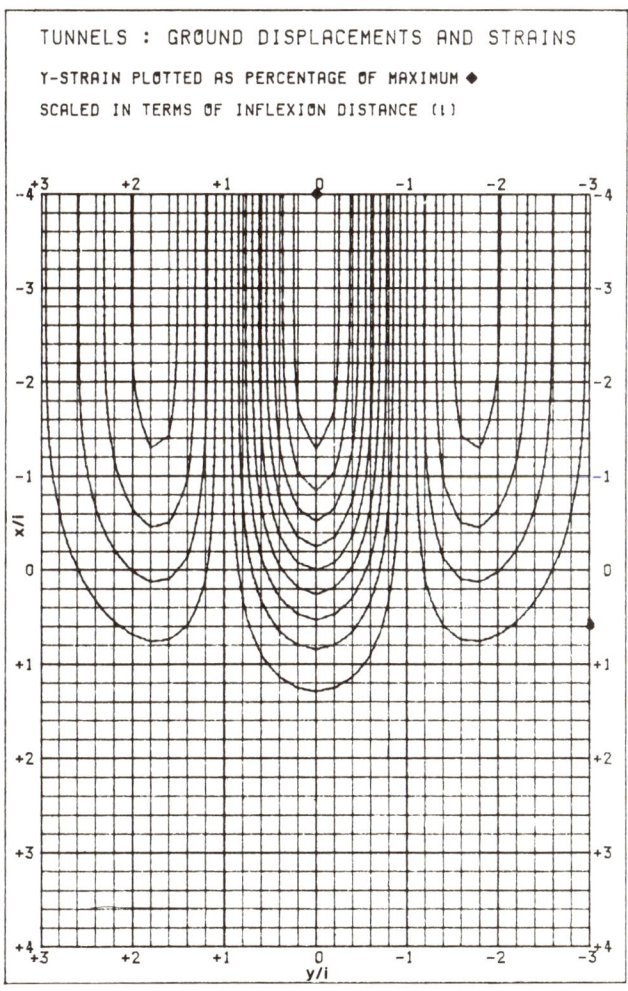

Figure 2.16 10% incremental ground strain (ε_y) curves, y-axis.

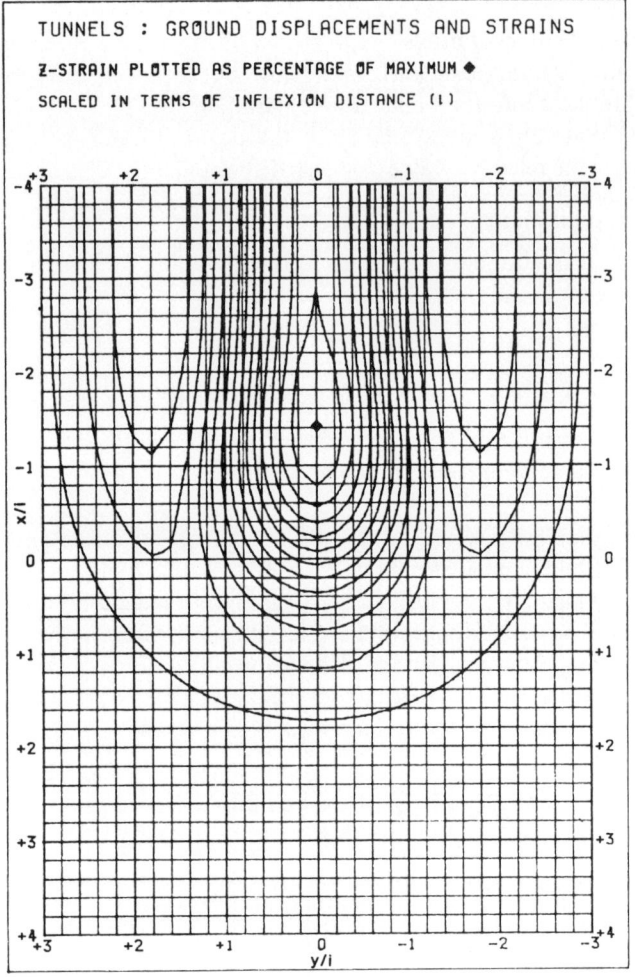

Figure 2.17 10% incremental ground strain (ε_z) curves, z-axis.

the following:

$$\frac{\partial u}{\partial y} = \frac{i_x^2 \gamma_{xy}}{i_x^2 + i_y^2} \sim \gamma_{xy}, \qquad \frac{\partial v}{\partial x} = \frac{i_y^2 \gamma_{xy}}{i_x^2 + i_y^2} \sim \gamma_{xy}. \tag{2.37}$$

$$\frac{\partial^2 u}{\partial x^2} = \left(\frac{x^2}{i_x^2} - 1\right)\frac{u}{i_x^2}, \qquad \frac{\partial^2 u}{\partial y^2} = \left(\frac{y^2}{i_y^2} - 1\right)\frac{u}{i_y^2}, \qquad \frac{\partial^2 u}{\partial x \partial y} = -\frac{y}{i_y^2}\varepsilon_x;$$

$$\frac{\partial^2 v}{\partial x^2} = -\frac{y}{i_x^2}\varepsilon_x, \qquad \frac{\partial^2 v}{\partial y^2} = \left(\frac{y^2}{i_y^2} - 3\right)\frac{v}{i_y^2}, \qquad \frac{\partial^2 v}{\partial x \partial y} = \left(\frac{y^2}{i_y^2} - 1\right)\frac{u}{i_x^2}$$

$$\tag{2.38}$$

It is noted in Reeves *et al.* (1983) that there is an argument for graphically displaying gradient magnitude $\sqrt{(\partial w/\partial x)^2 + (\partial w/\partial y)^2}$ and principal lateral strains/bending strains of the horizontal plane, but these curves depend on the ratio i_x/i_y, unlike contours for the above parameters.

Provided that values are available for the essential input parameters, the percentages of the reference values for each displacement and strain parameter are then assessed by interpolation between curves on the graphs. Numerical values for the displacement and strain parameters are calculated by defining the reference displacements and strains corresponding to absolute maxima defined as follows:

$$w_{max} = \frac{1}{\sqrt{2\pi}} \frac{V_s}{i_y} \quad \text{at } (-\infty, 0). \tag{2.39}$$

This is an important and well-used equation, expressed in the form

$$V_s = 2.5 i_y w_{max} \tag{2.39a}$$

which allows maximum settlement w_{max} to be estimated from an input value of V_s (itself usually estimated from the type of practical information given in section 2.4.1) and an input value of i_y (estimated from the information given in section 2.4.2). The volume of ground displaced by settlement at the surface (V_s) expressed per unit length of tunnel advance (usually in units of m^3/m = m^2) is obtained by integration of the area under the error curve assumed to define the form of the transverse settlement trough. Equation (2.39a) is derived mathematically in Attewell and Farmer (1974a), Appendix 1, pp. 379, 380.

$$v_{max} = -w_{max} \frac{n i_y}{z_0 - z} \exp\left[-\frac{1}{2}\right] \quad \text{at}(-\infty, 1), \tag{2.40}$$

$$u_{max} = -w_{max} \frac{1}{\sqrt{2\pi}} \frac{n i_x}{z_0 - z} \quad \text{at } (0, 0), \tag{2.41}$$

$$\varepsilon_{zmax} = w_{max} \frac{n}{z_0 - z} \left[\frac{\exp(-1)}{\sqrt{\pi}} + \frac{1 + \text{erf}(1)}{2}\right] \quad \text{at}(-\sqrt{2}, 0), \tag{2.42}$$

$$\varepsilon_{ymax} = -w_{max} \frac{n}{z_0 - z} \quad \text{at } (-\infty, 0), \tag{2.43}$$

$$\varepsilon_{xmax} = w_{max} \frac{n}{z_0 - z} \frac{\exp(-\frac{1}{2})}{\sqrt{2\pi}} \quad \text{at}(1, 0), \tag{2.44}$$

$$\gamma_{xymax} = \varepsilon_{xmax} \frac{i_x^2 + i_y^2}{i_x i_y} \quad \text{at } (0, 1). \tag{2.45}$$

(Note that 'erf' in equation (2.42) denotes 'error function'.)

Input parameters $i_x, i_y, z_0 - z$ are often specified in metres and V_s in cubic metres per metre. In that case the reference displacements expressed in millimetres are

$$w_{max} = 399 \frac{V_s}{i_y}, \quad v_{max} = -0.607 w_{max} \frac{n i_y}{z_0 - z};$$

$$u_{max} = -0.399 w_{max} \frac{n i_x}{z_0 - z}. \tag{2.46}$$

Similarly, if w_{max} is in millimetres, then the reference strains in microstrains are

$$\varepsilon_{zmax} = 1.129 \varepsilon_{ymax}; \quad \varepsilon_{ymax} = -1000 w_{max} \frac{n}{z_0 - z};$$

$$\varepsilon_{xmax} = 0.242 \varepsilon_{ymax}; \quad \gamma_{xymax} = \varepsilon_{xmax} \frac{i_x^2 + i_y^2}{i_x i_y}. \tag{2.47}$$

2.6 Ground movements and structures—use of design curves for preliminary assessment

Two matters must be emphasized initially. First, the ground–structure interaction is one of ground shear against the (assumed) stiffer contact surface of the foundation or buried pipe structure combined with an induced bending, the latter being simple or torsional. It is the ground displacements, not the strains, that are important, although, as stated earlier, the calculated ground strains provide upper-bound values for the structural strains and so are used here. Second, assumed limiting levels of structural deformation must be known, and these are specified in terms of the type of structure at risk (Attewell, 1978a; Norgrove et al., 1979). Although criteria suggested by Burland and Wroth (1975), and based on Skempton and Macdonald (1956) and others (see section 4.13 below), have been proposed for tunnelling-induced ground deformation, caution is required. Self-weight settlements and deformations are long-term developments, and much of their potential effect can be accommodated during the actual construction. Tunnelling-induced movements are imposed on a structure very quickly, and so the self-weight damage criteria will have no in-built conservatism with respect to dynamic movements.

2.6.1 Application of design curves—use of overlays

When a structure at risk has been identified, its foundation (in the case of a building) should be dimensionally-scaled on a transparent overlay and then superimposed on each of the design-curve graphs in turn. The distribution of the *10 per cent design curves* on the area of the structure is then noted and interpreted with respect to an adopted (usually empirical) damage criterion.

If the tunnel passes beneath a heavily built-up area it is recommended that the relevant portion of the national or state survey plan be photographed with national grid or other orientation feature marked. The scale of the plan is then noted and a further calculation performed to scale its dimensions by the factor i_s for further printing. Finally, a design-curve overlay transparency can be prepared from the print. Suppose that the scale of the survey plan is 1:S. The design-curve graphs, as originally drawn to scale by the microcomputer, reproduce a linear dimension scaled to i_s in units of $g = 25$ mm linear measure. If p is a dimension of a building measured (say in mm) on the plan, then the corresponding required length of the building on the photographic print overlay transparency is gpS/i_s. It is noted that this procedure is valid only for foundations in the same horizontal plane.

Example 2.3 Building foundation assessment

A sewer tunnel, having an excavated radius 1.105 m, is to be driven in a mainly granular soil at an axis depth of 10.5 m. The ground loss is estimated to be 5 per

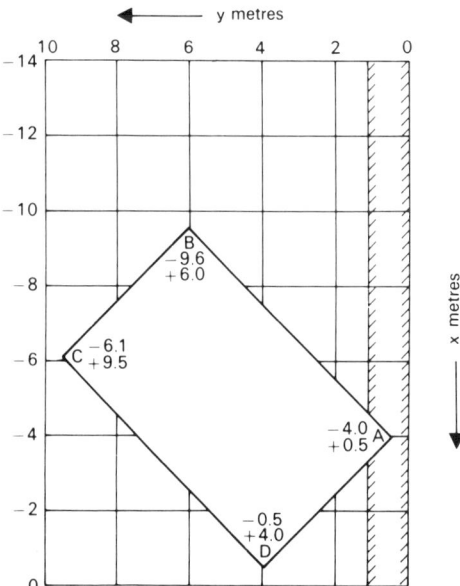

Figure 2.18 Plan location of building with respect to tunnel.

cent (i.e. $0.192 \text{m}^3/\text{m}$) and it is assumed that this loss is fully transferred to ground surface as settlement. A building, having a raft foundation of rectangular area at a mean depth of 0.5 m, is located with respect to the tunnel centre line as shown in Figure 2.18. Prediction of the effects on the building might proceed as follows:
$\overline{z_0 - z} = 10.5 \text{ m} - 0.5 \text{ m} = 10 \text{ m}$.
Let $i = 0.28(z_0 - z) - 0.1 \text{ m}$ (after O'Reilly and New (1982)).
Thus $i = 2.7 \text{ m}$. Let the n parameter be unity.
(Although the above equation is derived from UK tunnelling case histories, a check against a batch of international case histories suggests that it might be more generally relevant.)

The scaled coordinates of the foundation are:

x(m)	x/i	y(m)	y/i
− 4.0	− 1.481	+ 0.5	+ 0.185
− 9.6	− 3.555	+ 6.0	+ 2.222
− 6.1	− 2.259	+ 0.5	+ 3.518
− 0.5	− 0.185	+ 4.0	+ 1.481

The scaled foundation is redrawn in Figure 2.19 for superimposition on the design-curve graphs.

Lateral distortion and tensile failure. As the tunnel face advances, most of the foundation will eventually be subjected to virtually the whole deformation field on one side of the tunnel centre line. Tensile effects are most forcefully impressed on the foundation in two main regions:
$x > 0, y^2 < i_y^2$ and $x < 0, y^2 > i_y^2$. However, if $i_x = i_y$, wherever $\varepsilon_x \geqslant \varepsilon_y$ the more tensile principal strain in the horizontal plane is less than the overall maximum tensile $\varepsilon_y, \varepsilon_{yt\,max} = -0.466\varepsilon_{y\,max}$. The region ahead of the tunnel face $x > 0$, $y^2 < i_y^2$, is thus less dangerous than the region of permanent deformation $x < 0$, $y^2 > i_y^2$ transverse to the tunnel centre line. In this present example, $\varepsilon_{yt\,max} \simeq 0.127$ per cent.

A factor of some importance is the area of the foundation that is required to sustain a large tensile-strain field. Superimposing Figure 2.19 on Figure 2.16 shows that high permanent tensile strains would be experienced over a large area of the foundation, centred approximately on an x-axis just to the tunnel side of corner B. From Figure 2.13, -90 per cent $v_{max} \simeq 0.0065 \text{ m}$ displacement occurs over a horizontal distance of about 4.86 m, suggesting horizontal (y) distortion (see Figure 4.37) of approximately 0.096 per cent. At a superficial damage threshold of 0.05 per cent tensile strain the foundation would be vulnerable, even if a certain amount of interfacial slippage is allowed for.

Mid-planar bending. This can be analysed in a similar manner to horizontal strain. However, whereas in the case of horizontal strain it is the tensile directional component that is important, in a brittle material foundation, with bending the positive (convex up) and negative (concave up) components are of equal concern. Noting again the equivalence between ground curvature and direct horizontal strain curves, and superimposing Figure 2.19 overlay on Figure 2.16, it is seen that the more negative principal bending strain at corner A exceeds 80 per cent $(\partial^2 w/\partial y^2)_{max} = 80$ per cent $(w_{max}/i^2) \simeq 0.00389$ m^{-1}, and will eventually exceed 90 per cent $(\partial^2 w/\partial y^2)_{max}$ as the tunnel face advances a short distance further. This bending strain is clearly much more significant than the smaller, but near maximum and stabilized, positive principal bending strain already present at corner B.

If i_x is taken equal to i_y, the maximum ground slope, similar to the maximum horizontal displacement, is bounded absolutely in any region by the values

$$\left(\frac{\partial w}{\partial x}\right)_{max} = \frac{z_0 - z}{ni_x^2} u_{max} \simeq 0.42 \text{ per cent and}$$

$$\left(\frac{\partial w}{\partial y}\right)_{max} = \frac{z_0 - z}{ni_x^2} v_{max} \simeq 0.64 \text{ per cent.}$$

As with tensile failure, the average value may be more significant. From

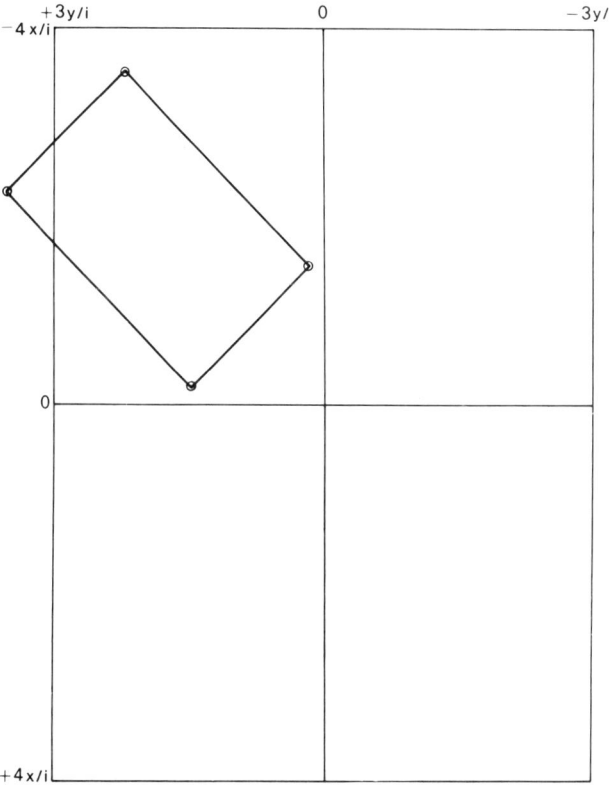

Figure 2.19 Overlay incorporating building plan dimensions scaled to i.

Figure 2.14 approximately 80 per cent w_{max} (0.0277 m) is developed over a distance of about 5 m, giving a y-direction *angular distortion* (see Figure 4.36) of approximately 0.454 per cent (1 in 220). Threshold values of angular distortion depend upon the actual construction of the building (Attewell, 1978a; Norgrove et al., 1979; see also Table 4.5). However, this predicted ground slope is sufficiently severe to create some concern for a building where the foundation is likely to follow the slope profile.

Example 2.4 Buried pipeline assessment

The general assessment of induced longitudinal pipe strain generally follows the above lines, predicted bending strains being additive to direct axial strains (the latter are the only strains experienced by the pipe neutral axis). For superimposition on the design-curve graphs the pipe can usually be represented by a line, since its diameter scaled to i is small. As earlier, it is assumed that the pipe deforms conformably with the predicted ground deformation that would be experienced without the presence of the pipe; the design curves effectively represent upper-bound predictions of pipe deformation. Any slippage at or near the pipe–soil interface implies that the actual direct and bending strains in the pipe may well be substantially less than those calculated on the basis of no interfacial slippage. Furthermore, the concept of an 'interface' may in many cases be invalid, since old cast iron pipes (for which assessments are likely to be made) may be severely corroded on the outside and so may have mechanical properties, over the zone of corrosion, intermediate between those of the uncorroded pipe material and those of the pipe bedding. Depending upon the type of pipe joint and its age, there may be a facility for rotation and direct translation at the joints which absorbs some of the potential strain in the pipe material.

For the example calculation, locked joints and an upper-bound strain condition have been conservatively assumed. Consider a 2ft (0.61 m) diameter pipeline buried at a depth of 1.5 m and lying parallel to—but 3.13 m from—the centre line of a tunnel, axis depth 13.5 m. The ground volume loss at the tunnel is estimated at 2.5 per cent or 0.103 m³/m, and the i-value is 6.26 m. It is assumed that n is close to unity.

An overlay incorporating a line distant $y/i = 3.13/6.26 = 0.5$ from and parallel to the tunnel centre line ($y/i = 0$) may be prepared (Figure 2.20) but this is really unnecessary since for this simple structural configuration all the information can be readily drawn by direct inspection of the design-curve graphs. From Figure 2.15 the maximum tensile strain along the line of the pipe $|y/i| = 0.5$ is 85 per cent ε_{xmax}, approximately $100\,\mu\varepsilon$. Maximum vertical and horizontal axial bending strains occur at the same point, since

$$\frac{\partial^2 w}{\partial x^2} \sim \varepsilon_x \quad \text{and} \quad \frac{\partial^2 v}{\partial x^2} = -\left(\frac{y}{i^2}\right)\varepsilon_x.$$

Consequently, maximum vertical axial bending is 85 per cent $(\partial^2 w/\partial x^2)_{max} \simeq 32\mu\varepsilon/\text{m}$, and maximum horizontal axial bending is approximately $9\mu\varepsilon/\text{m}$. The maximum resultant bending strain is approximately

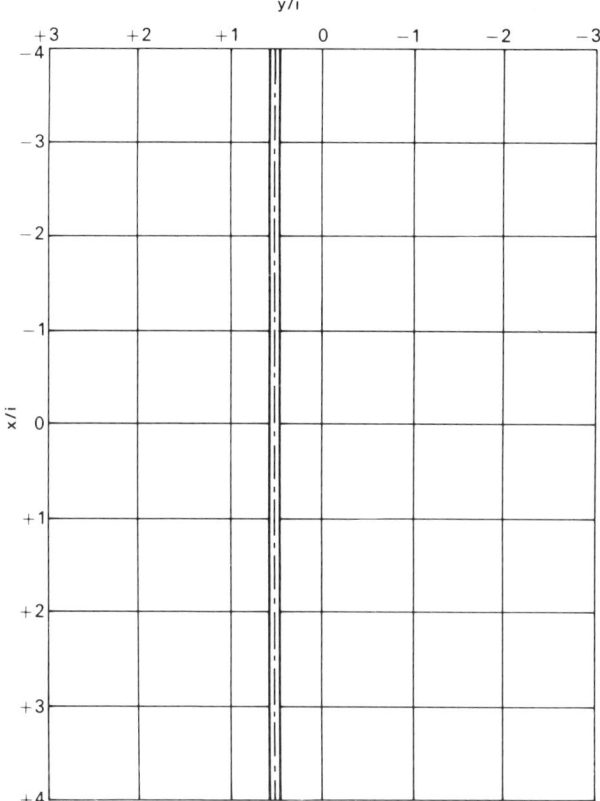

Figure 2.20 Example of pipeline lying parallel to the tunnel centre line.

$\sqrt{32^2 + 9^2} \simeq 33\mu\varepsilon/\text{m}$. Thus, assuming constant shear in a pipe of radius r, the maximum axial tensile strain is approximately $\varepsilon_x + r\partial^2 w/\partial x^2 \simeq 120\,\mu\varepsilon$.

Analyses of the above type would be applied to old pipes of traditional brittle material construction (cast iron or clay). Take, for example, an old pit-cast grey iron pipe of pre-BS 78 vintage, corroded on the outside, and unable to translate or rotate at the joints. Subject to all the assumptions discussed earlier, all of the potential strain calculated above would be experienced by the pipe material. With an allowable pipe stress of one-quarter the material ultimate tensile strength, an allowable strain of $400\mu\varepsilon$ would probably be applicable. It follows that the estimated upper-bound axial strain in the example pipe is less than one-third of its allowable strain. Had the estimated strains been much closer to the allowable strain, a criterion of failure based on the biaxial state of strain would have been applied, using the fact that for all pipe orientations the mean transverse strain is everywhere equal, but of opposite sign, to the axial strain under the assumption of no volume change. Two failure criteria could most conveniently be used: the Tresca maximum shear stress theory and the Huber–Hencky–von Mises distortional energy criterion. Examples of the application of these theories are not given in this book, but reference may be made to standard textbooks on material mechanics.

2.6.2 Pipes aligned obliquely to the direction of tunnel advance

Many of the major problems with underground services occur in urban areas when tunnel alignments and buried water and gas distribution mains are parallel to main roads, as in the example considered above. However, if a horizontal buried pipe is at an angle θ to the tunnel advance direction (x-axis), defined by direction cosines[†] $l, m, 0$, the following expressions apply for point values and mean values over a straight-line segment of length L between points 1 and 2:

$$\text{point axial strain} = \varepsilon_x l^2 + \gamma_{xy} lm + \varepsilon_y m^2 \qquad (2.48)$$

$$\text{mean axial strain} = [(u_2 - u_1)l + (v_2 - v_1)m]/L \qquad (2.49)$$

$$\text{point vertical axial bending} = \frac{\partial^2 w}{\partial x^2} l^2 + 2\frac{\partial^2 w}{\partial x \partial y} lm + \frac{\partial^2 w}{\partial y^2} m^2 \qquad (2.50)$$

mean vertical axial bending

$$= \left[\left\{ \left(\frac{\partial w}{\partial x}\right)_2 - \left(\frac{\partial w}{\partial x}\right)_1 \right\} l + \left\{ \left(\frac{\partial w}{\partial y}\right)_2 - \left(\frac{\partial w}{\partial y}\right)_1 \right\} m \right] / L \qquad (2.51)$$

point horizontal axial bending

$$= \frac{\partial^2 v}{\partial x^2} l^3 + \left(2\frac{\partial^2 v}{\partial x \partial y} - \frac{\partial^2 u}{\partial x^2}\right) l^2 m - \left(2\frac{\partial^2 u}{\partial x \partial y} - \frac{\partial^2 v}{\partial y^2}\right) lm^2 - \frac{\partial^2 u}{\partial y^2} m^3 \qquad (2.52)$$

mean horizontal axial bending

$$= [(\varepsilon_{y2} - \varepsilon_{y1} - \varepsilon_{x2} + \varepsilon_{x1})\sin 2\theta + (\gamma_{xy2} - \gamma_{xy1})\cos 2\theta]/2L. \qquad (2.53)$$

It is worth emphasizing that this section in general, and the above equations in particular, represent in effect upperbound solutions which assume that even a brittle pipe will follow the contours of the settlement trough. For specific solution of problems of buried pipes the reader's attention is directed to Chapter 3.

[†] $l = \cos\theta;\ m = \cos(90 - \theta) = \sin\theta$.

2.7 Relations between i_y and i_x

It has been assumed throughout the development of the ground movement equations in section 2.4 that the tunnel centre line settlement development profile may be described by a cumulative probability function based on the same statistical mean (w_{max}) and standard deviation (i) parameters as define the transverse normal probability settlement profile. In particular, it is assumed that $i_y = i_x$.

The easiest way of checking the reasonableness of this assumed and rather convenient equality is to compare a 'theoretical' cumulative probability centre line (xz-plane) settlement curve based on the i_y parameter with the actual 'best fit' recorded curve which defines i_x, the latter curve having been translated along the tunnel advance axis so that the 50 per cent maximum settlement (w_{max}) points on both curves match. Table 2.3 and Figure 2.21 demonstrate how this is done. This x-axis translation will usually be necessary because the

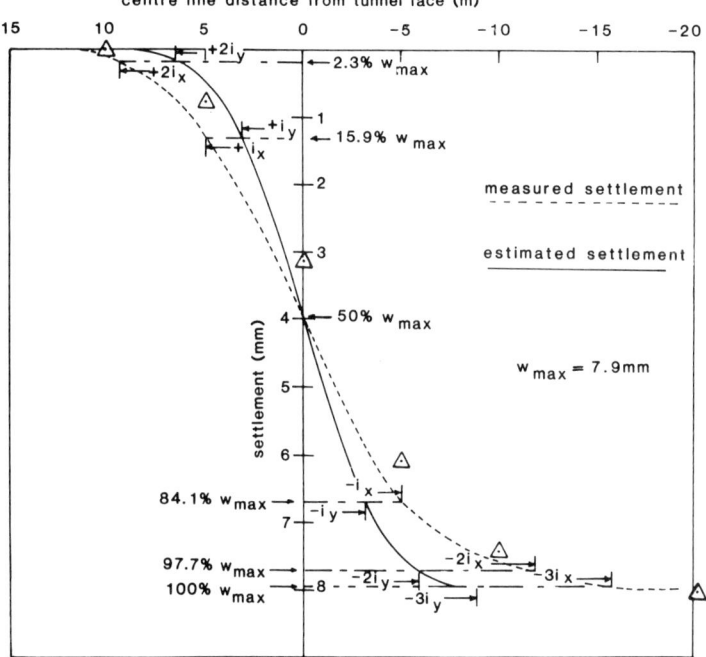

Figure 2.21 Centre line settlement distribution measured above a tunnel in stony/laminated clay at Hebburn, north-east England (Glossop, 1978), together with a distribution generated by the transverse settlement parameter i_y. The actual measurement points are marked by the triangular symbol and the measured curve has been translated forward by 1.3 m to set the 50% w_{max} point above the tunnel face.

Table 2.3 Percentage of maximum settlement as a function of position on the tunnel centre line settlement curve as expressed ideally in terms of the transverse settlement parameter i_y.

(Note: It is assumed that 50% w_{max} will have developed directly over the tunnel face on this centre line settlement profile; $+x$ distances are ahead of the face position and $-x$ distances are behind the face position.)

Location on centre line settlement profile	Approximate theoretical percentage of maximum settlement assuming that a cumulative probability function applies
$+3i_y$	$0\% w_{max}$
$+2i_y$	$2.3\% w_{max}$
$+i_y$	$15.9\% w_{max}$
0	$50\% w_{max}$
$-i_y$	$84.1\% w_{max}$
$-2i_y$	$97.7\% w_{max}$
$-3i_y$	$100\% w_{max}$

value of the ratio w/w_{max} at ground surface above the tunnel face is in almost all cases less than 50 per cent. For firm-to-stiff clays the range seems to be 30–50 per cent (Attewell and Woodman, 1982).

Obvious discrepancies between the two curves develop at low settlements and towards the maximum measured settlement. For such settlement development curves in general, these discrepancies can be quantified in terms of

$$i_y : \frac{+(3i_x)}{3}; \quad i_y : \frac{+(2i_x)}{2}; \quad i_y : +i_x; \quad i_y : -i_x; \quad i_y : \frac{-(2i_x)}{2}; \quad i_y : \frac{-(3i_x)}{2}.$$

The above relations between i_y and i_x are plotted in Figures 2.22–2.27 for the 14 tunnel case histories listed in Table 2.4 and carefully examined for this purpose. Original detailed measurements were available from most of these case histories.

Examination of measured centre line settlement development profiles indicated that the centre line settlement field almost always exceeds the length ($6i_y$ metres) of the cumulative probability curve generated from the transverse settlement trough i_y parameter. This is also shown by the plots on Figures 2.22–2.27, which lie generally above the equality line. These data points are also seen to lie closer to the estimated line ahead of $(+x)$ the face than they are behind $(-x)$ the face, where the incremental settlements are much attenuated. However, the symmetry of the estimated $(i_x = i_y)$ centre line settlement curves about the tunnel face position could only be satisfied if the

Table 2.4 Tunnel locations for centre line settlement estimations

Case history number on graphs (Figs. 2.22–2.27)	Tunnel location	Type of ground	Reference
1.	Belfast, N. Ireland	Soft silt with overlying fill	Glossop et al. (1979)
2.	Ouseburn, Newcastle upon Tyne, England	Recent fill (rubble, household waste and ash) in a soft clay matrix	Spencer (1978), Dobson et al. (1979)
3.	Howdon, Newcastle upon Tyne, England	Boulder clay	Glossop (1978)
4.	Willington Quay, Newcastle upon Tyne, England	Silty organic alluvial clay	Sizer (1976), Attewell et al. (1978)
5.	Hebburn, Newcastle upon Tyne, England	Stony laminated clay	Attewell and Farmer (1973)
6.	Green Park, London, England	London Clay	Attewell and Farmer (1974b)
7.	Grimsby B2, England	Soft alluvial clay (Marine 'warp')	Glossop (1980)
8.	Grimsby B1, England	as 7	as 7
9.	Grimsby C, England	as 7	as 7
10.	Thunder Bay South, Canada	Silt	Morton and Dodds (1979)
11.	Heathrow, London, England	Upper ground section of London Clay with 3.6 m of clay cover to tunnel under wet gravel	Wood and Gibb (1971), Smythe-Osborne (1971)
12.	Norton, Cleveland, N.E. England	Sandy, silty clay	Hurrell (1984a)
13.	San Francisco, USA	Slightly cemented dense silty fine sand dewatered by deep wells	Peck (1969)
14.	Washington DC, USA	Medium dense silty sand and gravel, interbedded with sandy silty clays	Hansmire (1975)

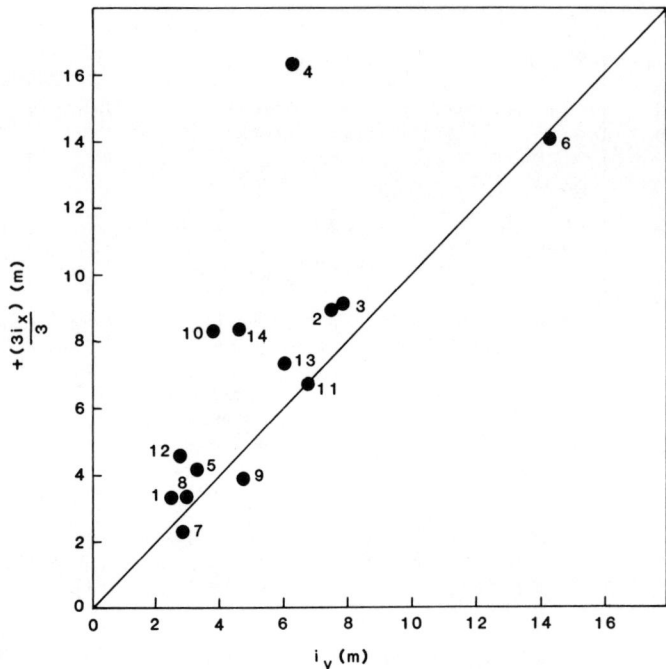

Figure 2.22 Quality of fit between estimated and measured tunnel centre line settlement curves just at the onset of settlement.

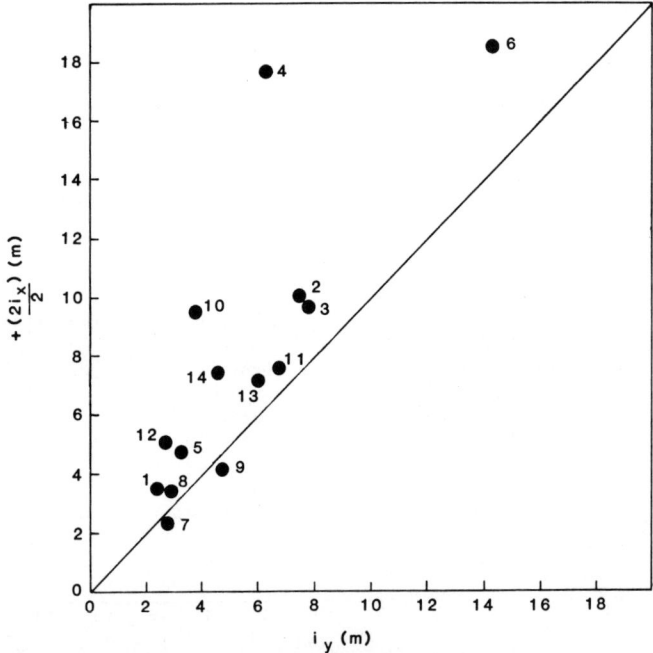

Figure 2.23 Quality of fit between estimated and measured tunnel centre line settlement curves at the point where the settlement is 2.3% maximum.

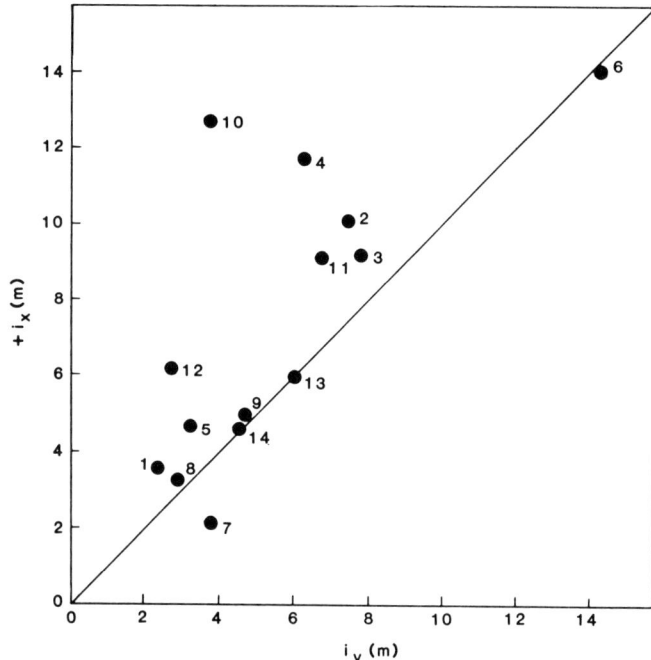

Figure 2.24 Quality of fit between estimated and measured tunnel centre line settlement curves at the point where the settlement is 15.9% maximum.

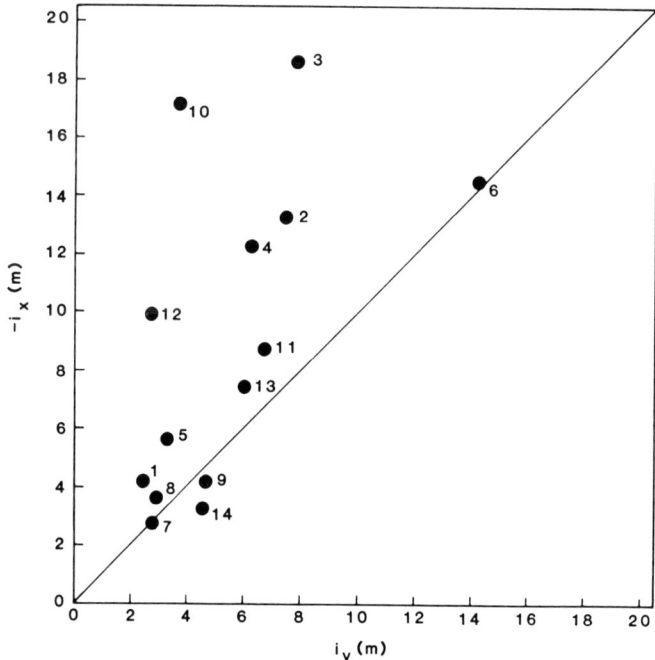

Figure 2.25 Quality of fit between estimated and measured tunnel centre line settlement curves at the point where the settlement is 84.1% maximum.

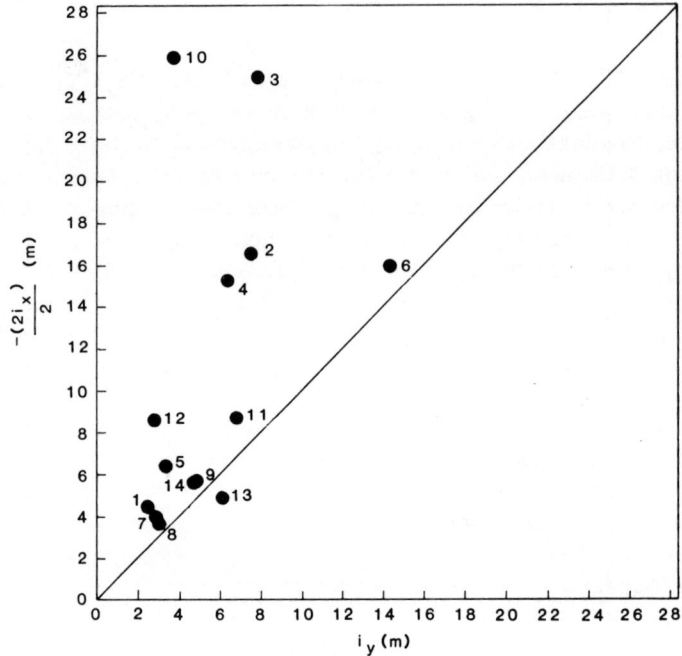

Figure 2.26 Quality of fit between estimated and measured tunnel centre line settlement curves at the point where the settlement is 97.7% maximum.

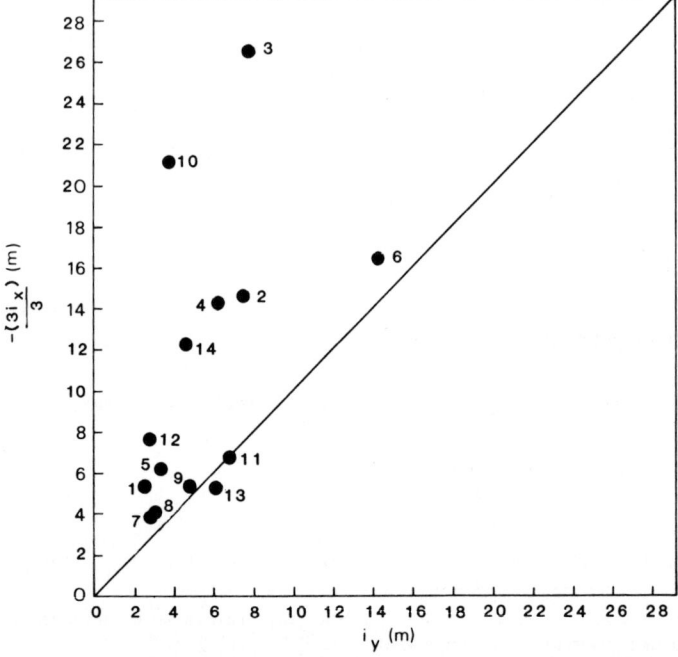

Figure 2.27 Quality of fit between estimated and measured tunnel centre line settlement curves at the point where the settlement has maximized.

tunnel actually behaved as the point source of linearly translating ground loss specified by Attewell and Woodman (1982) for their analyses and incorporated in equations (2.22) to (2.27). While it may be reasonable to idealize a tunnel face proper as a point source of loss for settlements ahead of the face, the continuing settlements above a lined tunnel owe their origins more to distributed radial losses at the tunnel, these losses being progressively inhibited and delayed as contact grout sets behind a segmental lining and the lining ring stiffness is fully mobilized. Because consolidation effects in clay soils, discussed in section 2.8, should be additive to ground loss settlements, the differences between measured and estimated centre line settlement curves behind the face could be expected to be less than actually indicated by the points on Figures 2.22–2.27.

The practical implications of the curve mismatches are not as serious as might at first be suspected. Since all points in the vicinity of the translating tunnel face experience the same wave of x-axis ground movement and strain, the preliminary translation in x on the graphs may be ignored in any practical appraisal. Ahead of the tunnel face the ground around buried pipes and building foundations lying along the centre line is strained such that they experience their worst longitudinal (x-axis) tensions and superimposed bending tensions. In this area where the curve mismatches are usually not large, any structural analysis based on i_y and an x-coordinate cumulative probability curve is reasonable. Behind the face, where the mismatches are greater, the cumulative probability centre line curve based on i_y estimates early postshield settlements and rates of settlement that are greater than those measured. Accordingly, the estimated (and temporary for any given element of ground) radii of curvature parallel to the tunnel centre line are usually greater than those actually measured, and any structural damage assessment based on the equation (2.22) generating the settlement curve, or the curve itself, would tend to be conservative.

It is concluded that the adoption of parameter $i = i_y = i_x$ for the estimation of settlement and its derivatives parallel to a tunnel line is generally valid for most practical design problems.

2.8 Time-dependent settlement

Having estimated ground losses at the tunnel (sections 2.1 and 2.3.1), for clay soils perhaps on the basis of an overload factor (see equation (2.28) and Figure 2.11), and then having decided on the degree to which those losses are transferred to form a settlement volume, there is still a need to predict additional consolidation[†] effects. Although consolidation actually begins at the excavation stage and with settlement trough formation, it is its superpo-

---

[†] The term 'consolidation' relates strictly to clay soils. Any long-term settlement effects in granular soils should be referred to as 'compression'.

sition on the shorter-term transverse settlement profile formed immediately after passage of the tunnel face that requires investigation. Deepening and (in many cases) widening of a ground-loss settlement trough can obviously change the response of a building or in-ground structure over the longer term of consolidation. *It is emphasized that the effects of consolidation must be added to those of ground loss in order to predict any terminal settlement magnitude and distribution.*

Two of the mechanisms promoting consolidation of the ground above a tunnel have been noted earlier. There is the phenomenon of direct gravitational drainage, under transient drawdown conditions, into the excavated void. Drainage may continue until the tunnel is sealed (lined, caulked and contact-grouted), although the sealing will rarely be absolute. Even very slight seepages can have a significant ground-volume reduction effect in view of the originally small volumes of porewater in clay soils. There is also the reduction of porewater-pressure effect in the soil around the tunnel. Soil dilates into the tunnel void until restrained by the lining and so encourages porewater migration towards the disturbed ground. In a similar manner stiff fissured clays may be locally sheared, weakened and dilated, any opening up of existing fractures creating a zone of higher permeability around the tunnel.

Authoritative guidance on this important subject is currently limited, since case history evidence is sparse. Some general comments may, however, be helpful for design purposes.

2.8.1 *Total maximum settlement estimation from simple overload factor*

Hurrell (1983, personal communication) has examined several case histories and, for cohesive soils below the water table, has *tentatively* suggested an empirical relation for the total (ground loss plus longer-term consolidation) maximum settlement w_{maxt} (mm) above the tunnel centre line:

$$w_{\text{maxt}} = (2w_{\text{max}})A \cdot \text{OFS} \tag{2.54}$$

where w_{max} is the maximum ground loss settlement (mm), A is a consolidation settlement coefficient to be determined, and OFS is the simple overload factor

$$\frac{\sigma_{z_0}(+q) - \sigma_i}{c_u} = \frac{\gamma z_0(+q) - \sigma_i}{c_u}.$$

The coefficient A is evaluated in Table 2.5 for the cases shown in Figures 2.28–2.32.

Figure 2.33 suggests that the lower the ground-loss settlement (that is, the stiffer the ground), the higher, proportionately, is the consolidation settlement contribution to the total settlement. The form of the relation between A and w_{max} can be used, albeit approximately, to predict the total maximum

ESTIMATION OF GROUND MOVEMENTS

Table 2.5 Some settlement case history data for the evaluation of the A coefficient.

Belfast (Glossop and Farmer, 1977)	Grimsby Array B1 (Glossop, 1980)	Grimsby Array C (Glossop, 1980)	Willington Quay (Attewell et al., 1978)
$w_{max} = 17$ mm	$w_{max} = 36$ mm	$w_{max} = 55$ mm	$w_{max} = 25$ mm
OFS = 3.64	OFS = 3.95	OFS = 5.62	OFS = 5.9
$w_{maxt} = 40$ mm	$w_{maxt} = 70$ mm	$w_{maxt} = 103$ mm	$w_{maxt} = 85$ mm
From eqn (2.54)	From eqn (2.54)	From eqn (2.54)	From eqn (2.54)
$A = 0.32$	$A = 0.25$	$A = 0.17$	$A = 0.29$

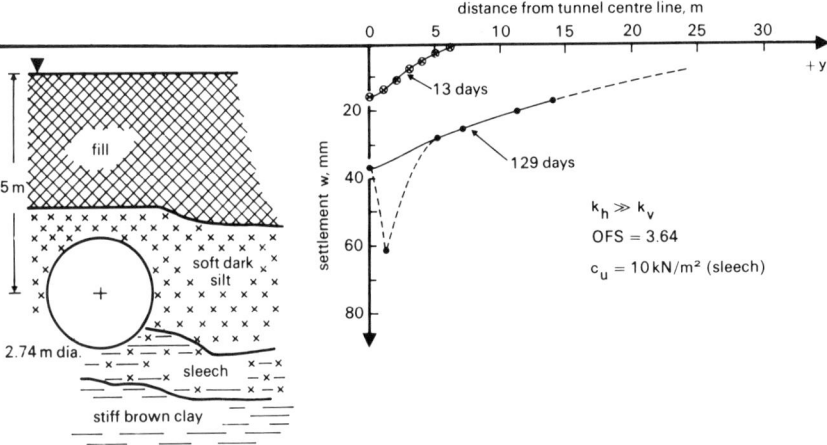

Figure 2.28 Transverse surface settlement distribution: Belfast, King George VI playing fields (after Glossop and Farmer, 1977, 1979).

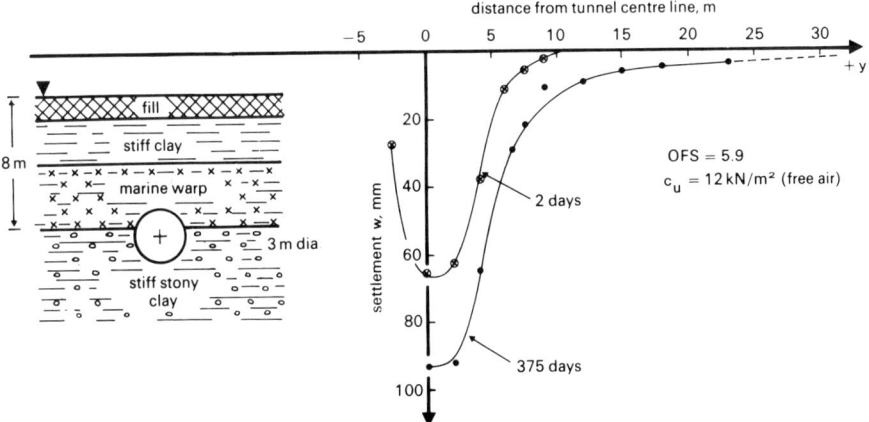

Figure 2.29 Transverse surface settlement distribution: Grimsby Array A (after Glossop, 1980; Glossop and O'Reilly, 1982).

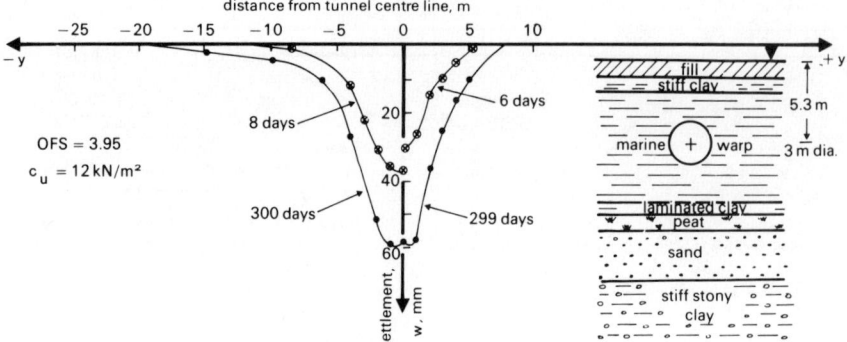

Figure 2.30 Transverse surface settlement distributions: Grimsby Array B1 (after Glossop, 1980, Glossop and O'Reilly, 1982).

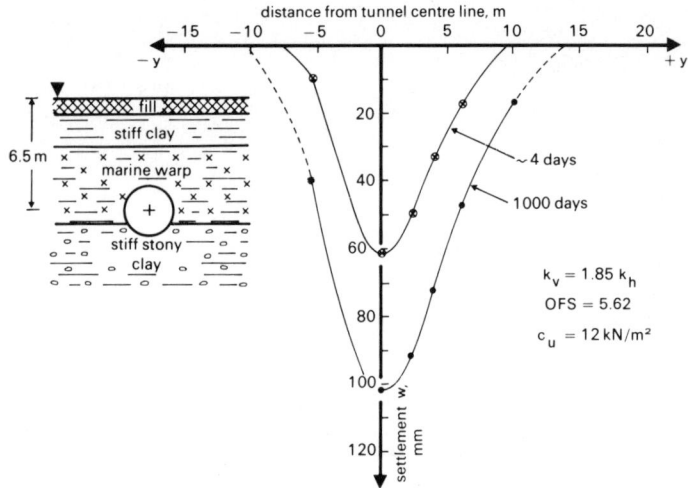

Figure 2.31 Transverse settlement distributions: Grimsby Array C (after Glossop, 1980; Glossop and O'Reilly, 1982).

settlement, w_{maxt}, for other case histories. This can be done by reading off the A value appropriate to a measured ground-loss settlement and then inserting it in equation (2.54) or by incorporating the A–w_{max} relation directly into equation (2.55):

$$w_{maxt} = 0.78 \, \text{OFS} \, (w_{max} - 0.01 \, w_{max}^2) \quad \text{for } 6\,\text{mm} \leqslant w_{max} \leqslant 63\,\text{mm}. \quad (2.55)$$

It is likely that this equation will be modified when more consolidation settlement data become available. Its application to some other case histories is shown in Table 2.6.

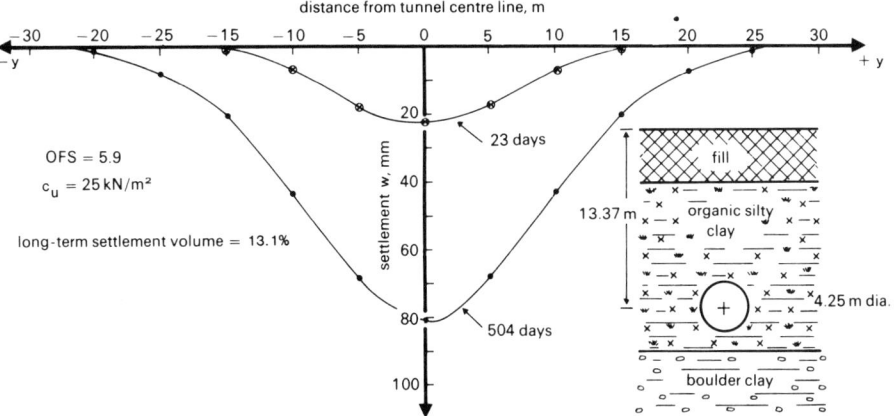

Figure 2.32 Transverse surface settlement distributions: Willington Quay (after Glossop, 1978).

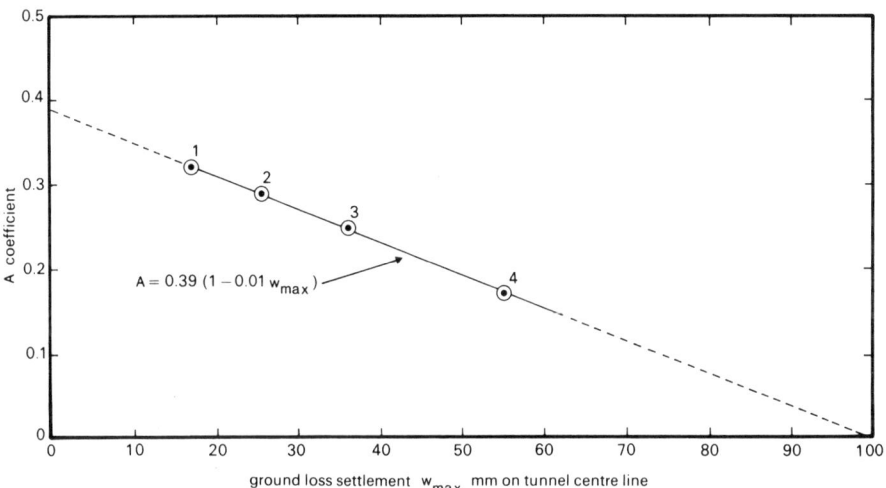

Figure 2.33 Variation of consolidation coefficient A with short-term ground loss settlement. For data points, see Figures 2.28–2.32 inclusive. 1, Belfast, OFS = 3.64; 2, Willington Quay, OFS = 5.9; 3, Grimsby B, OFS = 3.95; 4, Grimsby C, OFS = 5.62.

2.8.2 Total maximum settlement estimation using a compression index

This method, used by Attewell *et al.* (1978) for analysis on a Northumbrian Water Authority interceptor sewer at Willington Quay, northeast England (see Figure 2.32) requires measurement of the long-term porewater-pressure change by a piezometer installed above the tunnel crown. It takes no account of any possible consolidation below tunnel invert level. Maximum centre line consolidation settlement w_{maxc} is related to piezometric pressure change Δp by

Table 2.6 Comparison of observed and estimated settlements (based on Hurrell, 1984b).

Project	Tunnel diameter (m)	Depth z_0 (m)	OFS	Observed		Predicted		w_{maxt}^* (m)	Remarks
				w (m)	w_{maxt} (m)	w (m)	w_{maxt} (m)		
Willington Quay, Newcastle upon Tyne	4.25	13.375	5.9	25	85	75.4	85.4	85	Silty alluvial clay. Compressed air working. (Attewell et al., 1978)
Grimsby Array B1	3.0	5.3	3.95	36	70	44.8	76.2	70	Inorganic high plasticity clay. Compressed air working. (Glossop, 1980)
Grimsby Array C	3.0	6.5	5.62	55	103	66.1	98.2	103	
Belfast	2.74	5.0	3.64	17	40	27.4	56.1	40	Soft estuarine plastic clay. (Glossop and Farmer, 1977)
Green Park, London	4.15	29.5	2.10	6.17	12	5.5	8.5	9.5	Stiff fissured overconsolidated clay. (Attewell and Farmer, 1974a)
Regent's Park, London	4.15	34.0	2.90	5.0	10	5.9	12.6	10.8	Stiff, fissured overconsolidated clay. (Barratt and Tyler, 1975)
Hebburn, Newcastle upon Tyne	2.08	7.5	2.02	7.86	13	6.1	9.0	11.4	Stony/laminated clay. (Attewell and Farmer, 1973)
Howdon, Newcastle upon Tyne	3.68	14.18	1.50	8.90	11.2	5.4	6.0	9.5	Stiff, stony clay. (Glossop, 1978)

*denotes that the prediction is based on short-term response.

the equation

$$\Delta V = \varepsilon_z = \frac{\Delta H}{H} = \frac{\Delta e}{1+e_0} = \frac{C_c}{1+e_0} \log\left(\frac{p_0 + \Delta p}{p_0}\right) \qquad (2.56)$$

where ΔV is the volumetric strain, which is equal to the vertical strain ε_z for *one-dimensional* consolidation, ΔH is the change in thickness of soil element having initial total thickness H, Δe is the corresponding change in void ratio from an initial value of e_0, p_0 is the initial vertical pressure at the tunnel crown before measurable soil consolidation occurs, Δp is the change in ground pressure above the tunnel crown as a result of drawdown and soil consolidation (equivalent to the increase in effective stress); Δp ultimately is equal to $\gamma_w \Delta H$, where γ_w is the unit weight of water; and C_c is the compression index.

Note that the thickness, H, of the consolidating layer must be estimated from settlement measurements in the ground. Effective stress normal to ground surface then increases as a result of drawdown, and ΔH is equivalent to consolidation settlement w_{maxc}.

In the case history quoted by Attewell *et al.* (1978), values of 1, 0.3, and 4 m were assigned to e_0, C_c and H, respectively, for the soft silty alluvial clay. A p_0 value of 205 kN/m^2 was calculated from the tunnel depth to crown and the soil bulk unit weight. A recorded reduction in piezometric pressure Δp of 22 kN/m^2 was accompanied by a measured settlement of 25 mm. The calculated settlement based on the above equation was 27 mm.

Estimation of consolidation settlement above future tunnels may be based on knowledge of p_0 from the soil properties measured during the ground investigation and calculation of Δp from possible or allowable seepage rates into and along the tunnel as indicated by measured or by inferred ground permeabilities.

The method presented by Mitchell (1983) assumes consolidation settlement to occur through an entire soil profile from ground surface down to an impermeable stratum below tunnel invert. Mitchell (1983) takes the case of a tunnel, excavated diameter d, positioned with its centre at a distance z_2 above a stiff base layer which serves as a reference horizon. The axis depth of the tunnel is equal to $z_1 + h$, where h is height of the original groundwater table above axis before tunnelling and z_1 is the soil cover above that groundwater table. If γ is the unit weight of the soil, γ' is the submerged unit weight ($= \gamma - \gamma_w$). and $z_3 = h + d + 0.5z_2$, then the maximum consolidation settlement for a normally consolidated soil can be expressed as

$$\Delta H = \frac{hC_c}{1+e_0}\log\left(1 + \frac{\tfrac{1}{2}h\gamma_w}{z_1\gamma + \tfrac{1}{2}h\gamma'}\right) + \frac{z_2 C_c}{1+e_0}\log\left(1 + \frac{h\gamma_w}{z_1\gamma + z_3\gamma'}\right). \qquad (2.57)$$

Application of this equation relies on the identification, at the ground investigation stage before tunnelling, of a horizon, below tunnel level, that could be deemed 'impermeable'.

Table 2.7 A guide to values of compression index C_c of saturated soils (Lee et al., 1983, Table 5.1, p. 190)

Soil type	Index properties		C_c	Source
	Liquid limit	Plasticity index		(see Lee et al. for full references)
Normally consolidated estuarine silty clay (undisturbed)	100+	High	1 to 1.4	Lee et al. (1983)
Marine sediment, B.C., Canada	130	74	2.3	Finn et al. (1971)
Remoulded marine silty clay, Kyushu, Japan	70	43	1.1	Lee et al. (1983)
Deep-water brown marine clay	100 to 200	High	0.5 to 1	Noorany and Gizieski (1970)
Undisturbed organic silty clay, Delaware, USA	84	46	0.95	Schmidt and Gould (1968)
Undisturbed clay, New Orleans, USA	79	26	0.29	Lambe and Whitman (1969)
Stiff mottled clay	69	20	0.20	Lee et al. (1983)
Undisturbed Boston Blue clay	41	20	0.35	Lambe and Whitman (1969)

The Terzaghi and Peck (1967) equation

$$C_c \simeq 0.009(L_w - 10\%) \quad (2.58)$$

where L_w is the liquid limit, could be used for a very approximate value of the compression index. Alternatively, Table 2.7 (after Lee et al., 1983) could be used for guidance.

2.8.3 Form of the terminal (ground loss plus consolidation) transverse settlement profile

The complete long-term behaviour of superadjacent structures cannot be assessed without some knowledge of the form of the terminal settlement profile. Maximum settlement is only one element of ground movement composing this profile. Although there is a shortage of measurement evidence, some guidance can be given.

Since it is the soil drainage facility which controls its consolidation, the first approach considers the effect of permeability anisotropy. When the horizontal permeability k_h greatly exceeds the vertical permeability k_v the long-term settlement trough width would seem to increase substantially beyond the short-term ground-loss trough. As an approximation, the terminal transverse settlement curve may be constructed by dropping ordinates, equal in amplitude to the maximum consolidation settlement ($w_{maxt} - w_{max}$), from all points on the ground-loss curve and extrapolating laterally beyond the span of the latter. Thus, the terminal trough has a wider span than the ground-loss trough but maintains the same curvatures. In those cases where the soil permeability is more isotropic, the width of the transverse settlement trough will still tend to increase in the longer term but to a much lesser extent because of the dominance of vertical drainage. Since the potential for structural damage is a function of differential settlement it is suggested that for structures or parts of structures within the span of the ground-loss trough, this trough should be considered to be deepened but not widened by consolidation settlement. Thus, is the earlier predictive equations ((2.22) to (2.27) inclusive) for ground-loss settlement (and derivatives) distribution, w_{maxt} would be used instead of w_{max} and i_t would be the same as i, so inevitably increasing the ground curvatures and leading to a likely upper-bound pessimistic assessment of possible damage for structures within the ground loss settlement trough. For structures or parts of structures outside the span of the ground-loss trough, any differential settlements could be treated as being negligibly small.

A second approach to the prediction of consolidation settlement distribution (Hurrell, 1984b) considers the longer-term surface settlements to result from consolidation volume loss in the zone of disturbed ground adjacent to the tunnel springings. These loss sources are located at tunnel axis level and at distances plus and minus one tunnel diameter d from the tunnel centre line. They propagate normal probability-form settlement waves to the surface in a manner similar to that which generated the ground-loss settlement profile. The terminal transverse settlement profile is then the resultant superimposition of the short-term ground-loss settlement profile and the two consolidation settlement profiles, the normal probability form of the latter being defined by the same inflexion distance i-parameter determined empirically for the ground-loss profile. This condition is illustrated in Figure 2.34. Quantitatively, the component of maximum consolidation settlement w_{maxc}, at transverse distance $\pm d$, is

$$w_{\text{maxc}} = \frac{w_{\text{max}}(B-1)}{2\exp(-d^2/2i^2)} \qquad (2.59)$$

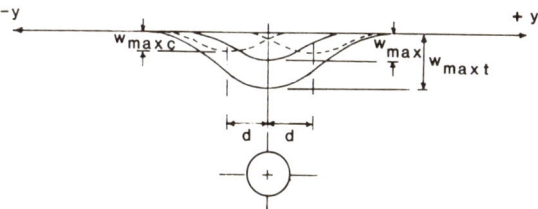

Figure 2.34 Development of long-term surface settlement.

where

$$B = 0.78\left(1 - \frac{w_{max}}{100}\right) \text{OFS for } 6\,\text{mm} \leqslant w_{max} \gtrsim 63\,\text{mm} \qquad (2.60)$$

after equations (2.54) and (2.55).

The long-term surface settlement profile is then defined by

$$w_t = w_{max}\exp(-y^2/2i^2) + w_{maxc}(\exp C_1 + \exp C_2) \qquad (2.61)$$

where

$$C_1 = -(y+d)^2/2i^2 \qquad (2.62)$$

and

$$C_2 = -(y-d)^2/2i^2. \qquad (2.63)$$

Thus, by this second method, the procedure for terminal transverse settlement profile definition is as follows:

Evaluate the short-term ground-loss trough parameters w_{max}, i, V_t. Parameter V_t is evaluated from equation (2.19) in section 2.1.3 and/or the information given in section 2.4.1. Parameter i may be evaluated from equations (2.31) or (2.32), or by reference to Norgrove et al. (1979) and Attewell and Woodman (1982). Parameter w_{max} is then evaluated from equation (2.22) in section 2.4 for $y = 0, G(\overline{x - x_i}/i) = 1$ and $G(\overline{x - x_f}/i) = 0$.

Calculate the value of the ultimate centre line surface settlement w_{maxt} from equation (2.55).

From equation (2.59), estimate the maximum component of consolidation settlement, w_{maxc}, at $\pm d$ from the tunnel centre line.

Use equation (2.61) above to define the complete long-term surface settlement w_t as a function of transverse distance $\pm y$ from the tunnel centre line.

2.9 Numerical methods

The finite-element method of analysis has direct application to the problem of tunnelling ground movements and the effect of those movements on structures. As shown in sections 2.5 and 2.6, any foundation or buried pipe within the zone of influence of the tunnel excavation will be subjected to a cycle of deformation. Accordingly, any finite-element modelling must take account of the three-dimensional character of the movements. It is not satisfactory to model two-dimensionally on a plane-strain basis solely for the permanent transverse deformation.

Finite-element modelling cannot be considered in any detail in this book. Three-dimensional finite-element programs are available commercially (e.g. PAFEC in the UK) and may purchased or rented by organizations such as universities for use on suitably powerful mainframe computers. Civil engineering consultants often have their own in-house facility. There are varying degrees of refinement on offer—mesh generation, plasticity, viscoelasticity and so on. It will be realized, however, that these programs are non-specific, and further programming may be needed to apply them to a particular ground engineering problem. For this reason it is often desirable to develop a program for the special problem of tunnelling ground movements and ground–structure interaction. In outlining the application of such a program, some of the problems and some of the methods of avoiding them are mentioned.

The relative dimensions of the structures in the ground create difficulties. A typical tunnel may be 2.5 m excavated diameter at an axis depth of, say, 13.5 m. Pipes likely to be affected by ground movements could be 0.5 m diameter at a depth of 1.5 m, but with a wall thickness of only a few millimetres. In most cases it is impractical to mesh the pipe wall(s) in continuity with, and as part of, the general mesh which also incorporates the tunnel. The same restriction applies to column footings.

Two methods, each with their own particular disadvantages, have been used in an attempt to overcome these problems. First, and in the case of buried pipes, the finite-element mesh in the vicinity of a pipe has been constructed so that a suite of nodal points creates a definable boundary line around the pipe at a distance of between 2.5 and 3 pipe diameters from it. The pipe is then removed from the mesh, the zone vacated by the pipe and surrounding soil remeshed to a scale compatible with that for the rest of the ground, and the program run for the tunnel in a pipe-less (and foundation-less) ground. Note is made of the orthogonal displacements at the nodes forming the above boundary. The limited zone of ground around the pipe is then expanded to a scale suitable for meshing the ground and the thickness of the pipe wall cross-section. The boundary node displacements for this limited zone, as previously determined in the absence of the pipe, are then applied. Output data with respect to total strain is then retrievable for the pipe with respect to position in advance of, or behind, the tunnel face. By noting the relative amplitude of strain, at pipe soffit and pipe invert, along the pipe length, it is then possible to resolve the bending components and direct components of strain which compose the total pipe strain at the different positions along the pipe, although clearly it is the total applied strain that will form the basis of any criticality assessment.

The second method again involves isolating a volume or volumes of ground around a pipe, but this time defining the movements u, v, w at the selected boundary nodes of the finite element mesh by the use of equations (2.24), (2.23), (2.22) in section 2.4. The BASIC computer program for the generation of the design curves in section 2.5 has been used by the present authors for this purpose. These displacements are then entered, as above, into the three-dimensional finite-element program for solution of the pipe strains.

It follows that the two-stage finite-element approach, and/or the analytical–

numerical hybrid approach, can be adopted for the solution of isolated foundation strains.

Modelling the geotechnical properties of the soil in a realistic manner presents problems. Anisotropy can often be handled rather more easily than can inhomogeneity. For the overconsolidated soil that nowhere approximates to a collapse condition at the tunnel, the deformation behaviour may be assumed to be elastic. Estimates of settlement should be based on effective stress parameters E' and v' for drained states of deformation and on total stress parameters E_u and v_u for undrained deformation. Values of v_u just less than— not equal to—0.5 would be adopted. For such numerical analyses, depth-dependent values of E', v', E_u, v_u would be required from a rather lengthy and expensive laboratory test program. The resulting values would be likely to show considerable scatter anyway, and there would be some difficulty in selecting representative values for input to the program. If the soil is only very lightly overconsolidated, or if the unsupported portion of the tunnel is close to collapse, the ground will behave plastically. The finite-element program must be able to accommodate such non-linear properties. It will also be realized that whereas soil deformation conditions in the vicinity of the tunnel face approximate quite closely to a state of plane stress undrained, further back down the tunnel near the newly lined section they become more nearly plane-strain and drained. Even further back, where contact grout has fully set and the transverse surface settlement trough has stabilized, the conditions will be those of plane strain and (if the tunnel has been properly sealed) undrained.

2.10 Site investigation for tunnels in soil

The term 'site investigation' incorporates the early desk studies and site walk-over studies, as well as the actual ground investigation. Knowing where to acquire information on ground conditions (including the geology of the area, geomorphology, hydrology and topography), together with the results from earlier ground investigations, is important at the desk-study stage. Sufficient information from the site investigation is needed for the construction work to be designed for maximum efficiency and safety, to enable contractors to submit realistically 'keen' bids for the work, and to allow employers to make sensible estimates of their financial commitments.

A site investigation that is 'inadequate' with respect to the ground conditions that are actually revealed and which adversely affect the contractor's planned rate of progress will incur financial penalties for the project promoter in the UK via Clause 12(1) of the ICE Conditions of Contract (Institution of Civil Engineers 1973, revised January 1979), which relates to the contractor's encountering physical conditions or artificial obstructions that as an experienced contractor he could not reasonably have foreseen, and/or perhaps via Clauses 44 (extension of time for completion), 51 (ordered variations) and 56(1) (increase or decrease of rate).

ESTIMATION OF GROUND MOVEMENTS 99

The following notes should in no way be regarded as a complete guide to site investigation for tunnels in soil! It is assumed that an employer who identifies and seeks to overcome some of the potential problems discussed in this book will engage the services of an experienced geotechnical engineer. For information on the technicalities of site investigation generally, readers are urged to refer to Dumbleton and West (1976a), to the British Standard Code of practice for site investigations (British Standards Institution, 1981), to Weltman and Head (1983) and to Joyce (1982). Site investigation specific to tunnels is covered in Dumbleton and West (1976b) and West et al. (1981). Contractual conditions for ground investigation are formulated in the document issued by the Institution of Civil Engineers (1983) and analysed in Cottington and Akenhead (1984). Contractual and cost benefit aspects of site investigation for tunnels are discussed in Attewell and Norgrove (1984, a, b, c,) and in Norgrove and Attewell (1984). Mathematical assessment of risk, decision-taking and decision reliability, often centred on Bayesian probability, have recently been discussed with respect to engineering geology and geotechnical engineering by Einstein and Baecher (1983) and by Whitman (1984), respectively.

Ground investigations are often phased so that, for example, later borehole locations, sampling and testing can be prescribed in the light of knowledge gleaned from earlier boreholes. This step-by-step approach is perhaps more easily implemented through a 'Schedule of Rates'-style term contract for the investigation. Retrieval of geological, geotechnical and groundwater information should continue during the construction phase. As a minimum requirement the exposed face should be mapped with respect to soil type each day, and the positions of any groundwater seepages noted on the sketch. This operation can be performed for smaller contracts by a competent clerk of works who has been instructed in the recognition of different soil types. If contractual claims based on unforeseen ground conditions are expected or have been submitted, it would then be sensible for the face to be mapped each shift and for the employer and contractor jointly to take soil samples. If there is any risk of inundation at the face the in-tunnel investigation would be extended to forward probing, with a cost penalty for production delays, but the need for any such probing should be anticipated when the Specification and Bill of Quantities are being written.

Although tunnelling is contractually and physiologically one of the most risky civil engineering construction operations with respect to ground conditions, special site investigation and test procedures are rarely adopted. Exploratory boreholes are normally put down at the centre of each access shaft location unless artesian pressures are anticipated, in which case the holes are offset from the shaft walls. Further exploratory boreholes are usually put down at 200 m or so intervals between shaft positions, these holes being offset about 1.5 tunnel diameters from the tunnel centre line. Care must be taken to record water strikes in each exploratory hole. The ground investigation

100   SOIL MOVEMENTS

contract documents must provide for drilling to cease as soon as water is encountered in a hole, and for time-related incremental readings of water-level increases to be recorded until a maximum head is achieved. Water levels should also be recorded at the beginning and end of each shift. Normal practice is to install standpipe piezometers in the holes for post-investigation readings of groundwater levels, but it should be remembered that smearing of clay soils at the sides of a borehole can render piezometer readings inaccurate. Great care should therefore be taken in the interpretation of readings for soil permeability. Boreholes not receiving piezometers should be suitably backfilled with an impermeable material and capped with a concrete slab.

There is often an argument for putting down large-diameter man-entry boreholes so that the ground can actually be inspected *in situ* through slots in

Figure 2.35 Ground investigation in Newcastle upon Tyne.

ESTIMATION OF GROUND MOVEMENTS   101

the borehole casing. These holes are expensive, however, and may not be easy to justify. For many tunnelling schemes access shafts are sunk at the beginning of the contract, and this should be encouraged purely for site investigation and geotechnical reasons.

Special problems can arise through lack of space in urban areas. Traffic flows may be disrupted, and locations for boreholes may be severely limited. Figure 2.35 shows a soft-ground percussive rig being operated in one of the narrow main streets of Newcastle upon Tyne, but even then with the hole and its casing passing through a vault beneath the pedestrian pavement—see Figure 2.36. Reinstatement after drilling must pay particular attention to sealing against future water ingress into such basements.

Most common *in situ* tests are pumping-in/out (borehole packers) for soil

Figure 2.36 Cased hole from the rig shown in Figure 2.35.

permeability assessment, standard penetration (SPT) for granular soils, and perhaps vane tests for cohesive soils. Ground investigation contractors would advise as to whether less common investigations, such as borehole photography/CCTV and pressuremeter tests, are justified. Simple rising-head tests are to be preferred because of their self-cleansing character.

Although the quantity of ground information may be limited, its quality and style of presentation are of major importance in tunnelling. The bid baseline might be expected to rise with reduced information density, as might the likelihood of claims based on unforeseen physical conditions.

The contract documents will include a factual site investigation report and perhaps an engineering report which expands this report. The factual report will contain the borehole logs which identify the soil type pictorially and descriptively. The logs will also identify water strike and rest water levels, type of sample (disturbed or 'undisturbed') and sample depth, nature and depth of any *in situ* test (e.g. SPT, dynamic cone penetration, vane, permeability). Borehole surface levels must be accurately surveyed from benchmarks or temporary benchmarks, and each borehole must be unambiguously identified on the log. Dates and drilling rates must also be recorded on the log. Within the factual report will be a large-scale plan or plans of the tunnel route and adjacent area. A 1:500 scale is often appropriate for the plan. Beneath the plan, often at a 1:100 scale, will be pictorial representations of the borehole logs, mutually aligned with respect to ordnance datum and showing both ground surface and proposed tunnel boundaries. Soil types and any rock head will be marked, as will water strikes and rest water levels. Such information, properly and unambiguously presented, is essential if a contractor is to price the job to the limit of his experience and if the employer is not to incur unnecessary expense in the form of contractual claims for unforeseen ground conditions that had, in fact, been foreseen.

For technically and geologically difficult work in an urban area where buried pipelines and high-value buildings may be at risk, and where some form of ground-improvement measure (perhaps compressed air) will almost certainly be needed, a more visual representation of the character of the ground can be beneficial. One such problem area is described in Norgrove *et al.* (1979). A simple three-dimensional representation of the ground is shown in Figure 2.37. It comprises a 1:500 area plan pasted to a baseboard, the board representing a certain level above ordnance datum. Quadrant dowelling is used to depict the tunnel ground investigation boreholes and also boreholes put down earlier by other clients for other developments. Rod lengths are scaled to the depths of the individual boreholes. The curved surfaces of the rods are coloured according to the different soil descriptions, and the flat surfaces of the rods are used for marking water strikes and rest water levels. Coloured cotton threads can be used to link stratifications between boreholes and so render interpretation a little easier.

The range of possible laboratory tests (grading, index properties, natural

Figure 2.37 Ground investigation borehole model.

moisture content, bulk density, strength, deformation modulus and chemical) on disturbed and 'undisturbed' samples is large, but those most generally used and useful for tunnel design are noted below. Ground investigation samples should be taken at tunnel face level, and between one tunnel diameter above soffit and one tunnel diameter below invert unless comprehensive finite-element analyses are to be performed, in which case the full depth profile to tunnel horizon should be sampled for testing (see section 2.9). Borehole log definition of the character of the soil at and just above face level is clearly important for estimating ground losses, as discussed in sections 2.1, 2.2, 2.3, 2.4.1, 2.4.2 and 2.8. Special attention must be directed to laminated clays in which silt intercalations containing water at sub-artesian or even artesian pressure have been known to create unexpected problems at the tunnel face. Water-bearing lenses in tills can create similar problems, and so the site investigation must aim, for example, to define any compressed-air or other ground-improvement requirements. It is obviously much less easy from borehole investigations to assess the presence of boulders—their spatial density and their possible size—but such information can be quite crucial in the case, for example, of slurry-shield tunnelling. If their presence is not revealed by the ground investigation (or if the contractor is not expressly cautioned about it in the contract documents), this could lead to substantial claims for extra payment. Borehole cores must be properly handled, labelled, logged, and then stored in an accessible place for inspection by the contractors bidding for the tunnelling work.

Granular soils

Grading curves. Careful note should be taken of the form of the particle-size distribution and the percentage of silt-size material. The amount of silt in the soil when assessed in the context of piezometric information from the borehole is an essential indicator of tunnel face stability and the choice of ground improvement measure(s). It should be remembered that there can be a loss of fines during the drilling process, leading sometimes to an underestimate of the silt contents of disturbed samples.

Cohesive soils

Index tests (liquid and plastic limits). From these values the soil should be defined according to the Unified Soil Classification System (see Casagrande, 1948; Terzaghi and Peck, 1967, pp. 39–42; Attewell and Farmer, 1976, pp. 36–42) and its location with respect to the A-line on the plasticity index–liquid limit modified plasticity chart (Terzaghi and Peck, 1967, p. 41) noted.

Natural moisture content. Great care must be taken throughout the sampling and testing procedures to preserve the true moisture content of the soil. When soil is percussively drilled and sampled above the groundwater table addition of water to the hole to assist penetration should not be allowed. Special note should be taken of how closely the natural moisture content approaches the liquid limit. Particularly with respect to tunnelling, it is useful to calculate the liquidity index from the natural moisture content and the Atterberg limits.

Bulk density. This soil property is invariably measured as part of a laboratory testing package, but in practice when used in calculations it is not normally a sensitive parameter. Soil unit weight is required for stability ratio estimation (see section 2.4.1).

Quick (UU) triaxial tests. There may sometimes be a shortage of material for testing of three or four specimens at different cell pressures. Multistage undrained triaxial tests on a single sample can then be used. The results, plotted as a Mohr diagram, give an approximate idea of soil shear strength, but it is not always easy to assess the strength contribution of any friction component. Undrained shear strength compared with the product (multiple) of tunnel depth and soil unit weight (and taking account of any air pressure temporary support at the face) produces a stability ratio from which the magnitudes of ground losses can be estimated (see section 2.4.1 and Figure 2.11). For overconsolidated soils and for those tunnelling conditions where the factor of safety against collapse is quite clearly high, the mechanical state of the soil will remain within the critical state boundary surface, its behaviour will be elastic and path-independent, and the ground movements will depend only on the initial and final states of the deformed soil. Simple triaxial testing may then be specified. If the soil is only lightly overconsolidated or mechanically could be close to collapse in an excavated, unsupported tunnel, then a high standard of sample acquisition must be specified and consideration given to implementing stress path testing.

All soils

Chemical analyses. Tests of pH, sulphates (Building Research Establishment, 1981*a*), and occasionally chlorides, in soil and water samples should be made. These tests are necessary for primary lining-concrete specification and perhaps extrados protection in chemical environments that are particularly aggressive. They are also needed for specifying worker protection in the tunnel.

Deformation modulus. The analysis of both pipeline deformation (Chapter 3) and structural foundation deformation (Chapter 4) requires data on foundation stiffness. Therefore, for these particular purposes, but not as a general ground investigation test procedure, laboratory stress–strain curves should be derived from specimens confined at pressures equivalent to overburden pressure at pipeline or foundation depth. It may difficult to select an appropriate modulus from a non-linear curve, and it must also be recognized that a laboratory sample can rarely be 'representative' of ground conditions generally. Furthermore, it is likely that a laboratory-derived modulus value will exceed an *in situ* modulus value and, if used in the design calculations could lead to underestimations in the pipeline or foundation deformation calculations.

It is important to note that the desk-study and walk-over phases of the investigation should give attention to the buildings and buried pipelines likely to be affected by ground movements. Operation in the UK of the Public Utilities Street Works Act 1950 (PUSWA) is noted with respect to buried pipelines in section 3.5.1, and the preparation by building surveyors of pretunnelling property schedules is discussed in section 4.14.

2.11 Measurement of ground movements

Measurement programmes serve three primary objectives: (1) to acquire settlement data in anticipation of claims for damage; (2) with some feedback to the resident engineer, and subject to contractual constraints, to suggest changes in construction method and perhaps progress of the works, leading to a reduction in contract costs and compensation claims; and (3) to conduct research into the fundamental causes of, and controls on, deformation and to increase the store of information for improving the quality of future ground-movement estimates.

Some degree of ground movement must be regarded as the unavoidable result of the construction operation in accordance with contract and in the sense of Clause 22(1) (b) (iv) of the ICE Conditions of Contract (Institution of Civil Engineers 1973, revised January 1979). It is tempting to infer that movements, and especially settlements, estimated empirically for the purposes of current project design from earlier case history measurement data, could form the bases from which an 'unavoidable' element could be specified. By implication, movements above those estimated for the particular geological, hydrological and geotechnical conditions pertaining on the current contract would be 'avoidable', and hence the contractor's responsibility on a contractor–client risk-sharing basis. However, problems would ensue from attempting to implement such an approach. It is preferable to use case history experience as a target for restricting ground-loss settlements and consolidation settlements by the adoption of the best possible workmanship in the tunnel, encouraged by a tight specification for the works and keenly overseen by the resident engineer and his staff.

Programme 1 involves precise levelling, by standard surveying methods, of stations secured to the ground surface within the expected zone of influence of the tunnel. For tunnels in urban areas it is not satisfactory simply to hammer studs into a tarmacadam surface and level to the studs. The stiffer membrane represented by a road or pavement construction resists settlement and will not therefore reflect the true soil settlement.

A typical measurement station, cut into a road surface, is shown in Figure 2.38. The 12.7 mm diameter stainless steel rod, cut to half-metre length, domed at the top and machined to a point at the bottom, is driven into the soil after breaking out the road surface. The soil is then excavated carefully by hand before concrete placement around it. The rod clears the top of the concrete by 25 mm and is then concealed by a manhole cover.

Subject to the constraints of space and the presence of adjacent buildings, several stations would normally constitute a transverse array, the likely span of the transverse settlement trough having been calculated previously. A

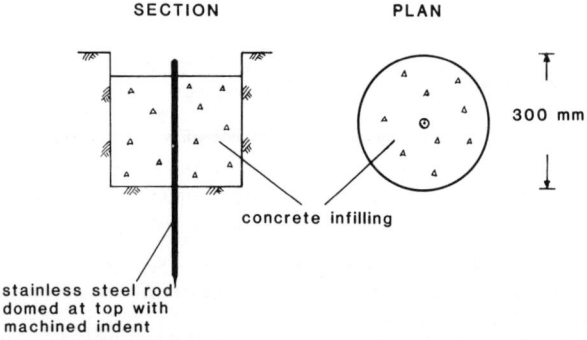

Figure 2.38 A typical surface levelling and banding station.

minimum of five stations, including a centre line station, is required for defining the form of half the settlement trough. There should be further measurement stations on the other side of the tunnel centre line, otherwise symmetrical movement about the centre line will have to be assumed.

Using repeated measurements as a tunnel passes beneath it, a single station on the tunnel centre line will define a settlement development profile. If possible, however, additional centre line stations should be prescribed in order

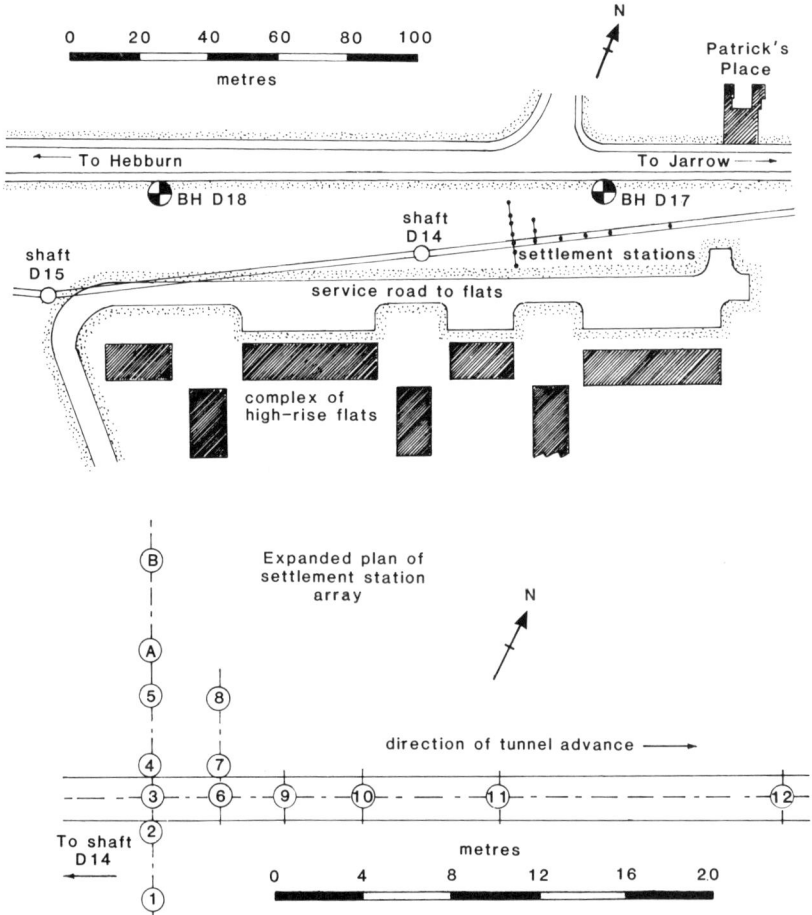

Figure 2.39 Instrumentation array for ground movement monitoring at Hebburn, north-east England. The tunnel forms part of the Northumbrian Water Authority's River Tyne South Bank Interceptor Sewer. Tunnel diameter and depth are 2.024 m (external) and 7.5 m (to axis), respectively, and the tunnel face was in stony/laminated clay. The tunnel, shield-driven in free air, runs approximately parallel to the river, and ground surface at the measurement array slopes at only about 2° northwards. Locations 1 to 12 inclusive represent boreholes containing inclinometer access tubes and magnetic monitors for in-ground settlement. Locations A and B are for surface settlement measurement only.

to extend the information density over a greater zone of tunnelling influence. A settlement station array used by Attewell and Farmer (1973) is shown in Figure 2.39.

Operationally, the base of the levelling staff rests on the domed head of the stainless steel rod. A temporary bench mark can be established from a permanent Ordnance Survey benchmark in close proximity to the settlement point array but outside the influence of any settlement trough. Because the changes in level above and to the sides of the tunnel centre line will often be very small, the demands for careful and precise measurement must be accompanied by equally careful choice of temporary bench mark location on an established rigid structure, minimally affected by environmental changes. A measurement accuracy of ± 0.2 mm must be aimed at.

As emphasized earlier, by the ground-loss equations in section 2.4 for example, there are horizontal movements and strains at ground surface. Such movements, integrated over the distances between levelling stations, have been measured as part of Programme 2 research by University of Durham research teams operating mainly on Northumbrian Water Authority tunnelling contracts in north-east England. Measurements have been taken by steel bands, tensioned (10 kg) by using a spring balance, between indents machined into the centres of the domed heads on the station rods. However, because there are inevitably few measurement stations and because the horizontal movements are much smaller than the vertical movements, the errors are higher than those related to surface levelling. The data are also insufficient to define horizontal movement and strain profiles, particularly in those places either side of the tunnel where the transverse strain changes from compression to tension. Corrections must be made for air temperature, and there may be slight banding errors caused by undulations on the ground surface. Under ideal conditions the limit of accuracy of such a measuring system is the sum of the observational errors, approximately ± 0.6 mm, but since in most cases the operational conditions are far from perfect the total error may be of the order of ± 1 mm.

Measurement of in-ground movement is accomplished first by sinking cased vertical boreholes 150 mm or 200 mm in diameter at specified measuring locations (see, for example, Figure 2.39). The base of each borehole is sunk to such a depth that it experiences no movement from the tunnel excavation (but movement due to any groundwater lowering cannot, unfortunately, be precluded). To extruded aluminium (or sometimes plastic) inclinometer access tubes, having flexibility normal to their long axes and each of assembled length equal to the depth of its particular hole, are attached magnetic settlement rings with spring 'spiders', each ring being located at a predetermined hole depth. The intention is for one of these magnetic ring monitors always to be positioned just above the tunnel crown so that it escapes excavation and continues to monitor the settlement there for comparison with the centre line settlement at ground surface (see section 2.3). It is wise to terminate this particular access tube just above the tunnel crown, for if the tube passes into the tunnel face section any vibration that occurs when the tube is sawn off in the tunnel might affect the accuracy of the crown settlement readings. After lowering each tube with its magnetic rings into its hole, the annulus between tube and casing is then filled with cement–bentonite grout designed to have a 28-day set strength equal to that of the surrounding undisturbed soil. Following grout pouring, the casing is then drawn and the grout topped up. One of the settlement rings is set to the bottom of each access tube line in order to act as a stable reference datum (see qualifying comment above). The authors have used a special removable 'dolly' insert to the inclinometer tubes for acting as the reference head upon which the staff for precise levelling can rest. A temporary bench mark and the deep settlement ring thus act as stable independent references for surface and in-ground settlement measurements.

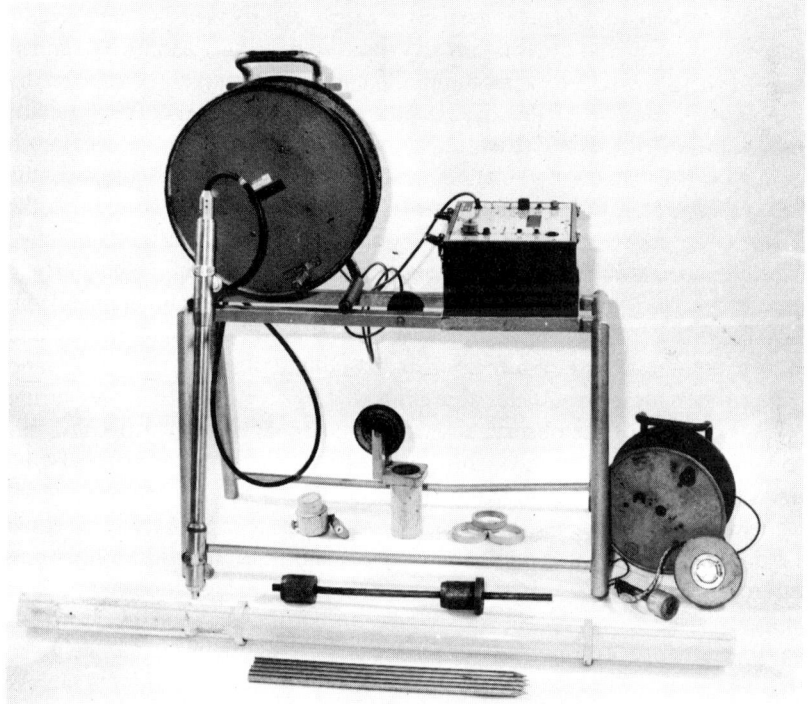

Figure 2.40 Some equipment used for measuring ground movements. Foreground: steel pins for ground surface movement (see Figure 2.38), inclinometer access tubing with two (unattached) magnetic settlement rings, 'dolly' for insertion into top of inclinometer tube and upon which the levelling staff is laid. Resting against the inclinometer support frame is an inclinometer torpedo which is lowered down the access tubing to measure horizontal ground movements. On top of the frame is the inclinometer cable drum and read-out system. Beneath the frame is a lockable top cap for access tubing and more magnetic settlement rings. To the right of the frame is the settlement ring monitor system.

Typical equipment used for measuring ground movements is shown in Figure 2.40. The principle of operation is that the magnetic settlement rings and the inclinometer access tube move according to the movements of the ground into which they are bedded. In spite of the lateral translations of the tubes to which they are attached, the magnetic rings are assumed always to measure vertical movement. Change of ring position is recorded by lowering a probe down the tube and noting the precise depths from the top of the tube at which the probe enters and leaves the ring's magnetic field, as indicated by an audible signal. Such probings require great care in order to achieve sufficient accuracy. The same person should always be designated to take the readings, and for each ring the magnetic field should be entered and exited on both an upward and downward transit. Access tubes contain orthogonal keyways along which a torpedo transducer is lowered for measuring horizontal movement. The torpedo produces an electrical signal, the strength of which is related to the inclination of the torpedo from the vertical. This signal is readily expressed as a horizontal displacement in a known direction, as determined by the orientation of the access tube keyways, and as an absolute magnitude related first to the stable base of the tube and secondly to some fixed surface datum via the top of the tube. Figure 2.4b shows an example of the deflections as measured by an inclinometer system. There is a steel measuring tape for monitoring torpedo depth, and in operation the torpedo is lowered and raised through the full depth of the access tube four times—once for each of the four keyways. Using a torpedo 1 m long, readings are taken at every 1 m increment of depth. Data from each incremental depth and for all four keyways are then reduced by computer to provide a suite of depth-dependent ground-movement vectors in a horizontal plane. These movements are then merged with the settlement ring data to provide three-dimensional ground-movement vectors at the locations of the access tubes.

Modern biaxial sensors record inclination simultaneously in the planes of the two orthogonal keyways, and so separate readings using both sets of keyways are not usually needed. Incremental data are displayed digitally and also recorded on cassette. A microcomputer/floppy disk/line printer system can then be used on-site to process the data and produce graphs of the form shown in Figure 2.4b.

In deep holes, summated twist over numerous sections of jointed inclinometer tube can lead to serious errors. Even with a manufacturer's tolerance on keyway straightness of within 1° per 3 m segment length, a 40 m deep hole could incur an integrated twist error of 13°. Inclinometer holes have been taken to four times this depth, and so it is sensible to incorporate a twist-correction procedure in the processing program. Such correction requires detailed measurement of changes of magnetic bearing with depth across two of the inclinometer tube keyways.

Clearly this type of measurement programme is lengthy, and very much of a research nature, aimed at establishing criteria upon which future predictions of ground movements can be based. The number of instrumented holes will never be deemed sufficient to present an acceptable three-dimensional picture of the movements. On the other hand, with a very large number of access tubes and settlement rings to probe very carefully as a tunnel face passes, it might prove difficult to accomplish a full data-gathering exercise in the available time. For such work it is sensible to have three inclinometer torpedo systems on site, each precalibrated in the holes before any tunnelling-induced ground movement occurs. One torpedo system can be used for measurement, and it should be presumed that one system could be off-site, perhaps at the manufacturer's works undergoing repair. The third system would then be on standby. Similarly, three magnetic ring probe systems should be assigned to each measurement job.

Figure 2.41 (a) Installing a horizontal ground anchor and magnetic ring system at the bottom of a shaft.

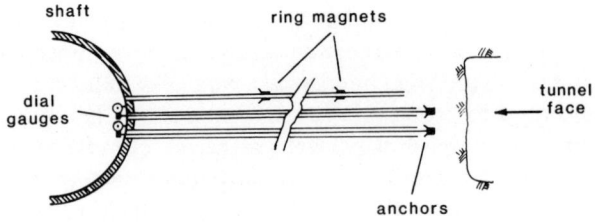

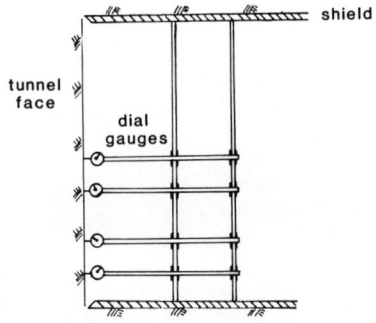

Figure 2.41 (*b*) Plan views of face movement measurements.

The importance of soil relaxation rate has been stressed in section 2.1.1. Laboratory measurements of deformation rate have been shown to accord with measurements taken in the ground. Deformation rates in London Clay have been measured using a ring magnet installed just above a tunnel crown. An alternative system of monitoring rates of inward face movement, which has been used by the authors, is shown photographically and diagrammatically in Figures 2.41*a* and *b*. If a shaft has been sunk in advance of an approaching tunnel face then the results from this system of monitoring may be directly compared with those from the model extrusion test. The system also has the advantage that the inward deformation profile over the tunnel face section may be described.

Adapting this method, three 50 mm diameter auger holes were drilled from the bottom of access shaft D 14 (see Figure 2.39) along the line of the approaching tunnel. Two of these holes were lined with metal tube for a distance of about 6 m. A steel rod having a ground anchor at the far end ran through two of these tubes. The ground anchors were so constructed that, on emerging from the end of the lining, three spring-loaded wings projected from the collar of the anchor. Although the anchor could be pushed into the clay soil, when slight tension was applied to the rods the wings were forced outwards and keyed the anchor into the ground. Where the rods projected into the access shaft they were equipped with dial gauges bearing on to a stainless steel plate which was rigidly fixed to the concrete lining of the shaft. The rods were supported in their tubes

ESTIMATION OF GROUND MOVEMENTS 113

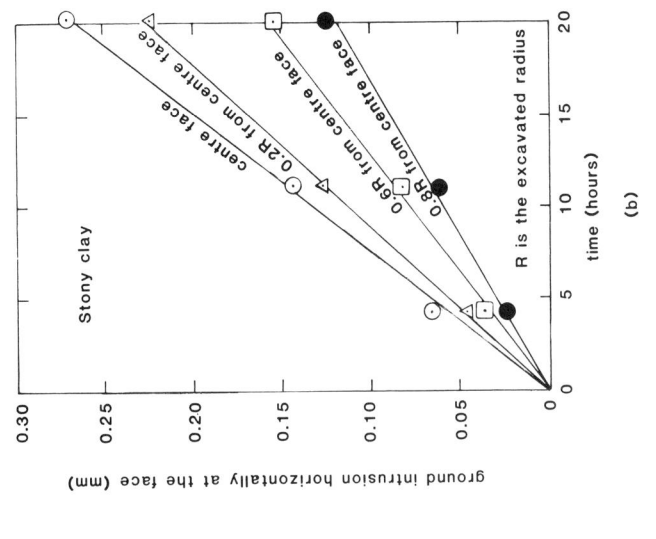

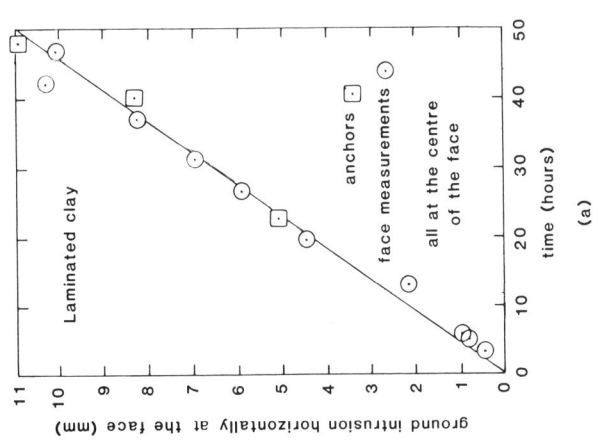

Figure 2.42 Rate of ground intrusion at the tunnel face.

by nylon bearings and the two ground anchors were installed at distances of 6.274 m and 6.223 m from the actual shaft wall. Time-dependent movement of the anchors into the tunnel face as it approached was directly monitored by the dial gauges in the shaft. The third hole contained a small-bore plastic tube with magnetic rings around its circumference at distances of 2 m and 5 m from the shaft wall. An audible reed switch of the type used for monitoring the same form of settlement ring installed on vertical access tubing for an inclinometer was for this work mounted on a stainless steel rod.

Measurements of clay soil intrusion of the face have also been made using dial gauges mounted on extension rods within the shield itself (Figure 2.41b). The extension rods were clamped to bars passing across the mouth of the shield and the gauges bore on to aluminium plates wedged into the tunnel face immediately following the end of excavation for the week.

Figure 2.42a is an example plot of the magnitude of soil inward movement against time from two ground-anchor experiments and two direct-face experiments performed in a tunnel in laminated clay. The points relate to the centre of the face where the intrusion rate is highest; they produce a good straight-line fit having a gradient (intrusion rate) of 0.221 mm/h. This figure agrees closely with the extrusion rate value of 0.218 mm/h derived from laboratory extrusion tests made on undisturbed samples of this same clay from the same location at the same overburden pressure.

Figure 2.42b is a plot of soil intrusion against time related to four points on a tunnel face in stony clay. In this case, data were obtained from four dial gauges located across the horizontal radius of the face (Figure 2.41b). Although the data are limited, the increase in intrusion rate towards the centre of the face is clear, indicating that the clay soil has tended to develop a dome-like form at the face of the shield rather than shearing around the cutting edge and intruding as a plane-ended cylinder. This feature is quite consistent with observations made on laboratory extrusion tests (Figures 2.3a, b) where the face 'domes' until failure, at which point the clay begins to extrude as a cylindrical plug by shearing around the circumference of the aperture. There was no evidence of direct shear at the tunnel face proper. The intrusive doming effect may be accommodated in any ground-loss analysis by the application of equation (2.1) in section 2.1. It is noted that the dial gauge at the centre of the face in stony clay indicated a maximum soil intrusion rate of 0.0134 mm/h, which was much lower than the intrusion rate for laminated clay.

2.12 Measurements on structures

Pipelines. Movements in the surrounding ground will be transferred in part to the pipe bedding material and thence to the pipe itself. Pipe strains may be resolved into direct (axial) compressive or tensile strain, to longitudinal bending strains having components in the vertical and horizontal planes, and to ring bending. Vibrating-wire strain gauges seem to be the most suitable for

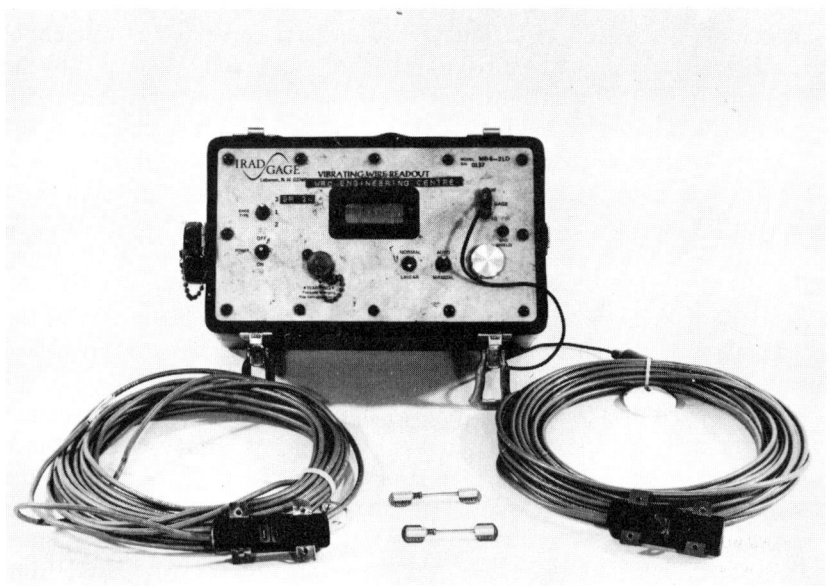

Figure 2.43 Vibrating wire strain gauges for resolving induced strain distribution in buried pipelines.

measuring pipe strains, having long-term stability and inherent robustness. IRAD and Gage Technique vibrating wire acoustic gauges have been used for this purpose (see Figure 2.43).

Before the approach of the tunnel face, the ground at the chosen location is excavated down to the crown of the pipe and just down to the pipe axis on one side. The minimum of ground is disturbed to allow man-entry for fixing the gauges. It may be necessary to cut away corrosion protection applied to the extrados of the pipe in order to attach the gauges. A firm screwed attachment of gauge to pipe is ideal, but in many cases a statutory undertaker will not permit this, and a strong epoxy-type adhesive must be used. In order to resolve an imposed state of strain a minimum requirement is for two gauges mutually at right angles in both the crown and axis locations. One gauge of each pair is aligned axially along the pipe and the other is aligned circumferentially. After mounting and attaching lead-out wires for the conditioning unit, any exposed pipe must be reprotected, as must be the complete gauge assembly. The access trench is then carefully backfilled and recompacted.

Buildings. Two systems of measurement have been used by the authors. First, tiltmeters have been installed in basements of buildings likely to experience some rotation as a result of nearby tunnelling. Two meters are most

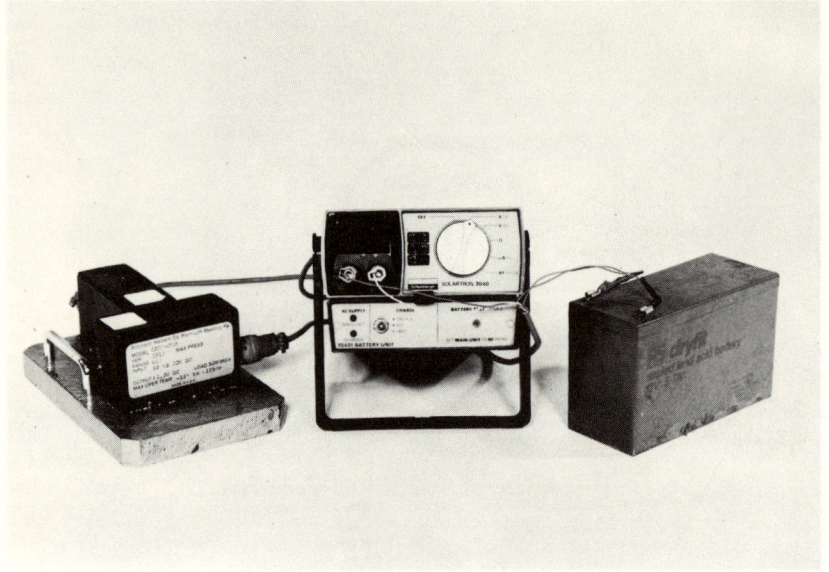

Figure 2.44 Two tiltmeters, set at right angles, for recording rotational movements in buildings.

conveniently arranged at right angles to one another (Figure 2.44) so that the magnitude and direction of changes in slope can be resolved in relation to the tunnel face position. These meters are powered by a stable DC voltage and the output, precalibrated in terms of angular rotation (tilt), may be read on a millivoltmeter or recorded either on a chart recorder or on magnetic tape.

The second system involves the fixing of Demec points, usually on the inside or outside walls of buildings likely to be affected by tunnelling and on both sides of pre-existing cracks. Demec points are small stainless steel discs, 6.6 mm in diameter and 1.64 mm thick, containing a drilled-out centre hole into which the locating pins both of the standard setting gauge and of the measuring gauge fit accurately (Figure 2.45). The setting gauge is used to locate pairs of pins an exact predetermined distance apart, the pins being fixed to the wall by means of epoxy resin. Any change in distance between a pair of points preset to standard is then indicated on the dial of the measuring gauge and quantified as a strain via a manufacturer's (W. H. Mayes and Son (Windsor) Ltd) calibration factor. Readings down to 20 microstrains are theoretically possible, but it is perhaps inadvisable to quote to an accuracy of more than 100 microstrains. In some instances, information on relative *displacement* between points can be rather more valuable, in which case the resolved strain should simply be multiplied by the standard gauge length. It is advisable to correct readings for any temperature variations that may occur on the wall at the different times and dates of reading.

Figure 2.45 Demec gauge and points.

It is likely that future developments in laser technology will allow remote monitoring of structural deformation (Wilson, 1984). It is less likely that strain gauges of the wire or foil resistance type, in say a 45° rosette configuration, will be used for deformation measurements on buildings, but there may be occasions when strain-gauge-based load cells can be inserted between structural members in order to measure changes in tunnelling-induced loads.

2.13 Shafts

Although this book is concerned with ground movements and tunnelling, the fact that shaft construction for access to tunnels, for manholes and for ventilation purposes is an essential part of the overall tunnelling operation suggests the inclusion of some brief notes on the most usual shaft-construction methods.

Underpinning. Bolted segments are sequentially hung from an initially erected ring of segments which are secured against sinking by grouting or by use of a suitable collar at ground surface. During excavation at the base of the shaft only sufficient ground is taken out for the assembly of a complete ring of segments before lifting and bolting to the ring above. This method is attractive for construction in poor ground since the unsupported excavation area can be restricted to that sufficient to erect and bolt one segment to the ring above.

Excavation continues in this manner, ring by ring, until formation level is reached. It is important for the rings to be grouted frequently, otherwise downdrag forces on the outside of the segments can cause high tensile forces to develop around the rings, leading to cracking of the segments and fracturing of the bolts. This is a quick, traditional method of construction suitable in cohesive soils and dry granular soils.

Permeable strata and/or lenses of water-bearing granular soil may be encountered during sinking. Groundwater must be removed, most conveniently by pumping from a sump dug in the base of the excavation, so that construction can proceed in sensibly dry conditions. If this proves to be inadequate, then vacuum ejectors may have to be installed around and outside the perimeter of the deepening shaft.

The shaft base must be sufficiently thick so that, together with the weight of the lining rings (adjusted for wall friction), the total weight exceeds the maximum likely hydrostatic uplift forces on the base. If a base thinner than that calculated must be used, bleed wells are normally incorporated into the construction to allow the hydrostatic pressures to be continuously dissipated.

Caisson method. Precast concrete rings are sunk under their own weight or by the application of kentledge, the soil either being grabbed out by bucket in the dry or under water. Skin friction may be reduced by the application of bentonite mud maintained in an annulus above a choker ring, itself above the cutting edge.

Problems with water inflow and blowing at the excavation base can be removed by using mechanical excavation under water. Calculations for base heave, involving shaft-wall weight, should take no account of frictional resistance if a bentonite lubricant is used.

Construction within a sheet-piled cofferdam. Lateral transmission of groundwater can be cut off by driving a sheet-piled cofferdam to a suitable distance below the base of the excavation. If the piles are to be incorporated into the permanent works the base slab will be fastened to them. On the other hand, the shaft can be constructed within the cofferdam and the sheet piles withdrawn when the base slab is cast. In either case, control of the hydrostatic pressure beneath the base slab must be maintained until the weight of the shaft exceeds the maximum hydrostatic uplift force. Any friction between sheet piles forming part of the permanent works and the surrounding soil can be used to resist hydrostatic uplift pressure and should be entered into the calculations.

Ideally, and if available at a convenient depth, the sheet-pile toes will be seated in a less permeable and firmer horizon. If not, then a suitable cut-off depth can most simply be estimated by flow-net analysis involving the hand-sketching of flow lines and equipotential lines.

Diaphragm wall method. Instead of sheet piling, a diaphragm wall—say of hexagonal or pentagonal shape—can be used to support the excavation, and the wall can be incorporated into the permanent works. Construction of such a wall, using bentonite for temporary support with fluid concrete being tremied in at the base to surround a reinforcement cage while displacing the bentonite upwards and out of the trench, is usually a specialist operation.

Jet grouting method. For shaft construction, jet grouting involves the formation *in situ* of contiguous panels of replacement material, usually a cement slurry, for soil removed (see Figure 2.46). Initially, two guide holes, nominally 150 mm diameter, are sunk to the depth at which the panel wall is to be formed. A monitor containing two jet orifices is lowered to the base of one hole, the other hole serving as a disposal route for the soil slurry that is replaced. The upper, soil-excavating, jet relies for its power on water expelled under very high pressure through a very fine nozzle directed towards the spoil-removal hole. The water jet is surrounded by a concentric collar of compressed air which concentrates the jet, particularly below the water table. Cement slurry is injected through the lower jet into the space created by the removal of the soil fines as the monitor is progressively lifted. Adjacent panels of set cement slurry are linked to form continuous walls of low permeability.

Shaft base. A shaft base should be cast on to a clean gravel blanket and contain bleed wells which must be maintained in an effective condition until such time as the weight of the base and the walls exceeds the total hydrostatic uplift forces. The excavation will create an immediate, quasi-elastic heave of the

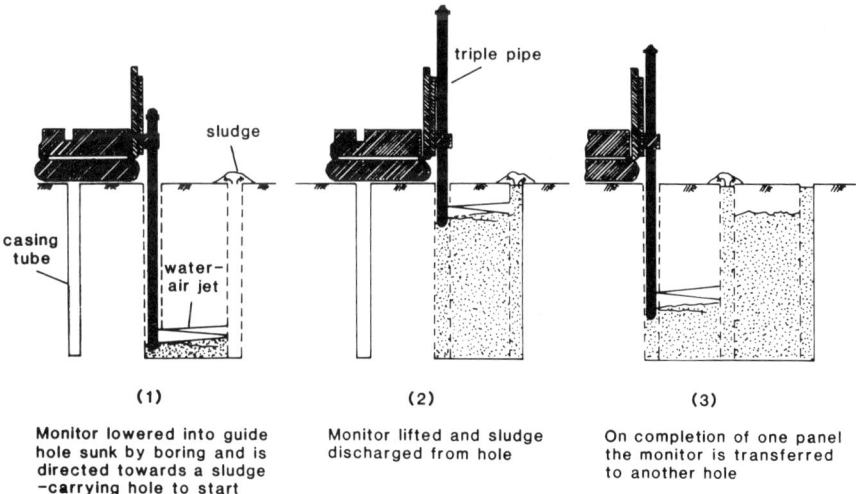

Figure 2.46 GKN Keller Jet Grouting for panel construction.

base—more in the centre than at the sides, which are contained by the shaft walls. Added to this heave there is usually swelling over a period of time in a clay soil. Progressive excavation will obscure the immediate heave and some of the swelling will be resisted by early placement of the base slab, itself to be reduced by subsequent consolidation settlement. The consolidation settlement cannot exceed the degree of swell, and the side friction along the walls of the shaft will tend to reduce load applied by the base. When the calculated weight of the emplaced construction material exceeds the total hydraulic uplift, the wells can be shut off. The original groundwater table will then be progressively restored, resulting in upward elastic and time-dependent movements at the shaft. Side-wall friction at the shaft will restrict this movement, the magnitude of which will in any case be less than the total settlement experienced by the shaft construction before bleed-well closure. Connections between tunnels and shaft should either be of rigid design to resist differential movements between the two or should be flexible enough to accommodate them. It is, in any case, advisable to complete the shaft and base construction some time before making the tunnel connections so that most movement will have ceased.

Buried services. There are several options, including the following.

Isolation: Excavation around an old cast iron main in a carefully staged operation and replacement of the original, compacted backfill with a more deformable material such as blast-furnace coke. If the trench excavation is in a busy street, then the trench must be bridged by cast iron sheeting. After ground movements have ceased, the temporary backfill material must be removed and replaced by specified backfill recompacted to standard. For very old mains, where considerable corrosion may have taken place, this isolation operation may not be feasible, and in any case the movements accompanying exposure could exceed those caused by tunnelling. Almost certainly, the opportunity would be taken to replace the main with a new polyethylene pipe.

Replacement: Replace with less brittle, polyethylene pipe.

Insertion: Renovate old cast iron pipelines of 75 mm and 100 mm diameter by forced insertion of thermoplastic polyethylene. A machine is drawn through the old main, expanding and fracturing it progressively and pulling through a replacement PVC sleeve. A polyethylene pipe is then placed within this protective sleeve. Longitudinal slip movement can be accommodated between pipe and soil.

2.14 Pretunnelling protection of structures

There are several methods whereby building foundations can be isolated, or partially isolated, from ground that is expected to deform around them. These include diaphragm walling, underpinning (with provision for progressive jacking to overcome settlements), grouting (to render soil adjacent to the structural foundation less prone to deform), sheet piling (although perhaps with the penalty of some vibration damage and weakening of the foundation soil) and contiguous bored-pile walls.

Such methods may be only partially successful and, in view of their expense, should be given very careful consideration before adoption. Most of the methods, and particularly underpinning, are really only applicable with vertical movements in view, whereas lateral distortions can be potentially severe. As noted above (particularly with respect to sheet piling, although it applies to all protective measures) the protective engineering operations could well induce more structural damage than might have been caused by the tunnelling-induced ground movements without the protection.

3 Ground movements and buried pipelines

3.1 Introduction

It has always been recognized that ground movement can cause longitudinal bending stress in buried pipelines. By the 1970s research had shown that the primary causes of fractures in grey iron distribution mains are corrosion and system disturbance due to ground movement (for example, DoE, 1977; Roberts and Regan, 1974, 1977; Needham and Howe, 1979). It has been estimated that the UK gas and water distribution system consists of over 500 000 km of buried pipeline, excluding service connections (DoE, 1977; National Water Council, 1977). In 1972 more than 90 per cent of both the gas and water distribution network comprised grey iron pipework (Collins *et al.*, 1973). Similar usage of grey iron pipework is reported in a survey of the 100 largest US cities (Sears, 1968). By 1984 the proportion of grey iron had been reduced to approximately 60 per cent of the UK gas system and 85 per cent of the water system, due to the large-scale introduction of medium-density polyethylene and ductile iron. This still represents over 400 000 km of distribution mains known to be vulnerable to the effects of ground movement.

Pipeline failure has been defined by Anderson and Misund (1983) as an incident leading to a significant leak or otherwise requiring immediate repair. In terms of engineering structural design, failure would also include excessive deformation associated with yield or buckling in ductile materials. For brittle grey iron pipelines the failure modes can be classified as:

Transverse fracture associated with longitudinal bending
Longitudinal split associated with ring bending
Blow-out, hole or perforation associated with long-term corrosion
Leakage at pipeline joint
Leakage at service connection joint
Damage caused by direct impact.

In low-pressure distribution systems joint leakage is commonplace, due to about 40 per cent of the network having lead–yarn joints and the large number of service connections. The consequences of joint leakage are not usually as severe as other types of failure, and for old iron pipework this is almost accepted as a routine maintenance requirement. Excluding leakage and direct impact, corrosion is the main cause of failure in metal pipes in 25–50 per cent of cases, depending on pipe diameter and ground conditions. The other principal cause of failure is transverse fracture caused by differential ground movement. This differential ground movement arises from traffic loading, ground temperature and moisture changes, and ground movement associated with adjacent excavation such as trenching or tunnelling.

Since 1975 the British Gas Corporation Engineering Research Station has sponsored a comprehensive research programme to quantify the causes of stress in buried pipelines and to develop damage-control procedures for the distribution system. The UK water industry is a major promoter of deep trenches and tunnels in urban areas, primarily for sewer construction. Such excavations inevitably cause ground movements, and a possible conflict of interest between a pipeline operator's responsibility for the safety and security of the distribution system and the water industry's need to construct deep trenches and tunnels without extensive and costly diversions of other services. In 1979 the UK water industry started research into the effects of ground movement on buried pipelines. The main areas of interest are:

Form of ground movement
Magnitude of ground movement
Interaction of pipeline with ground movement
Deformation properties and strength of iron pipes
Deformation properties and strength of pipe joints
Deformation properties and strength of soil around pipes
Interaction of pipe networks with ground movement including the effect on service branches.

Howe (1985a) and Rumsey and Dorling (1985) give a summary of the effects of deep trench excavation on adjacent buried pipelines. Research for the UK water industry on the effect of tunnelling-induced movement has been carried out mainly by the University of Durham and the Northumbrian Water Authority. Broad agreement has now been reached with the gas industry on the areas of interest listed above. The methods given in Chapter 2 are used to estimate the form and magnitude of ground movement. The present chapter deals with the interaction of a pipeline with these tunnelling movements. The linear elastic methods that are used have been found to be suitable for practical application. For the interaction analysis to be of any practical use, equal importance is given to adequately defining the strength and deformation properties of pipes, pipe joints and the soil around the pipeline. Although the methods of analysis were developed for tunnelling movements and grey iron pipelines, the approach is equally applicable to other forms of ground movement and other types of pipeline. It is now clear that ground movement can cause significant longitudinal bending stress in all small-diameter pipelines.

The near-surface ground movement that occurs as a result of tunnelling in soil depends, amongst other things, on the stiffness of the soil above the tunnel.

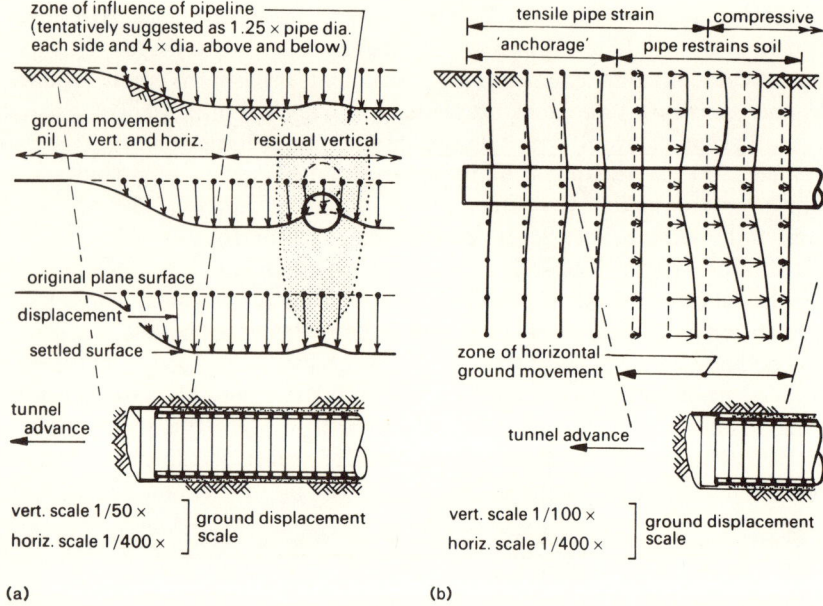

Figure 3.1 Typical restraint to ground movement due to buried pipeline. (*a*) Ground movement transverse to pipeline. (*b*) Ground movement parallel to pipeline.

In the vicinity of a buried iron pipeline the 'predicted ground movement' is modified, since the pipe stiffness is of the order of 1000 to 3000 times the soil stiffness. This restraint on ground movement is illustrated in Figure 3.1. The principal effect to be evaluated is longitudinal bending of the pipeline associated with transverse ground movement (that is, ground movement at right angles to the longitudinal axis of the pipeline). Reference to Figures 2.10*a*, *b*, *c* shows that bending in the sagging mode will tend to develop above the tunnel centre line. Also, bending in the hogging mode will tend to occur offset from the centre line at a transverse distance of about $2i$. Transverse to the tunnel line a continuous rigidly-jointed plain pipeline will be subjected to a bending moment of the form shown in Figure 3.2*a*. The magnitude of this moment depends, amongst other things, on the longitudinal flexural rigidity of the pipe and the stiffness of the soil around the pipe. The more rigid the pipe then the greater is the restraint to ground movement and hence the greater is the bending moment. This does not necessarily imply higher stresses in more rigid pipes, since increasing rigidity may be accompanied by increasing bending strength. Reference to Figure 2.10*a* shows that hogging and sagging bending moments tend to develop in a parallel pipeline. For a continuous rigidly-jointed plain pipeline the bending moment is of the form shown in Figure 3.2*b*.

Clearly, the restraint to ground movement and the bending moment induced in a pipeline will be much affected by any joint flexibility and the

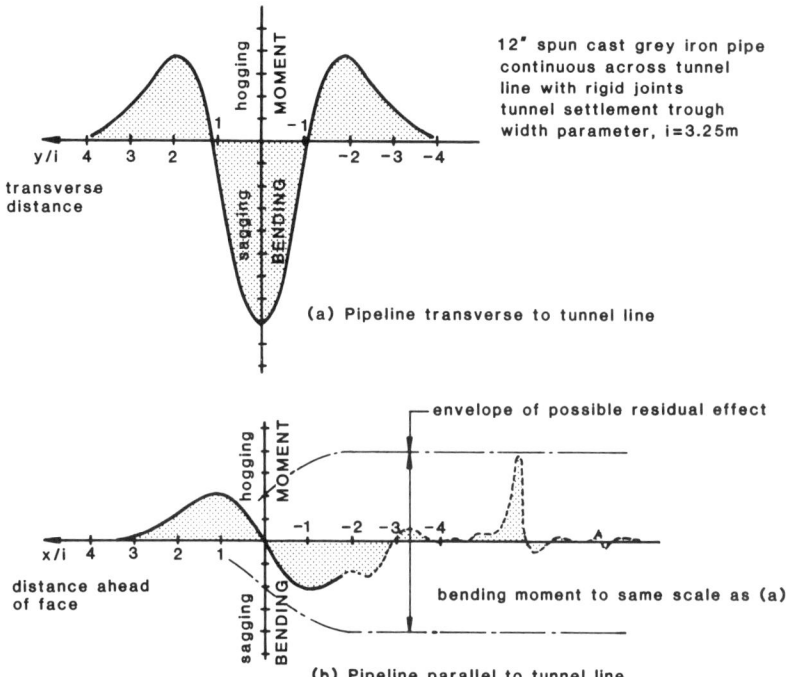

Figure 3.2 Typical form of bending moment induced by tunnelling beneath buried pipeline.

position of these joints. For small-diameter grey iron pipelines and shallow tunnels, flexible joints in the pipeline may have only a small reducing effect at the joint spacings commonly used, if these joints happen to be in the least helpful position. This position is equally spaced each side of the tunnel centre line for transverse pipelines, and approximately equally spaced each side of the point of maximum bending in parallel pipelines. As a tunnel advances parallel to a pipeline, then at some stage the joints are in this least helpful position. There are no practical cases where regularly spaced joints (that is, 5.5 m centres or less) cause an increase in the bending moment compared with a continuous rigidly-jointed pipeline. The analysis in section 3.2 therefore starts with the simple case of a continuous rigidly-jointed pipeline. Relaxation of bending moments due to rotation at pipe joints need be considered only if significant stress is induced in a continuous pipeline. Even then further detailed analysis may not be required for small-diameter pipelines since relaxation at the joint may not affect the stress in the pipe midway between the joints.

As an alternative to bending associated with the overall pattern of ground movement, longitudinal bending in a pipeline may result from small local variations in the ground movement or soil–pipe stiffness. These stresses will be apparently randomly distributed along the pipe. This applies particularly to parallel pipelines where there may be significant residual stresses after the settlement wave has passed. Also, local restraint to pipe movement may occur in pipeline networks at branches. These effects are of particular importance in relatively flexible pipes (that is small-diameter metal pipes and all plastic pipes) and are considered in detail in sections 3.2.7 and 3.4.6.

In addition to bending stress associated with lateral ground movement, axial tensile and compressive forces are induced by parallel ground movement (that is, ground movement parallel to the longitudinal axis of the pipeline). Load is transferred by the shear stress at the soil–pipe interface as illustrated in Figure 3.1. This shear stress is due to the relatively stiff pipe locally restraining soil movement in the zone of tunnel-induced ground movement, and the soil restraining pipeline movement in the zone of pipeline 'anchorage'. The mechanism is analogous to the transfer of load to the ground by friction piles as described by Cooke (1975). For a continuous rigidly-jointed pipeline, axial compressive forces tend to occur above the tunnel centre line in transverse pipelines and behind the tunnel face position in parallel pipelines—coinciding approximately with the position of maximum sagging moment. Tensile forces develop offset from the tunnel centre line in transverse pipelines and ahead of the tunnel face position in parallel pipelines—coinciding approximately with the position of maximum hogging moment. Fixity conditions are crucial in determining whether tensile or compressive forces are induced, and the magnitude also depends on the soil–pipe bond and soil–pipe shear (adhesive) strength. For these reasons the possible compressive forces associated with parallel ground movement cannot be relied on to *reduce* tensile bending stresses associated with transverse ground movement. Measurements on pipelines have shown that axial forces are certainly a secondary effect compared with longitudinal bending. Residual axial forces in pipelines parallel to a tunnel are negligible compared with the generally small axial forces that arise through seasonal temperature variations. The analysis in section 3.3 starts with the simple case of a continuous pipeline with rigid (non-extendable) joints. The axial forces that can develop are limited by the soil–pipe shear strength. If necessary, allowance can be made for the significant reduction in axial forces that occurs in jointed pipelines. Rubber gasket joints transfer insignificant forces and lead–yarn joints transfer only very small forces.

Analysis forms only a part of the design problem, solutions to most problems in practice being based on engineering judgement. Analysis is carried out by the simplest approximate procedure that will give the required solution to the desired accuracy in a reasonable time. The purpose of the soil–pipe interaction analysis in sections 3.2 and 3.3 is to assess an upper limit for the possible increase in pipe stress associated with ground movement. It must not be expected that the actual stress distribution will always correspond closely with the calculated value. A particular difficulty with soil–structure interaction problems is being able to define small enough limits for the soil deformation properties. The results of the analysis may be very sensitive to the assumptions made, and it is usual to test the sensitivity to changes in these properties. Section 3.4 discusses these important parameters. It has been found that for pipeline problems this is not now a major difficulty, and useful solutions as well as an overall understanding can be obtained. Continuing research into soil and pipe-joint deformation properties may lead to possible refinements in the soil–pipe model.

The final section of this chapter outlines consultative procedures for pipeline operators and promoters of tunnels. It should now be possible for these parties to avoid problems and later disputes through early consultation at the design stage of tunnel works. The main difficulties now lie in the apportionment of costs, should renewal or diversion of a pipeline be necessary in advance of tunnelling works. Inevitably the major share of any costs is borne by the promoter of the tunnel works, whereas the risk of seriously damaging a pipeline usually arises because the pipeline is already significantly overstressed. Unfortunately, many of the old grey iron distribution mains are already in this condition and have little capacity for additional load. A wider understanding of the significance of system disturbance as a cause of stress in buried pipelines should lead to fewer pipeline failures in the future.

3.2 Ground movement transverse to a pipeline

The horizontal and vertical ground movements associated with tunnelling in soil can be estimated by the methods detailed in Chapter 2. As outlined in that chapter, the expressions for ground curvature, for example

$$\frac{\partial^2 w}{\partial y^2} = \frac{1}{i^2}\left(\frac{y^2}{i^2} - 1\right) w_{max} \exp\left[\frac{-y^2}{2i^2}\right]$$

may be equated with the effect on a buried pipeline by assuming that the pipeline has zero flexural rigidity. Clearly $w = f(y/i)$ can only be differentiated if this is a smooth function. In reality, all case history data consist of discrete values of w against y/i (in many cases at wide intervals) and there is no suggestion that actual settlement troughs precisely follow a smooth curve. In this context the expressions for curvature may be meaningless. In practical terms this amounts to evaluating $\partial^2 w/\partial y^2$ numerically from a set of discrete values of w. At best this is a dubious numerical process. If, however, the flexural rigidity of the pipeline is taken into account, no limitations are placed on the form of the settlement trough. Any shape of settlement trough can be specified, including the smooth forms referred to in Chapter 2. The flexural rigidity of the pipeline has the effect of smoothing out the discontinuities in the settlement trough. The following analysis has been developed to assess the effect of ground movement on iron pipelines where the ring deformation of the pipeline is negligible compared with its longitudinal deformation. A study of the very small stresses induced in the circumferential direction has been reported by Valliappan and Raja-Sekar (1985). For large-diameter, thin-walled flexible pipelines some modification to the approach that follows may be necessary (see Gumbel and Wilson, 1981, for a definition of rigid and flexible pipes).

Longitudinal bending in a pipeline is caused by differential *slope* on the ground-movement profile. Depending on the flexural rigidity of the pipeline, the maximum bending stress may be associated either with the overall form of movement or with discontinuities in slope on the ground movement profile. As the flexural rigidity decreases then discontinuities and other non-uniformities in the soil–pipe system may become more important than the overall form of movement. It is important that this is recognized at the outset. For example, Nath (1983) modelled the effect of trench-induced movement on a pipeline; since the soil model was homogeneous it could generate only smooth forms of ground movement. Inevitably the conclusion was that as the flexural rigidity of the pipeline decreased, then the bending curvature was determined solely by this smooth ground movement. In this case the soil–pipe interaction modelling amounted merely to a numerical differentiation of this movement profile, irrespective of the methods actually used. With decreasing flexural rigidity, the pipe stress was proportional to the pipe diameter. Unfortunately this is inconsistent with the measurements and fracture records which indicate that, for this relatively short wave of ground movement, stress *increases* with decreasing pipe diameter. With decreasing flexural rigidity, the effect of a smooth ground-movement profile is only a lower limit for pipe stress. Discontinuities in slope and other non-uniformities give the upper limit which tends to the relationship of stress *inversely* proportional to pipe diameter. This is consistent with the fracture records—for example, Roberts and Regan (1977). Also, the empirical relationship for pipe stress given by Rumsey and Dorling (1985) agrees with the theoretical relationship if allowance is made for reasonable local variations in ground movement or pipeline restraint.

Theoretical approaches to this problem are similar to the methods available for piles subjected to lateral loading or lateral soil movement. Three broad categories of analysis are described by Poulos and Davis (1980).

(1) The subgrade reaction approach where the reaction between the pipeline and the soil at a point is related to the deflection at that point (Winkler foundation).

(2) Methods based on the theory of elasticity that employ the equations of Mindlin (1936) for subsurface loading within a semi-infinite mass.

(3) Numerical methods, in particular finite-element methods.

The limitations of these methods have been extensively discussed (for example, Poulos and Davis 1980, pp. 163 and 164, and Elson, 1984) and it is shown in section 3.2.1 that the subgrade reaction model is practically satisfied provided that certain restrictions are placed on its use. Whereas for piles it is generally necessary to allow for the variation of soil properties with depth along the pile, this condition does not apply to pipelines. It is important to recognize that the soil stiffness immediately adjacent to the pipeline is of primary importance, just as for laterally loaded piles the soil stiffness near the surface—to a depth of only a few pile diameters—has a dominant influence on the behaviour of the pile. Irrespective of the method of analysis, the main factors determining the behaviour of a pipeline are the magnitude and distribution of soil movement, the soil–pipe stiffness (including the effect of joints) and the yield pressure of the soil. The precise nature of the elastic response of the soil is of secondary importance compared with determination of the appropriate soil moduli.

3.2.1 Subgrade reaction analysis

The subgrade reaction foundation model, first introduced by Winkler (1867), is characterized by the assumption that the pressure in the foundation is proportional at every point to the deflection occurring at that point and is independent of pressures or deflections produced elsewhere in the foundation. This model has been widely used in foundation engineering, and there is considerable experience in applying the theory to practical problems. For the case of 'long' beams on an elastic foundation, the use of the model is not merely a device to simplify the analysis, although it undoubtedly does.

An exact theoretical solution can be found for the bending, under a concentrated load P, of an infinite beam resting on an elastic solid. Biot (1937) gives the solution for the maximum bending moment occurring under the load as

$$M_{max} = 0.332 Pb \left[C(1 - v^2) \frac{E_b I_b}{E b^4} \right]^{0.277} \tag{3.1}$$

where $C = 1.00$ if the distribution of pressure is uniform across the width of the beam and $1.00 < C < 1.13$ if the deflection is uniform across the width of the beam, $2b$ is the beam width, E_b and I_b are elastic modulus and second moment of area, respectively, of the beam, and E and v are respectively the elastic modulus and Poisson's ratio of the elastic foundation.

In the interval $0.01 < b/c < 1$, the error in the expression for M_{max} is of the order of 2–3 percent where

$$c = \left[C(1 - v^2) \frac{E_b I_b}{E} \right]^{1/3} \tag{3.2}$$

Biot's solution has been extended by Vesic (1961) to give the full solution for bending moment, shear force and deflection for both a concentrated load and a couple. Vesic has investigated in detail the reliability of the subgrade reaction approach and the magnitude of error induced by its application compared with the modelling of an elastic continuum. For beams of infinite length the

Winkler foundation model is practically satisfied. For an infinite beam,

$$k_\infty d = K_\infty = 0.65 \sqrt[12]{\frac{E_g d^4}{E_p I_p} \left[\frac{E_g}{1 - v_g^2}\right]} \qquad (3.3)$$

where k is the *coefficient* of subgrade reaction (dimensions FL^{-3}), K is the *modulus* of subgrade reaction (dimensions FL^{-2}), d is the pipe outside diameter (i.e. beam width) $E_p I_p$ is the flexural rigidity of the pipe, and E_g, v_g are elastic properties of the foundation.

Throughout the remainder of this chapter the subscript p is used for the pipe and g for the soil around the pipe. For a pipe, the second moment of area about the centre is

$$I_p = \frac{\pi}{64}[d^4 - (d - 2t)^4],$$

where t is the pipe wall thickness.

Using equation (3.3) for K and the Winkler foundation model, the maximum bending moment is overestimated by 4 per cent at $b/c = 0.01$, 1 per cent at $b/c = 0.03$ and 12 per cent at $b/c = 1.0$. The deflection is underestimated between zero and 15 per cent, and the contact pressure is underestimated between zero and 20 per cent. The recommended procedure for analysis is shown in Table 3.1, and according to Vesic it has been substantiated by large-scale model tests.

From equation (3.3), K_∞ can be calculated if E_g is known. Over the range of the variables applicable to pipeline problems, K_∞ is in the range 0.4 to 0.7

Table 3.1 Recommended procedures for analysis

Size of beam	Free length (L) beyond loaded length[c]	Recommended procedures		
		For rough estimates	For refined analysis	
Long	$\lambda L > 2.5$	Conventional analysis assuming infinite beam and K from equation (3.3)[a]		
Moderately long	$1 < \lambda L < 2.5$	Conventional analysis assuming finite beam and K from equation (3.3)[a]		
Moderately short	$0.4 < \lambda L < 1$	Conventional analysis[a]		[b]
Short	$\lambda L < 0.4$	Treat as perfectly rigid[a], i.e. linear contact pressure		[b]

[a] After Vesic (1961).
[b] Use of the equations of Mindlin (1936) or finite-element methods.
[c] $\lambda = \sqrt[4]{\dfrac{K}{4 E_p I_p}}$

times E_g. Poisson's ratio, v_g, has a minor effect and a reasonable value can be assumed. Alternatively, the value of K_∞ can be calculated from $\bar{k}$ for a square plate of width B. Denoting $\bar{k}B$ by $\bar{K}$, and knowing that for a rigid square plate subjected to normal pressure p

$$\frac{E_g}{1-v_g^2} = 0.815 \frac{pB}{w} = 0.815\bar{K}$$

then

$$K_\infty = 0.52 \sqrt[12]{\frac{\bar{K}d^4}{E_p I_p}} \bar{K}. \qquad (3.4)$$

This compares with the formula given by Terzaghi (1955) which for overconsolidated clays is conservatively suggested to be

$$K_\infty = 0.67\bar{K}$$

irrespective of beam flexibility value. Equation (3.4) yields

$$K_\infty = (0.20 \text{ to } 0.65)\bar{K}$$

for

$$\sqrt[12]{\frac{\bar{K}d^4}{E_p I_p}} = 0.40 \text{ to } 1.25,$$

where the lower value is for very rigid beams and the upper value is for very flexible beams.

In the case of a beam below the surface of an elastic solid, the foundation modulus is the sum of the moduli on each side of the beam. At infinite depth, and for Poisson's ratio of 0.5, the effective modulus is

$$K_{\text{eff}} = 2K_\infty$$

where K_∞ is given by equation (3.3). For horizontal soil movement transverse to a buried pipeline, the presence of a free boundary (ground surface) above the pipeline can, for practical purposes, be ignored and is conservative. For vertical ground movement the effective modulus depends on the depth of burial and pipe diameter, and c given by equation (3.2). Again, it is suggested that for practical purposes, $K_{\text{eff}} = 2K_\infty$, which affects the soil–pipe stiffness parameter by at most 19 per cent and is within the precision of estimating the soil modulus E_g. The differences between the soil above and below the pipeline will tend to overestimate sagging bending moments.

3.2.2. Basic theory

In the foundation model, the pressure (p) in the foundation is proportional at every point to the deflection occurring at that point, that is,

$$p = kw_p$$

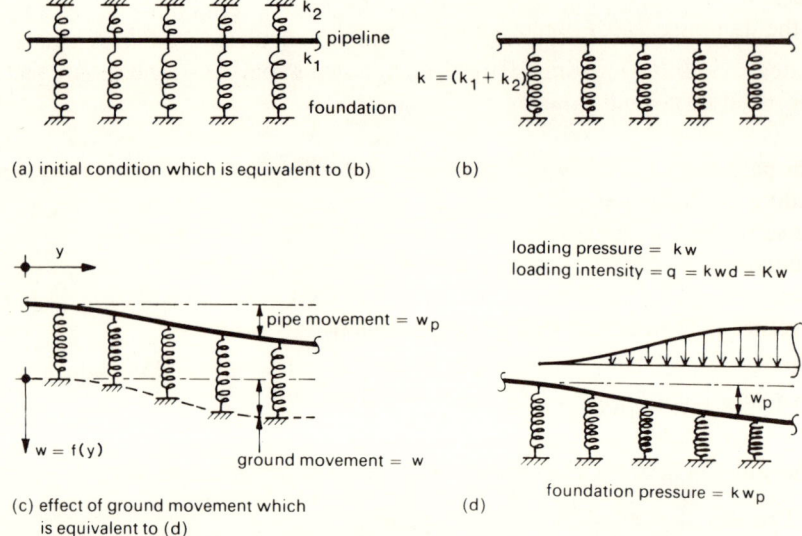

Figure 3.3 Beam on elastic foundation model. Note: representation of foundation reaction as discrete springs is diagrammatic only—reaction in foundation model is continuous.

where k is the *coefficient* of subgrade reaction (FL^{-3}), and w_p is the deflection of the pipe.

Initially, the pipeline is in equilibrium with soil pressure p_1 and p_2, respectively below and above the pipeline, as illustrated in Figure 3.3a. The foundation coefficient is equivalent to $k_{eff} = k_1 + k_2$. The ground-movement profile, $w = f(y)$, is the ground movement that occurs remote from the pipeline or if the pipeline were not present. The effect of this movement is shown in Figure 3.3c and causes pipeline movement w_p. Adjacent to the pipeline, soil and pipeline movement are the same, since it is assumed that the soil and pipe remain in contact. The increase in soil pressure is

$$p = k(w_p - w).$$

This is equivalent to a foundation pressure kw_p acting upwards plus an applied loading pressure kw acting downwards, that is

$$p = kw_p + (-kw)$$

as shown in Figure 3.3d. Putting the foundation modulus $K = kd$, where d is the pipe outside diameter, the applied loading intensity is

$$q = Kw.$$

The equation of the elastic line for the pipeline is

$$\frac{\partial^4 w_p}{\partial y^4} + 4\lambda^4 w_p = 4\lambda^4 f(y) \tag{3.5}$$

where w_p is the pipeline movement, $f(y) = w$ is the ground-movement profile, λ is the damping factor (dimensions L^{-1}), and

$$\lambda = \sqrt[4]{\frac{K_{\text{eff}}}{4E_p I_p}}. \tag{3.6}$$

The parameter $1/\lambda$ will be called the soil–pipe stiffness. Thus, the procedure for finding the deflection, shear force and bending moment in the pipeline due to the settlement, w, of the foundation is to apply a loading to a beam on an elastic foundation where the loading intensity is $q = Kw$. Hetenyi (1946) gives a comprehensive treatment of the theory of beams on elastic foundations together with solutions that are applicable to this problem.

3.2.3. General method of analysis

The soil/pipeline model is linear elastic and the assumptions are:

The Winkler foundation model applies
The soil is already precompressed and always remains in contact with the pipe
The pipe material is linear elastic, homogeneous and isotropic
The soil around the pipe is linear elastic and homogeneous
The pipeline is homogeneous (that is, a plain pipeline with rigid joints) and continuous for a distance of $3/\lambda$ beyond the point at which the slope of the ground-movement profile is negligible.

Using the procedure outlined in section 3.2.2, a typical load distribution parallel to the tunnel drive is shown in Figure 3.4. Where the loading varies with distance x, it is divided into, say, 10 equal length elements and a linear variation of load (i.e. settlement) is assumed across each element. The load element (n) to $(n + 1)$ is trapezoidal and is equivalent to the sum of two triangular loads of maximum intensities q_n and q_{n+1}. The element length can be chosen as a multiple of $1/\lambda$, for example $3/\lambda$, $2/\lambda$, $1/\lambda$, $1/2\lambda$ and so on, since this reduces the numerical work if manual calculation methods are used. The bending moment (and deflection and shear if required) at each point $1, 2, \ldots, n$, and so on, is the sum of the bending moments caused by the, say, 19 triangular load elements and the uniform load (if any). Since the pipeline is assumed continuous for at least $3/\lambda$ past the point of load variation, analysis is for an infinite beam on an elastic foundation. The relevant formulae for bending moments, shears and deflections are given in Appendix A, together with influence values for bending moment which can be used for more rapid calculation by manual methods. If required, the deflection can be calculated by a similar method or by a twofold numerical integration of the pipeline curvature diagram. The integrals are theoretically infinite, but rapidly converge and can be terminated at $3/\lambda$ past the point of load variation.

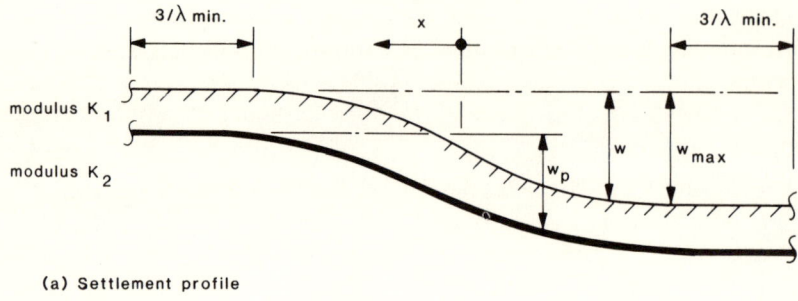

(a) Settlement profile

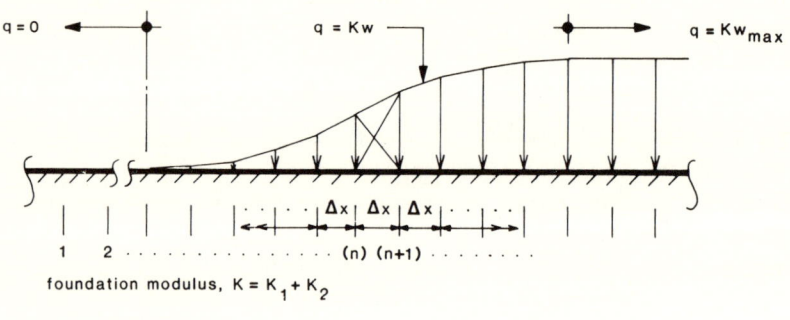

(b) Equivalent loading

Figure 3.4 Typical loading on pipeline due to ground movement.

This analysis for a continuous pipeline also represents a rigidly-jointed pipeline where the bending moment is continuous at the joints. The solution for a pipeline with perfectly flexible joints is obtained by relaxing the fixing moments at the joint positions in a manner similar to moment distribution in framed structures. A free end to a pipeline is obtained by relaxing the bending moment and shear. Pipe networks can be cut into the individual line elements, and the required solution obtained by restoring the geometrical and fixity conditions at the joints between the elements. The relevant formulae for moment or load applied at the end of a semi-infinite beam on an elastic foundation are given in Appendix B.

In the application of the analysis to practical problems, the following should be noted:

(1) w is the ground-movement profile at *pipeline level*, and it is generally convenient to put $w_{max} = 100$ and *pro rata* for the required w_{max} value at the end of the calculation.

(2) λ is given by equation (3.6) and has dimensions L^{-1}.

(3) $K_{eff} = (K_1 + K_2)$ and has dimensions FL^{-2}. Generally take $(K_1 + K_2) = 2K_\infty$ where K_∞ is given by equation (3.3). Approximately then $K_{eff} = 0.8$ to 1.4 times the soil modulus E_g.

(4) In equations (3.3) and (3.6), E_p and E_g are appropriate *secant* values corresponding to the range of stress and are values applicable *adjacent* to the pipeline, not for the soil as a whole.

(5) A range of E_g, and where appropriate I_p, should be substituted to test sensitivity to chosen values.

(6) Possible ground-movement profiles, in addition to those described in Chapter 2, should be considered to test sensitivity to smoothness of profile.

(7) The increase in soil pressure below the pipeline is $K_1(w_p - w)/d$, and the decrease above it is $K_2(w_p - w)/d$. These expressions involve a small difference which makes the calculation of soil pressure approximate. This is not a defect in the method, but merely a reflection of the physical reality of the problem.

(8) The method also applies to horizontal soil movement lateral to a pipeline which would have to be combined with vertical bending to give the maximum resultant bending.

3.2.4 Solutions for a rigidly-jointed pipeline transverse to a tunnel drive

The maximum soil movement transverse to a pipeline occurs when the pipe is transverse (at right angles) to the tunnel drive and the settlement trough is fully developed. The pipeline is in the plane $x = $ constant and $4i$ or more behind the

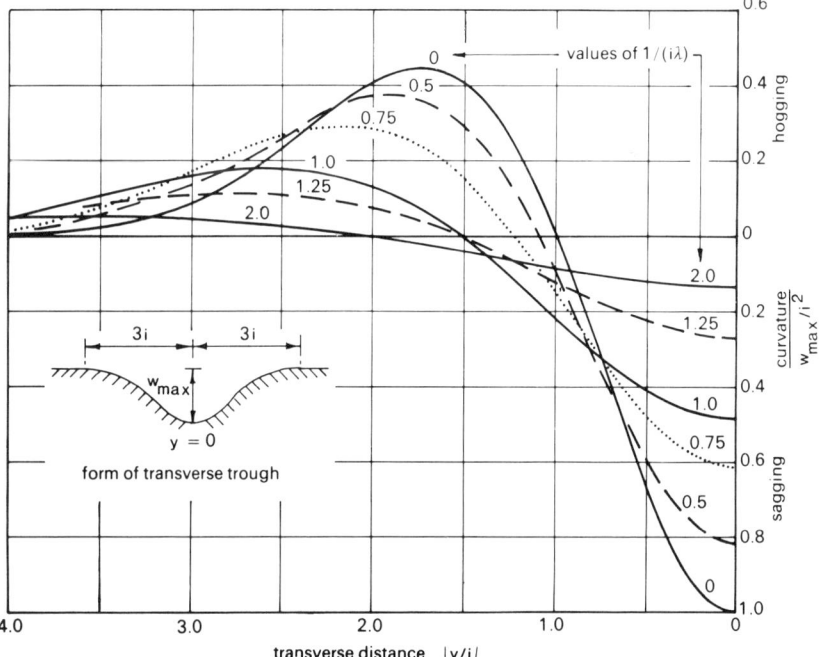

Figure 3.5 Pipe curvature variation with soil–pipe stiffness.

tunnel face, where i is the trough width parameter. This transverse soil movement is wholly vertical. For a pipeline with rigid joints extending $3/\lambda$ into undisturbed ground ($1/\lambda$ being defined by equation (3.6)), dimensionless curves are shown in Figure 3.5 which relate pipe bending curvature $\partial^2 w_p/\partial y^2$ to transverse distance y/i and soil–pipe stiffness $1/\lambda$. The bending moment is given by

$$M = E_p I_p \frac{\partial^2 w_p}{\partial y^2} \qquad (3.7)$$

and the extreme fibre bending strain is

$$\varepsilon_{yp} = \frac{d}{2} \frac{\partial^2 w_p}{\partial y^2} \qquad (3.8)$$

The assumed ground-movement profile is the normal probability form. Interpolation on Figure 3.5 is facilitated by Figures 3.7 and 3.8 which give the maximum sagging and hogging curvatures as a function of soil–pipe stiffness and the position of zero and maximum hogging moments. The maximum sagging moment is always at $y = 0$. Sensitivity to the assumed settlement trough is shown in Figure 3.7. The area of the settlement trough in Figure 3.6

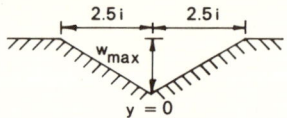

Figure 3.6 Worst transverse discontinuous slope profile.

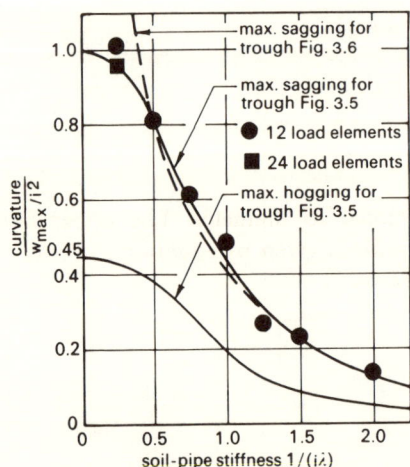

Figure 3.7 Maximum pipe curvature variation with soil–pipe stiffness.

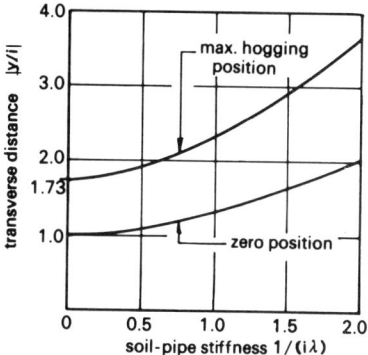

Figure 3.8 Position of maximum and zero bending moment.

is the same as for the normal probability form, that is, both troughs reflect the same settlement volume. For practical purposes there is no difference in maximum bending moment for $1/\lambda > 0.5i$. The solutions for *maximum* sagging bending can be made identical by adjusting the length of slope in Figure 3.6 from $2.5i$ to $2.3i$. Thus the maximum sagging bending curvature is for $1/i\lambda > 0.7$,

$$\frac{\partial^2 w_p}{\partial y^2} = \frac{i\lambda}{2.3}\left[1 - A(2.3i\lambda)\right]\frac{w_{max}}{i^2} \tag{3.9}$$

where the function

$$A(i\lambda) = \exp(-i\lambda)\left[\cos(i\lambda) + \sin(i\lambda)\right]$$

and in the trigonometric functions the parameter $i\lambda$ is in radians.

For $1/i\lambda < 0.7$, the bending moment depends on the precise form of the settlement trough. This generally cannot be adequately defined. The lower limit is given by the perfectly smooth settlement trough, Figure 3.5. As $1/i\lambda$ tends to zero, this smooth trough gives a maximum of

$$\varepsilon_{yp} = \frac{d}{2}\frac{w_{max}}{i^2}$$

that is, stress proportional to diameter. The upper limit for the worst discontinuous slope profile is given by equation (3.9). As $1/i\lambda$ tends to zero this reduces to, for $1/i\lambda < 0.7$,

$$\frac{\partial^2 w_p}{\partial y^2} = \frac{i\lambda}{2.3}\frac{w_{max}}{i^2} \tag{3.10}$$

Combining equations (3.8) and (3.10) with equations (3.3) and (3.6) gives approximately

$$\varepsilon_{yp} \propto \frac{d}{2}\frac{w_{max}}{i}\left[\frac{E_g}{E_p I_p}\right]^{0.25}$$

For a constant ratio of pipe diameter to wall thickness, I_p is proportional to d^4. Thus, stress is approximately independent of diameter for the upper-limit case.

For the settlement trough shown in Figure 3.5, the relationship between maximum sagging curvature and soil–pipe stiffness shown in Figure 3.7 can be read as a reduction factor associated with soil–pipe stiffness. For negligible stiffness ($1/\lambda = 0$), the curvature factor is 1.0, reducing to 0.8 at $1/\lambda = 0.5i$, 0.45 at $1/\lambda = 1.0i$, and so on. Hogging curvatures are reduced by a similar amount. This reduction factor due to soil–pipe stiffness is greatest where the 'span of the pipeline' is least. The fully-developed transverse settlement trough is the plane in the three-dimensional ground movement field which has the least 'span'. Thus irrespective of the orientation of the pipeline relative to the tunnel drive, the factor (bending curvature for actual soil–pipe stiffness)/(bending curvature for negligible soil–pipe stiffness) cannot be less than the value given by Figure 3.7.

Example 3.1 Calculation of bending strain in large-diameter transverse pipeline

Tunnel excavated diameter $2R = 1.5\,\text{m}$.
Volume of ground loss/unit advance (in granular soil at tunnel level) $V_t = 5$ per cent of face area $= 0.0884\,\text{m}^3/\text{m}$.
Tunnel depth, ground level to axis level $z_0 = 5.0\,\text{m}$.
Pipeline depth, ground level to axis level, $z_p = 1.5\,\text{m}$, thus $z_0 - z_p = 3.5\,\text{m}$.
Settlement trough width parameter (based on predominantly cohesive soil above tunnel and equation (2.31a), $i = 0.43\,(z_0 - z_p) + 1.1\,\text{m}$ whereby $i = 2.6\,\text{m}$.
Maximum ground displacement at pipeline level (i.e. at $z_0 - z_p = 3.5\,\text{m}$) is

$$w_{max} = \frac{V}{\sqrt{2\pi}\,i} = \frac{0.0884}{\sqrt{2\pi} \times 2.6} = 0.0136\,\text{m}$$

$$= 13.6\,\text{mm}.$$

Associated surface settlement (at $z = 0$) is 10.9 mm.
Pipe outside diameter $d = 0.50\,\text{m}$.
Pipe wall thickness $t = 0.018\,\text{m}$.
Pipe second moment of area,

$$I_p = \frac{\pi}{64}\left[d^4 - (d - 2t)^4\right]$$

$$= \frac{\pi}{64}\left[0.50^4 - 0.464^4\right]$$

$$= 7.93 \times 10^{-4}\,\text{m}^4.$$

Pipe secant modulus $E_p = 70\,\text{GN/m}^2 = 7 \times 10^{10}\,\text{N/m}^2$.
Soil secant modulus (value adjacent to pipe) $E_g = 10\,\text{MN/m}^2 = 10 \times 10^6\,\text{N/m}^2$.
Soil Poisson's ratio $\nu_g = 0.5$.

GROUND MOVEMENTS AND BURIED PIPELINES

Modulus of subgrade reaction, K_∞, from equation (3.3),

$$K_\infty = 0.65 \sqrt[12]{\frac{E_g d^4}{E_p I_p}} \left[\frac{E_g}{1 - v_g^2} \right]$$

$$= 0.65 \sqrt[12]{\frac{10 \times 10^6 \times 0.50^4}{7 \times 10^{10} \times 7.93 \times 10^{-4}}} \left[\frac{E_g}{0.75} \right] = 0.596 E_g$$

$$= 5.96 \times 10^6 \text{N/m}^2.$$

Effective K value (due to depth of burial etc.)

$$K_{eff} = 2 \times K_\infty = 11.9 \times 10^6 \text{N/m}^2.$$

Damping factor, λ, from equation (3.6)

$$\lambda = \sqrt[4]{\frac{K_{eff}}{4 E_p I_p}} = \sqrt[4]{\frac{11.9 \times 10^6}{4 \times 7 \times 10^{10} \times 7.93 \times 10^{-4}}} = 0.481 \text{ m}^{-1}.$$

Soil–pipe stiffness parameter $1/\lambda = 2.078$ m

$$= \frac{2.078}{2.6} i = 0.8 i.$$

For the pipeline transverse to the tunnel drive there are permanent bending curvatures given by Figures 3.5 and 3.7. Above the tunnel centre line

$$\frac{\partial^2 w_p}{\partial y^2} = 0.56 \frac{w_{max}}{i^2} = \frac{0.56 \times 0.0136}{2.6^2} = 1.127 \times 10^{-3} \text{m}^{-1}.$$

The extreme fibre bending strain

$$\varepsilon_{yp} = \frac{d}{2} \frac{\partial^2 w_p}{\partial y^2} = \frac{0.50}{2} \times 1.127 \times 10^{-3} = 2.82 \times 10^{-4}$$

$$= 282 \text{ microstrain}.$$

The maximum bending stress

$$\sigma_{yp} = \varepsilon_y E_p = 282 \times 10^{-6} \times 7 \times 10^{10} = 19.7 \times 10^6 \text{ N/m}^2$$

$$= 19.7 \text{ N/mm}^2.$$

This represents approximately 10 per cent of the rupture strength of a low-grade grey iron in longitudinal bending (see section 3.4.2).

At a transverse distance of approximately $2.2 i = 5.7$ m, there is a maximum hogging curvature of

$$\frac{\partial^2 w_p}{\partial y^2} = 0.27 \frac{w_{max}}{i^2} = 5.43 \times 10^{-4} \text{m}^{-1}.$$

The extreme fibre bending strain

$$\varepsilon_{yp} = \frac{d}{2} \frac{\partial^2 w_p}{\partial y^2} = 136 \text{ microstrain}.$$

The maximum hogging bending stress is

$$\sigma_{yp} = \varepsilon_{yp} E_p = 9.5 \, \text{N/mm}^2.$$

The effect of *soil drainage* can be allowed for by substituting drained soil parameters in the expression for K. Taking $E'_g = 2/3(1 + v'_g) E_g$ and $v'_g = 0.3$, then

$$K'_{\text{eff}} = 0.486 E'_g = 4.21 \times 10^6 \, \text{N/m}^2 \text{ and } \lambda' = 0.441$$

$$1/\lambda' = 2.266 \, \text{m} = 0.87 i.$$

The bending strains are reduced from $282 \mu\varepsilon$ to $259 \mu\varepsilon$ and from $136 \mu\varepsilon$ to $121 \mu\varepsilon$. Soil and pipe material creep would give similar small reductions in pipe strain with time.

Example 3.2 Calculation of bending strain in small-diameter transverse pipeline

The tunnel details, pipeline depth and soil stiffness are the same as in Example 3.1.

$$i = 2.6 \, \text{m}, \quad w_{\max} = 13.6 \, \text{mm}, \quad E_g = 10 \, \text{MN/m}^2, \quad v_g = 0.5.$$

Pipe outside diameter $d = 0.10 \, \text{m}$.
Pipe wall thickness $t = 0.010 \, \text{m}$.
Pipe second moment of area

$$I_p = \frac{\pi}{64} [d^4 - (d - 2t)^4] = 2.90 \times 10^{-6} \, \text{m}^4.$$

Pipe secant modulus $E_p = 70 \, \text{GN/m}^2$
$$= 7 \times 10^{10} \, \text{N/m}^2.$$

Modulus of subgrade reaction, K_∞, from equation (3.3)

$$K_\infty = 0.65 \sqrt[12]{\frac{E_g d^4}{E_p I_p}} \left[\frac{E_g}{1 - v_g^2} \right] = 0.557 E_g = 5.57 \times 10^6 \, \text{N/m}^2.$$

$$K_{\text{eff}} = 2 \times K_\infty = 2 \times 5.57 \times 10^6 = 11.1 \times 10^6 \, \text{N/m}^2.$$

$$\lambda = \sqrt[4]{\frac{K_{\text{eff}}}{4 E_p I_p}}$$

$$= \sqrt[4]{\frac{11.1 \times 10^6}{4 \times 7 \times 10^{10} \times 2.90 \times 10^{-6}}} = 1.923 \, \text{m}^{-1}.$$

$$1/\lambda = 0.520 \, \text{m} = \frac{0.52 i}{2.6} = 0.20 i.$$

From Figures 3.5 and 3.7 and for a smooth settlement trough as Figure 3.5, above the tunnel centre line the extreme fibre bending strain is

$$\varepsilon_{yp} = \frac{d}{2} \frac{\partial^2 w_p}{\partial y^2} = \frac{0.10}{2} \times 0.97 \times \frac{0.0136}{2.6^2} = 98 \text{ microstrain};$$

at a transverse distance of approximately $1.8 i = 4.7 \, \text{m}$, the maximum hogging bending strain is

$$\varepsilon_{yp} = \frac{d}{2}\frac{\partial^2 w_p}{\partial y^2} = \frac{0.10}{2} \times 0.44 \times \frac{0.0136}{2.6^2} = 44 \text{ microstrain.}$$

Since $1/\lambda < 0.7i$, consideration must be given to more severe effects being caused by changes in slope on the ground movement profile, as illustrated by Figure 3.6. The upper limit for these effects is given by equation (3.10). The maximum extreme fibre sagging bending strain is

$$\varepsilon_{yp} = \frac{d}{2}\frac{\partial^2 w_p}{\partial y^2} = \frac{d}{2}\frac{i\lambda}{2.3}\frac{w_{max}}{i^2}$$

$$= \frac{0.10}{2} \times \frac{1}{0.20 \times 2.3} \times \frac{0.0136}{2.6^2} = 219 \text{ microstrain.}$$

The maximum hogging strain is half this value

$$= 109 \text{ microstrain.}$$

A reasonable design case is to allow for a bending strain midway between the upper and lower limit cases, i.e.

maximum sagging bending strain

$$= 158 \text{ microstrain,}$$

maximum hogging bending strain

$$= 77 \text{ microstrain.}$$

3.2.5 Solutions for a rigidly-jointed pipeline parallel to a tunnel drive

For a pipeline parallel to a tunnel, the maximum soil movement transverse to the pipe occurs when the pipeline is directly above the tunnel drive. The pipeline is in the plane $y = 0$, and this transverse soil movement is then wholly vertical. In a similar manner to section 3.2.4, Figures 3.9–3.12 relate the pipe bending curvature to distance ahead of the tunnel face, x/i, and soil–pipe stiffness. The term 'distance ahead of tunnel face' is interchangeable with 'distance ahead of point of maximum slope on parallel settlement trough' without affecting the validity of the diagrams. For the cumulative normal probability form of ground movement shown in Figure 3.9, the sagging bending at x/i behind the tunnel face is numerically equal to the hogging bending at x/i ahead of the face. Sensitivity to the assumed settlement trough is shown in Figure 3.11. The area displaced by the settlement trough in Figure 3.10a is the same as that displaced by the cumulative normal probability form and reflects the same total load on the pipeline. Figure 3.10b is based on case history data for bolted segmentally-lined tunnels in firm-to-stiff cohesive soil where the forward part of the settlement trough is approximately 80 per cent of the form shown in Figure 3.9 and postgrouting deformation makes a significant contribution, extending the development of the trough to $4i$ behind the face. Figure 3.10b is also applicable to the initial

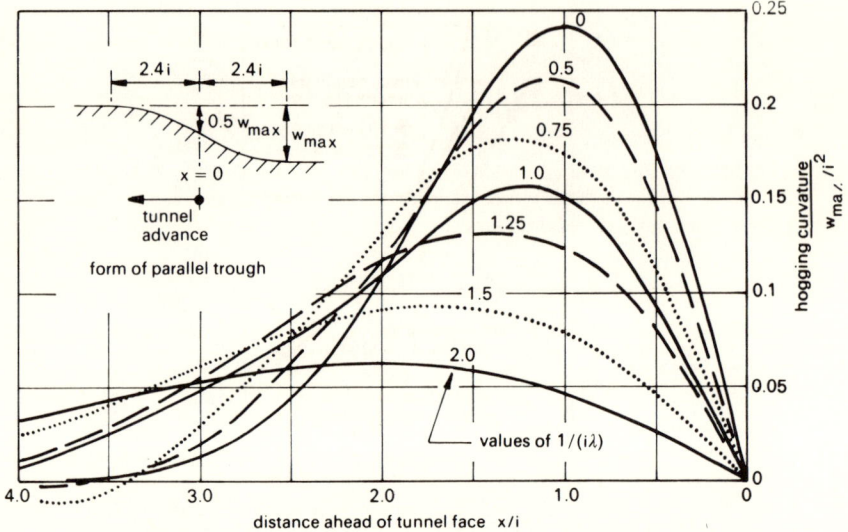

Figure 3.9 Pipe curvature variation with soil–pipe stiffness.

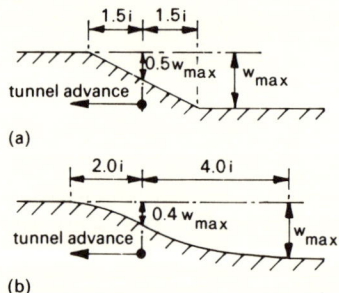

Figure 3.10 Parallel settlement troughs. (*a*) Worst parallel discontinuous slope profile. (*b*) Case-history data for tunnels in firm/stiff clay.

ground-loss settlement in soft, silty clays, but in this case the origin of the x-axis is shifted to give a trough starting approximately $1.0i$ ahead of the face.

For practical purposes, the maximum hogging bending associated with Figure 3.10*b* is merely 80 per cent of the value for Figure 3.9. For $1/\lambda > 1.0i$ there is no practical difference between the maximum bending for the movement profiles of Figures 3.10(*a*) and 3.9. The solutions for *maximum* bending can be made identical by adjusting the length of slope in Figure 3.10*a* from $3.0i$ to $3.3i$. Thus the maximum bending curvature is:

for $1.0 < 1/i\lambda < 2.0$

$$\frac{\partial^2 w_p}{\partial x^2} = \frac{i\lambda}{6.6}[1 - A(3.3i\lambda)]\frac{w_{max}}{i^2} \qquad (3.11)$$

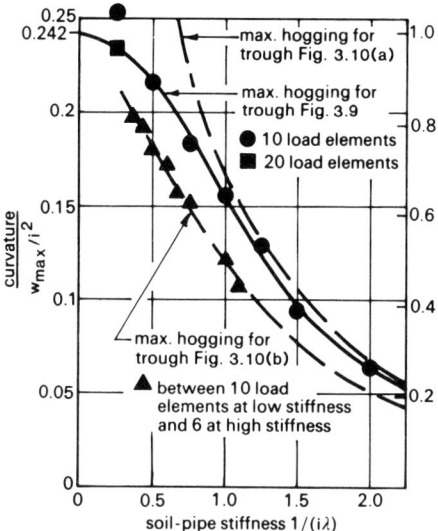

Figure 3.11 Maximum pipe curvature variation with soil–pipe stiffness.

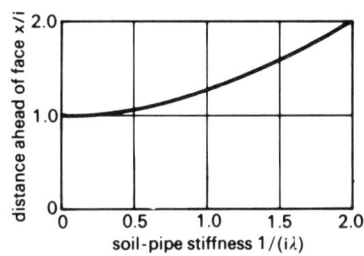

Figure 3.12 Position of maximum bending moment.

and for $1/i\lambda > 3.0$

$$\frac{\partial^2 w_p}{\partial x^2} = 0.322\lambda^2 w_{max} \qquad (3.12)$$

where the function $A(i\lambda)$ is defined as in equation (3.9).

For $1/i\lambda < 1.0$, the bending moment depends on the precise form of the settlement trough. As $1/i\lambda$ tends to zero, the perfectly smooth settlement trough gives the lower limit for maximum bending curvature of

$$\frac{\partial^2 w_p}{\partial x^2} \to 0.242 \frac{w_{max}}{i^2}. \qquad (3.13)$$

The upper limit for the worst discontinuous slope profile is given by equation (3.11). As $1/i\lambda$ tends to zero this reduces to

$$\text{for } 1/i\lambda < 1.0, \frac{\partial^2 w_p}{\partial x^2} = \frac{i\lambda}{6.6} \frac{w_{max}}{i^2}. \qquad (3.14)$$

For constant ratio of pipe diameter to wall thickness, equation (3.13) implies stress proportional to diameter. Equation (3.14) implies stress independent of diameter, whereas, for very stiff pipes or a very short wave of ground movement, equation (3.12) implies stress inversely proportional to diameter.

As in section 3.2.5, the relationship between maximum curvature and soil–pipe stiffness can be read as a reduction factor associated with soil–pipe stiffness. Reduction factors are shown on the right-hand scale of Figure 3.11, taking the relative curvature for $1/\lambda = 0$ as 1.0. The curvature factors are 1.0 at $1/\lambda = 0$, 0.89 at $1/\lambda = 0.5i$, 0.65 at $1/\lambda = 1.0i$, and so on. As mentioned in section 3.2.4, the reduction is less than for pipelines transverse to the tunnel drive.

Example 3.3 Calculation of bending strain in large-diameter parallel pipeline

The tunnel details and pipeline details are the same as in Example 3.1. The pipeline is parallel to and directly above the tunnel drive.
Maximum ground displacement at pipeline level, $w_{max} = 13.6$ mm.
Pipe outside diameter $d = 0.50$ m.
Soil–pipe stiffness parameter $1/\lambda = 2.078$ m

$$= 0.8i.$$

From Figures 3.9 and 3.11 and for a smooth settlement trough as in Figure 3.9, ahead of the tunnel face the maximum hogging curvature is

$$\frac{\partial^2 w_p}{\partial x^2} = 0.18 \frac{w_{max}}{i^2} = \frac{0.18 \times 0.0136}{2.6^2} = 3.62 \times 10^{-4} \text{m}^{-1}.$$

The extreme fibre bending strain is

$$\varepsilon_{xp} = \frac{d}{2} \frac{\partial^2 w_p}{\partial x^2} = \frac{0.50}{2} \times 3.62 \times 10^{-4} = 9.1 \times 10^{-5}$$

$$= 91 \text{ microstrain}.$$

Since $1/\lambda < 1.0i$, consideration must be given to more severe effects caused by changes in slope. The upper limit given by equation (3.14) is

$$\varepsilon_{xp} = \frac{d}{2} \frac{i\lambda}{6.6} \frac{w_{max}}{i^2} = \frac{0.50}{2} \times \frac{1}{0.8 \times 6.6} \times \frac{0.0136}{2.6^2}$$

$$= 95 \text{ microstrain}.$$

Since in this example the tunnel is overlain by predominantly cohesive soil, the settlement wave is more likely to be of the form illustrated in Figure 3.10(b). In this case a reasonable design is to allow for a maximum bending strain of

$$\varepsilon_{xp} = 80\% \frac{(91 + 95)}{2} = 74 \text{ microstrain}.$$

Example 3.4 Calculation of bending strain in small-diameter parallel pipeline

The tunnel details and pipeline details are the same as in Example 3.2. Pipeline is parallel to and directly above the tunnel drive.

$$i = 2.6\,\text{m}, \quad w_{max} = 13.6\,\text{mm}, \quad d = 0.10\,\text{m}, \quad 1/\lambda = 0.20i.$$

For a smooth settlement trough, ahead of the tunnel face the maximum hogging bending strain is

$$\varepsilon_{xp} = \frac{d}{2}\frac{\partial^2 w_p}{\partial x^2} = \frac{0.10}{2} \times 0.236 \times \frac{0.0136}{2.6^2} = 24\ \text{microstrain}.$$

The upper limit for slope discontinuities is

$$\varepsilon_{xp} = \frac{d}{2\,6.6}\frac{i\lambda}{i^2}\,w_{max} = \frac{0.10}{2} \times \frac{1}{0.20 \times 6.6} \times \frac{0.0316}{2.6^2} = 76\ \text{microstrain}.$$

The design case midway between these lower and upper limits is $\varepsilon_{xp} = 50$ microstrain.

The analysis outlined so far envisages hogging bending ahead of the tunnel face and a similar (or possibly reduced) sagging bending behind the tunnel face. When the tunnel has advanced a distance of $4.0i + 3/\lambda$ past the point in question then these stresses theoretically reduce to zero. It is not thought that this necessarily represents the true situation. Possible residual stresses are considered in section 3.2.7.

Pipeline parallel to but offset from the tunnel drive

Consider a pipeline in the plane $y = y_p$. In this case the ground movement transverse to the pipeline causing longitudinal bending is the sum of vertical movement, w, and lateral movement, v. The axial movement, u, has associated axial pipe movement only. The expressions for ground movement can be reduced to the following.

The settlement along the tunnel centre line, $y = 0$, is denoted by $w_{y=0}$. Thus at $y = y_p$ the vertical movement is

$$w_{y=y_p} = \exp\left[\frac{-y_p^2}{2i^2}\right] w_{y=0}$$

and the lateral movement is

$$v_{y=y_p} = \frac{-n y_p}{z_0 - z_p} \exp\left[\frac{-y_p^2}{2i^2}\right] w_{y=0}.$$

These expressions for ground movement offset from the tunnel centre line apply irrespective of the form of movement along $y = 0$. Thus the effect on a pipeline offset from the tunnel drive is the same as on the centre line except that the magnitude of movement and the magnitude of the effect are reduced. The maximum movement on the centre line is w_{max}. At $y = y_p$ the maximum

SOIL MOVEMENTS

movement is

$$\text{vector sum } (v+w)_{y=y_p} = \left[1 + \left(\frac{ny_p}{z_0 - z_p}\right)^2\right]^{1/2} \exp\left[\frac{-y_p^2}{2i^2}\right] w_{max}$$

$$= \text{(reduction factor)} \; w_{max}$$

Table 3.2 gives the variation of this reduction factor with the geometric and ground movement parameters. Because the form of ground movement offset from the tunnel drive is identical with the form directly above the tunnel, the reduction factors apply equally to the bending induced in rigid or flexibly-jointed pipelines and to transient or permanent effects. The pulling effect on branches from offset parallel pipelines is considered in section 3.4.6.

Table 3.2 Reduction factors for longitudinal bending in offset parallel pipelines

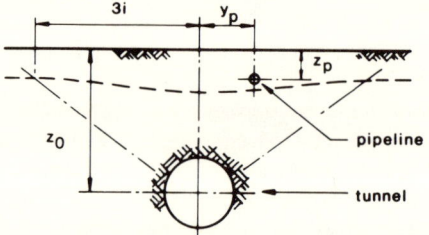

Offset distance $\frac{y_p}{z_0 - z_p}$	Reduction factor for values of $i/(z_0 - z_p)$									
	Granular soil above tunnel $n = 0.9$					Cohesive soil above tunnel $n = 1.0$				
	0.23	0.24	0.25	0.26	0.27	0.4	0.5	0.6	0.7	0.8
0	1.0	1.0	1.0	1.0	1.0	1.0	1.0	1.0	1.0	1.0
0.1	0.91	0.92	0.93	0.93	0.94	0.97	0.99	0.99	0.99	0.1
0.2	0.70	0.72	0.74	0.76	0.77	0.90	0.94	0.96	0.98	0.99
0.3	0.44	0.47	0.50	0.53	0.56	0.79	0.87	0.92	0.95	0.97
0.4	0.23	0.27	0.30	0.33	0.35	0.65	0.78	0.86	0.91	0.95
0.5	0.10	0.13	0.15	0.17	0.20	0.51	0.68	0.79	0.87	0.92
0.6	0.04	0.05	0.06	0.08	0.10	0.38	0.57	0.71	0.81	0.88
0.7	0.01	0.02	0.02	0.03	0.04	0.26	0.46	0.62	0.74	0.83
0.8	0	0	0.01	0.01	0.02	0.17	0.36	0.53	0.67	0.78
0.9			0	0	0	0.11	0.27	0.44	0.59	0.72
1.0						0.06	0.19	0.35	0.51	0.65
1.2						0.02	0.09	0.21	0.36	0.51
1.4						0	0.03	0.11	0.23	0.37
1.6							0.01	0.05	0.14	0.26
1.8							0	0.02	0.08	0.16
2.0								0.01	0.04	0.10

Example 3.5 Calculation of effect of ground movement on parallel offset pipeline

The tunnel details and pipeline details are the same as in Example 3.3 except that the pipeline is offset from the tunnel drive.

Offset distance, tunnel centre line to pipeline centre line, $y_p = 2.0$ m

$$z_0 - z_p = 3.5 \text{ m}, \quad i = 2.6 \text{ m}, \quad w_{max} = 13.6 \text{ mm}.$$

For predominantly cohesive soil above the tunnel, the ground movement parameter $n = 1.0$.

$$\frac{y_p}{z_0 - z_p} = \frac{2.0}{3.5} = 0.57, \quad \frac{i}{z_0 - z_p} = \frac{2.6}{3.5} = 0.74, \quad n = 1.0.$$

From Table 3.2, the reduction factor $= 0.85$.

Maximum ground movement transverse to pipe $= 0.85 \times w_{max}$
$= 11.6$ mm.

Design maximum bending strain $= 0.85 \times \left[80\% \frac{(91 + 95)}{2} \right]$.

(assuming still that $K_{eff} = 2K_\infty$)

$$\varepsilon_{xp} = 63 \text{ microstrain}.$$

Note: the final horizontal movement inwards towards the tunnel centre line is

$$v = \frac{-ny_p}{z_0 - z_p} \exp\left[\frac{-y_p^2}{2i^2} \right] w_{max} = 5.8 \text{ mm}.$$

This may generate axial tension or joint extension in a branch of the main pipeline.

3.2.6 Solutions for pipe deflection and soil pressure

The change in pressure at the soil–pipe interface due to settlement of the ground is:

Increase below pipeline

$$p_1 = K_1 (w_p - w)/d \quad (3.15)$$

Decrease above pipeline

$$p_2 = K_2 (w_p - w)/d \quad (3.16)$$

where K_1 and K_2 are the foundation modulus values below and above the pipeline (dimensions FL^{-2}) as discussed in sections 3.2.1 to 3.2.3. The maximum changes in soil pressure, that is maximum differential soil–pipe movements, coincide with those points where the differential slope on the ground movement profile is a maximum. Transverse to the tunnel, this is at $y = 0$ and $y = \pm 1.73i$ for Figure 3.5 movement profile and at $y = 0$ and $y = \pm 2.5i$ for Figure 3.6 profile. Parallel to the tunnel the maxima are at $x = \pm i$ for Figure 3.9 profile and at $x = \pm 1.5i$ for Figure 3.10a profile. Typical distributions of differential movement are illustrated in Figures 3.14d and 3.15d. The foregoing applies to rigidly-jointed continuous pipelines and would be considerably modified by joint flexibility.

For the rigidly-jointed pipeline transverse to the tunnel line, the maximum numerical value of $(w_p - w)$ occurs directly above the tunnel at $y = 0$.

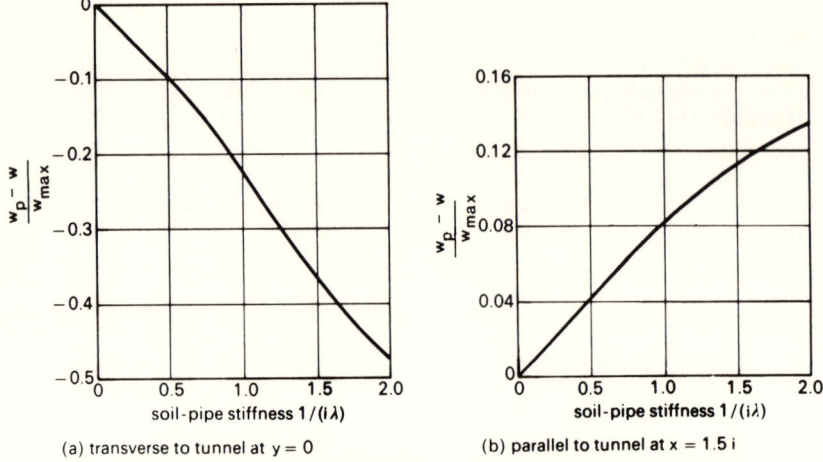

Figure 3.13 Maximum differential soil–pipe movement.

Figure 3.13a gives the value of this maximum pipeline restraint to ground movement for the ground-settlement profile shown in Figure 3.6. At $y = 0$ and for $1/\lambda < 0.5i$,

$$w_p = w_{max}\left(1 - \frac{1}{5i\lambda}\right) \to w_{max} \quad \text{as} \frac{1}{i\lambda} \to 0.$$

Parallel to and directly above the tunnel drive, and taking Figure 3.10a movement profile, the maximum value of $(w_p - w)$ is shown in Figure 3.13b. This maximum occurs at $x = 1.5i$; thus $w = 0$ at this position for the ground-movement profile adopted. The restraint to ground movement behind the tunnel face is similar to that ahead of the face position. At $x = 1.5i$ and for $1/\lambda < 1.0i$,

$$w_p = \frac{1}{12i\lambda} w_{max} \to 0 \quad \text{as} \frac{1}{i\lambda} \to 0.$$

The discontinuous-slope ground-movement profiles give the upper limit for differential soil–pipe movement. The 'normal probability' curve of ground movement produces significantly lower values. In the following examples, the pipe deflection for this latter case was calculated by the alternative methods mentioned in section 3.2.3. For the method involving a twofold numerical integration, the error in the calculated *differential* movement was less than 5 per cent starting from curvature values to four significant figures at intervals of 0.25i. As mentioned in section 3.2.3, there is no practical merit in carrying out such calculations to this apparent order of accuracy, since the results are inevitably very approximate. Calculation of change in soil pressure is carried out to give a rough check on the assumptions concerning soil secant modulus and pipeline remaining in contact with the soil. Pipeline movement is required

GROUND MOVEMENTS AND BURIED PIPELINES   149

where a network of connected pipes is to be analysed, and the load–deformation equations are required at each connection.

Example 3.6 Calculation of deflection and soil pressure for large-diameter pipeline

The tunnel and pipeline details are the same as in Example 3.1 and the pipeline is transverse to the tunnel line.

$$i = 2.6 \text{ m}, \quad w_{max} = 13.6 \text{ mm}, \quad d = 0.5 \text{ m}, \quad 1/\lambda = 2.078 \text{ m} = 0.8i.$$

From Figure 3.13a, the pipe restraint directly above the tunnel is

$$\frac{w_p - w}{w_{max}} = -0.17 \text{ and } w_p - w = -0.17 \times 13.6 = -2.3 \text{ mm}$$

and since $w = w_{max}$ at $y = 0$

$$w_p = (1 - 0.17) w_{max} = 0.83 \times 13.6 = 11.3 \text{ mm}.$$

For the 'normal probability' form of ground movement, the pipe slope and deflection have been calculated by a numerical integration of the bending diagram with the conditions of slope being zero at $y = 0$ and deflection being zero at $y = 3i + 5/\lambda$. The bending curvature, slope and deflection diagrams for $1/i\lambda = 0.8$ are shown in Figure 3.14a to c. The differential soil–pipe movement is given in Figure 3.14d and is equivalent to the loading on the pipeline. From

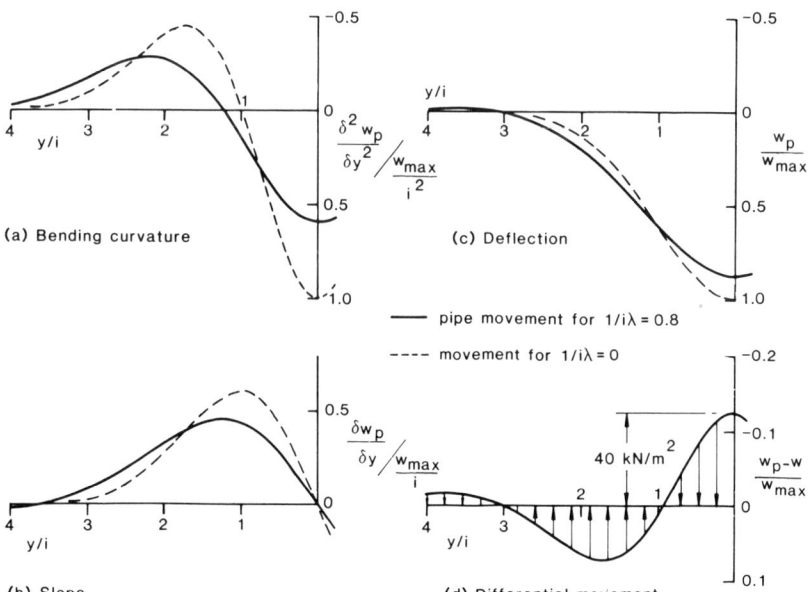

Figure 3.14 Movement of large-diameter pipeline transverse to tunnel.

Figure 3.14d:

at $y = 0$, $w_p - w = -0.124 w_{max} = -1.69$ mm (compared with -2.3 mm above)

$$w_p = (1 - 0.124) w_{max} = 11.91 \text{ mm}$$

and at $y = 1.73i$, $w_p - w = +0.074 w_{max} = 1.01$ mm.

$$w = w_{max} \exp\left[\frac{-y^2}{2i^2}\right] = 3.03 \text{ mm}$$

$$w_p = 1.01 + 3.03 = 4.04 \text{ mm}.$$

In Example 3.1, $K_1 = K_2 = 5.96 \times 10^6 \text{N/m}^2$. From equation (3.15) the increase in pressure below the pipeline at $y = 0$ is

$$p_1 = K_1 \frac{(w_p - w)}{d} = 5.96 \times 10^6 \frac{(-0.124 \times 0.0136)}{0.50} \text{N/m}^2,$$

$p_1 = 20 \text{ kN/m}^2$ *decrease* below the pipeline,

and $p_2 = 20 \text{ kN/m}^2$ *increase* above the pipeline.

The overburden pressure is $\gamma z_p = 19 \text{kN/m}^3 \times 1.5 \text{ m} = 28.5 \text{kN/m}^2$, and so the pipeline remains in contact with the soil. The total break-out load is 40kN/m^2. The assumed soil modulus of 10 MN/m^2 relates to a firm clay of shear strength $c_u = E_g/200 = 50 \text{kN/m}^2$. For the pipe depth-to-diameter ratio of $z_p/d = 3.0$, and assuming the soil–pipe adhesion is $0.5 c_u$, then the ultimate break-out resistance is 150 to 200 kN/m^2 at a differential displacement of 10 to 20 mm. Major yielding is not therefore envisaged, and the assumption of linear behaviour is reasonable.

For the pipeline parallel to and directly above the tunnel, as in Example 3.3, then from Figure 3.13b ahead of the tunnel face

$$\frac{w_p - w}{w_{max}} = 0.068, \quad w_p - w = 0.068 \times 13.6 = 0.9 \text{ mm}.$$

For the 'cumulative normal probability' form of ground movement, the bending curvature, slope and deflection diagrams are shown in Figure 3.15a–c. The differential movement at $x = i$ is

$$w_p - w = 0.0284 w_{max} = 0.39 \text{ mm (compared with 0.9 mm above)}.$$

The total bearing pressure is $p_1 + p_2 = 9 \text{kN/m}^2$ compared with a bearing capacity of about 400kN/m^2.

3.2.7. *Effect of non-uniformity in soil–pipe system*

In the preceding sections, the assumption of homogeneity in the model gives zero residual bending stress in pipelines parallel to the tunnel drive when the settlement wave has passed. Also, for an assumed smooth transverse settlement trough, theoretical bending stresses in small-diameter pipelines are small, since the bending is proportional to $w_{max}/i^2 \times d/2$. It is not thought that this always represents the true situation. Additional bending stresses may result from the small local variations in the overall pattern of ground movement or from local restraint to pipe movement.

Denoting the overall pattern of ground movement transverse to the pipeline as w, and assuming that local variations of up to ± 15 per cent w may occur, then adapting equation (3.12), the maximum extreme fibre bending strain is

$$\varepsilon_{pmax} = 0.097\lambda^2 \frac{d}{2} w. \tag{3.17}$$

Alternatively, assuming that the pipeline movement may be locally restrained by ± 5 per cent w, then from Hetenyi (1946)

$$\varepsilon_{pmax} = 0.100\lambda^2 \frac{d}{2} w. \tag{3.18}$$

Combining equations (3.17) and (3.18) with equations (3.3) and (3.6) gives approximately

$$\varepsilon_{pmax} \simeq 0.055 \frac{d}{2} w \left[\frac{E_g}{E_p I_p} \right]^{0.5}. \tag{3.19}$$

For pipelines parallel to the tunnel, w may be replaced by w_{max} in equation (3.19), since all points are eventually subjected to the maximum movement. The reduction factors in Table 3.2 apply to offset parallel pipelines. Equation (3.19) is thus comparable with an empirical formula that has been suggested for predicting the effect of ground movements caused by deep trenching on adjacent parallel pipelines. Rumsey and Dorling (1985) give the formula for this case which is equivalent to

$$\varepsilon_{pmax} = 0.058 \frac{d}{2} w_{max} \left[\frac{E_g}{E_p I_p} \right]^{0.4}. \tag{3.20}$$

(E_g here is the elastic secant modulus for the soil adjacent to the trench excavation (which influences the magnitude of ground movement) and is *not* the elastic modulus for the soil adjacent to the pipeline (which influences the soil–pipe interaction).) The formula is the result of linear-elastic finite-element modelling coupled with the results of measurements on full-scale trench construction. In equation (3.20), E_g and E_p are in units of kN/m^2, I_p is in metres^4 and d and w_{max} are in metres. Equation (3.20) is said to have a typical range of possible error of ± 25 per cent and is applicable to iron pipelines up to 300 mm (12 inches) in diameter. For this case, pipeline joint flexibility does not significantly affect the maximum pipe bending strain. Over the intended range of application, equation (3.20) gives a pipe strain approximately 50 per cent higher than does equation (3.18) if it is assumed that the soil modulus next to the trench is three times the soil modulus around the pipeline.

152   SOIL MOVEMENTS

Example 3.7 Calculation of allowance for bending strain in large-diameter pipeline associated with non-uniformity

The tunnel and pipeline details are the same as in Example 3.1 and the pipeline is transverse to the tunnel line.

$$w_{max} = 13.6 \text{ mm}, \quad d = 0.5 \text{ m}, \quad \lambda = 0.481 \text{ m}^{-1}.$$

If w is the ground movement at the particular point where the effect of non-uniformity occurs, then from equation (3.18)

$$\varepsilon_{p\,max} = 0.100 \lambda^2 \frac{d}{2} w$$

$$= 0.100 \times 0.481^2 \times \frac{0.50}{2} \times w = 5.78\, w \text{ microstrain},$$

where w is in mm.

It is unlikely that the effect of non-uniformity and the maximum ground movement on the transverse settlement trough both occur at the same point. A reasonable design case is to take $w = 0.75 w_{max}$ which covers 75 per cent of the possibilities. Thus $\varepsilon_{p\,max} = 0.75 \times 5.78 \times 13.6 = 59$ microstrain. Referring to Example 3.1, the bending strain calculated for immediately above the tunnel centre line is 282 microstrain. No further allowance is therefore required for other possible effects.

With the pipeline parallel to the tunnel, as in Example 3.3, the residual

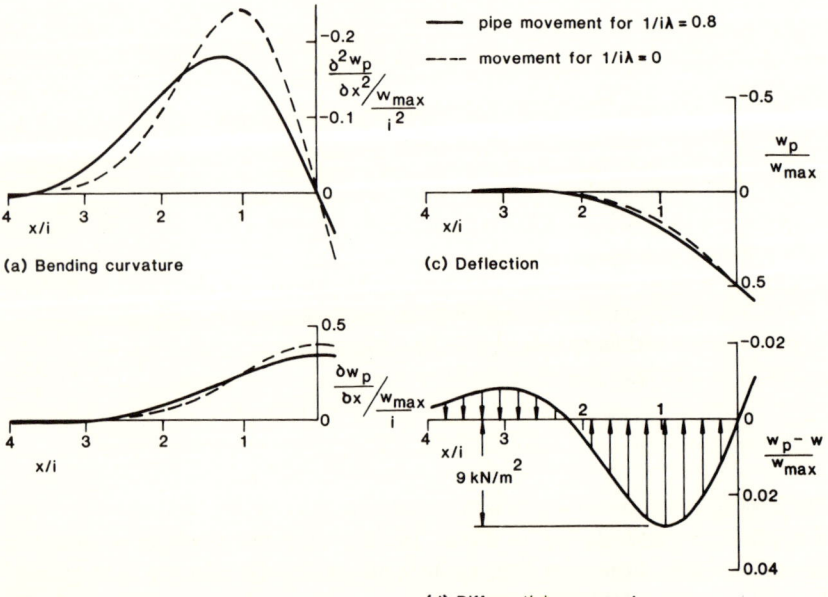

Figure 3.15 Movement of large-diameter pipeline parallel to tunnel.

GROUND MOVEMENTS AND BURIED PIPELINES 153

settlement of the pipeline at all points is w_{max}. Any effects associated with non-uniformity are bound to coincide with this maximum ground movement. Thus $\varepsilon_{p\,max} = 5.78 \times 13.6 = 79$ microstrain. This compares with the transient bending ahead of the tunnel face position of between 74 and 95 microstrain, depending on the precise form of ground movement. For a pipeline in good condition, the residual bending strain of up to 79 microstrain would be of more significance than the transient effect of up to 95 microstrain. However, if the pipeline was already in a highly overstressed condition then the wave of bending ahead of the tunnel face might be of greater importance. As the tunnel advances then the maximum bending is bound to coincide with points of weakness in the pipeline. For example, a top-entry service hole of 50 mm diameter in a 500 mm grey iron main reduces the hogging bending strength by a factor of 2.0. The 95 microstrain is thus equivalent to 190 microstrain temporary load. The possible residual strain of up to 79 microstrain is not bound to coincide with these points of weakness.

Example 3.8 Calculation of allowance for bending strain in small-diameter pipeline associated with non-uniformity

The tunnel and pipeline details are the same as in Example 3.2 and the pipeline is transverse to the tunnel line.

$$w_{max} = 13.6 \text{ mm}, \quad d = 0.10 \text{ m}, \quad \lambda = 1.923 \text{ m}^{-1}.$$

From equation (3.18), $\varepsilon_{p\,max} = 0.100 \lambda^2 \dfrac{d}{2} w$

$$= 18.49 w \text{ microstrain},$$

where w is in mm.

As in Example 3.7, transverse to the tunnel, take the value of $w = 0.75 w_{max}$. Thus $\varepsilon_{p\,max} = 0.75 \times 18.49 \times 13.6 = 189$ microstrain. Referring to Example 3.2, the bending strain allowance for the overall form of ground movement is 158 microstrain. This should now be increased to 189 microstrain to allow for the magnitude of ground movement.

With the pipeline parallel to the tunnel, as in Example 3.4, the maximum residual bending strain is $\varepsilon_{p\,max} = 18.49 w_{max} = 253$ microstrain. This compares with the transient bending ahead of the tunnel of up to 76 microstrain. The possible residual effects thus exceed the transient effects when $w/w_{max} > 0.3$ for this particular case, that is, $x < 0.5i$.

3.2.8 *Effect of pipe-joint rotation*

The longitudinal bending effects on a pipeline have so far been restricted to a continuous rigidly-jointed pipeline. In this case the bending moment is continuous across the joints and there is no differential rotation at the joints. The moment/rotation characteristics of the common joint types are given in section 3.4.3. Welded, flanged, cement- or mortar-caulked and turned-and-

Table 3.3 Joint rotation for bending moment of 10 per cent of rupture strength in 8"(200 mm) spun cast grey iron main

Joint type	Short-term loading (not applicable to tunnelling-induced movements)		Medium-term loading (transverse and parallel pipelines)	Long-term loading (transverse pipelines)
	rotation (degrees)	time to apply loading (minutes)	rotation (degrees)	rotation (degrees)
Lead yarn with hardened jute packing (gas main)	0.3	6	0.5	1.0
Lead yarn with soft jute packing (water main)	1.7	6	3.0	4.0 metal binding in joint at 7% of the rupture strength
Bolted-gland mechanical joint (gas main)	4.0	—	4.0	4.0
	metal binding in joint at a bending moment of 0.13% of the rupture strength			
Rubber gasket push-in joint (water main)	5.0	—	5.0	5.0
	metal binding in joint at a bending moment of 0.16% of the rupture strength			

bored joints can be classified as perfectly rigid for the purpose of analysis. Rubber gasket push-in joints, bolted-gland mechanical joints and rubber gland mechanical couplings can be classified as perfectly flexible, that is, incapable of transferring significant bending moment, and can be modelled as hinged. The lead–yarn joint is classified as semi-rigid at least for short-term loading, that is, a hinge with rotation proportional to applied bending moment. Taking the data in sections 3.4.2 and 3.4.3, then for a bending moment at the joint in a 200 mm diameter spun-cast grey iron main of 10 per cent of the longitudinal rupture strength, the required differential rotation across the joint is given in Table 3.3. For the lead–yarn joint, creep increases the rotation for this externally applied bending moment. The allowable joint movement for long-term loading (transverse pipelines) is 2.5° for rubber gasket joints, 1.5° for lead–yarn joints in a water main, and between zero and 1.0° for lead–yarn joints in a gas main. For rubber gasket joints the allowable rotation is set so as to prevent metal bending in the joint. For lead–yarn joints the limitations on rotation are to prevent joint leakage. It is clear from the foregoing comparison between applied stress and joint rotation that limits on joint movement can be an important consideration.

For a preliminary assessment of the longitudinal bending of a flexibly-jointed pipeline, the joints can be taken as being perfectly rigid for the purpose of calculating possible bending stresses. There are no practical cases where

GROUND MOVEMENTS AND BURIED PIPELINES 155

relaxation of regularly-spaced joints (18 feet (5.49 m) centres or less) causes an increase in maximum bending moment. Only if significant stresses are induced in a rigidly-jointed pipeline need consideration be given to the possible reduction associated with joint movement. Similarly, a quick check on possible joint rotation can be obtained by assuming that the pipe between the joints is perfectly rigid and that the joints accommodate all the differential ground movement. The joints then lie on the ground movement profile with the pipe being straight between the joints.

Example 3.9 Calculation of potential pipe-joint rotation

The pipeline is parallel to the tunnel drive, and the pipe and tunnel details are the same as in Example 3.3 and 3.4. The joint spacing is 18 feet (5.49 m). For a pipeline parallel to the tunnel centre line it is only necessary to consider the joints in the worst position since, as the settlement wave advances, then at some stage the joints are in this position.

$$w_{max} = 13.6 \text{ mm}, \ i = 2.6 \text{ m}.$$

Joint spacing, $L_J = 5.49 \text{ m}; \quad L_J/i = 2.11$.

For the 'cumulative normal probability' form of ground movement shown in Figure 2.10a, the worst position for the joints is at $x_J = 10.98$ m, 5.49 m, zero, -5.49 m and so on.

$$\text{At } x = 5.49 \text{ m}, \quad x/i = 2.11, \quad w/w_{max} = 0.0174$$

$$\text{at } x = \text{zero}, \quad w/w_{max} = 0.5000$$

$$\text{Differential settlement (in 5.49 m)} = 0.4826 \ w_{max}$$

The small differential settlement between $x = 5.49$ m and $x = 10.98$ m is 0.0174 w_{max}. The potential rotation at the joint ahead of the tunnel face position is

$$\theta_J = \frac{(0.4826 - 0.0174)}{L_J} w_{max}$$

$$= \frac{0.4652 \times 13.6 \times 10^{-3}}{5.49} = 1.15 \times 10^{-3} \text{ rad} = 0.07°.$$

For the form of ground movement shown in Figure 3.10a the potential rotation for $L_J \leqslant 3i$ is simply

$$\theta_J = \frac{w_{max}}{3i} = \frac{13.6 \times 10^{-3}}{3 \times 2.6} = 1.74 \times 10^{-3} \text{ rad} = 0.10°.$$

The ground movement transverse to a pipeline offset from the tunnel line is of the same form but of reduced magnitude. The reduction factors in Table 3.2 apply equally to joint rotation. Thus, for the pipeline offset by 2.0 m, as in Example 3.5, the potential rotation is merely 85 per cent of the above values.

Where the pipeline is transverse to the tunnel, as in Examples 3.1 and 3.2, then the potential joint movement depends on the position of these joints relative to the tunnel line. For these examples the worst position is at $y_J = 0, 5.49$ m, 10.98 m and so on. The best position (as far as limiting joint movement) is $y_J = -2.745$ m, 2.745 m, 8.235 m and so on. For the worst case,

$$\text{at } y = 5.49 \text{ m}, \quad y/i = 2.11, \quad w/w_{max} = 0.1075$$

$$\text{at } y = 0, \quad w/w_{max} = 1.0000$$

$$\text{Differential settlement (in 5.49 m)} = 0.8925\, w_{max}$$

A similar differential settlement occurs between $y = 0$ and $y = -5.49$ m. The potential rotation at the joint immediately above the tunnel is

$$\theta_J = \frac{2 \times 0.8925 \times 13.6 \times 10^{-3}}{5.49} = 4.42 \times 10^{-3} \text{ rad} = 0.25°.$$

For the best position, $\theta_J = 0.08°$. A reasonable design case is to assume joints at $y_J/L_J = -0.875, 0.125, 1.125$ and so on. This includes 75 per cent of the possibilities; that is, there is a 0.75 probability of not being exceeded. For this case $\theta_J = 0.24°$. (Note that the form of ground movement is such that the 0.75 probability does not lie 75 per cent of the way between $\theta_J = 0.08°$ and $\theta_J = 0.25°$.) For the form of ground movement shown in Figure 3.6 the potential rotation for $L_J \leqslant 2.5i$ is simply

$$\theta_J = \frac{w_{max}}{1.25i} = 4.18 \times 10^{-3} \text{ rad} = 0.24°.$$

Reduction in longitudinal bending due to joint movement. The approaches to the solution of longitudinal bending in a jointed pipeline correspond to the methods used for the analysis of continuous beams in linear framed structures —see, for example, Gregory (1966). In the 'force' method the unknowns to be solved are the internal generalized forces in the structure, namely, bending moments and shears acting across 'releases' or 'cuts'. The final equations specify geometrical continuity on each side of the cuts. The physical picture of the iterative solution is to imagine pairs of men with levers and jacks stationed each side of the cuts. They begin with zero forces and successively apply equal and opposite sets of forces across their cuts until geometrical continuity is everywhere achieved. For a jointed pipeline, the pipeline would be cut into its constituent pipe lengths on elastic foundations and the ground movement applied to this cut structure. Equal and opposite shears would then be applied at the pipe ends to restore displacement continuity at the joints. The process is frequently ill-conditioned, and starts with 'short' beams for which the subgrade reaction analysis does not apply.

The alternative approach is the 'deformation' method. Here the unknowns are the joint rotations (that is, differential joint rotations for a pipeline). The final equations specify static equilibrium in the jointed pipeline. The physical picture of the iterative solution is men with levers and jacks stationed at the pipeline joint positions. Starting with the forces in a continuous pipeline they are successively instructed to release their constraining forces in turn so that these forces are reduced to negligible magnitudes. It is not necessary to work in terms of end rotations and displacements; end moments and shears may be used instead. This technique for the solution of the unknown deformations is moment distribution. The process is usually extremely well-conditioned and starts with the solution for a continuous pipeline for which the subgrade reaction analysis is comparable with the elastic continuum model (at least in the solution for bending moment). A further advantage is that for relatively flexible pipelines the release of moment at a joint does not even carry over to adjacent joints. In the following analysis only moment release will be considered. Moment *and* shear release are required at a free end to a pipeline. Partial release of moment is required for stiff joints. Other continuity conditions apply, for example, where the end of a pipeline is built-in to a relatively rigid structure.

The bending moment profile for a release at a hinged joint is shown in Figure 3.16. The distribution is given by

$$M_x = M_J A(\lambda x) \tag{3.21}$$

where the function $A(\lambda x)$ is defined in equation (3.9). The full set of load–

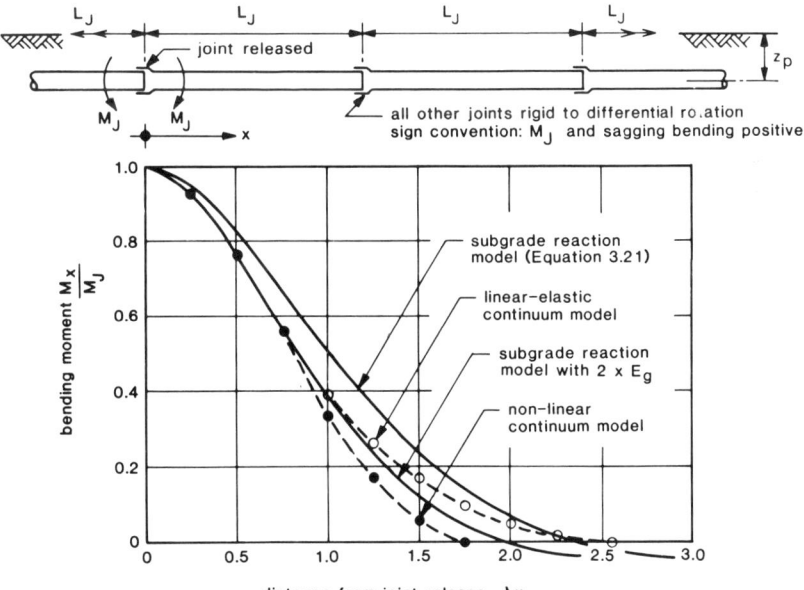

Figure 3.16 Bending moment profile for moment release at hinged joint.

deformation equations is given in Appendix B. In equation (3.21), M_x is zero or negligible beyond $\lambda x = 3\pi/4 \simeq 2.5$ (as noted in Table 3.1). For the subgrade reaction analysis the critical length for moment release is thus

$$x = \frac{0.75\pi}{\lambda} = 2.0d \left[\frac{4E_p I_p}{E_g d^4} \right]^{0.271} \quad (3.22)$$

for Poisson's ratio between 0.3 and 0.5. Equation (3.22) is practically identical to the equation given by Randolph (1981). For moment at the end of a pile in a linear elastic continuum, this finite-element analysis gives the equivalent of

$$\text{critical pile length } l_c = 2.0d \left[\frac{4E_p I_p}{E_g d^4} \right]^{0.286} \quad (3.23)$$

for Poisson's ratio between 0.3 and 0.5.

As noted in section 3.2.1, the subgrade reaction foundation model, with K_∞ given by equation (3.3), slightly overestimates bending moment compared with the elastic continuum model. This is indicated in Figure 3.16 where the distribution for a linear elastic continuum is given. For a jointed pipeline this has the effect of *underestimating* the final bending moment. Allowance for this factor can be made by either using the elastic continuum distribution for moment or by substituting a value for E_g of twice the value used previously for the continuous pipeline calculation. The effect of doubling the value for E_g in the subgrade reaction model is indicated in Figure 3.16. Potentially more important than the precise nature of the elastic soil–pipe model is the non-linear response of the soil. Increasing soil stiffness with reducing strain means that the extent of the effect of a moment release may be even less than in the linear elastic continuum model. An approximate indication of the possible nonlinear response is also shown in Figure 3.16. It is a matter of judgement as to how extensive an allowance should be made for moment release. Any reasonable distribution may be adopted based on the comparisons in Figure 3.16.

If the bending moment reduction due to joint flexibility is less than 10 per cent, this will be classified as 'no effect'. From Figure 3.16 together with Figures 3.5 and 3.9 the following preliminary observations can be made.

(1) If $L_J \lambda \geqslant 4.0$ the maximum bending moment in a *parallel* pipeline is not affected by joint flexibility.

(2) $L_J \lambda \geqslant 4.0$ the maximum bending moment in a *transverse* pipeline is not affected by joint flexibility if these joints happen to be in the worst position, that is, equally spaced each side of the tunnel line.

(3) If $L_J \lambda \geqslant 4.6$ the maximum bending moment in a *transverse* pipeline is not affected by joint flexibility with the joints at $y = -0.375\,L_J, 0.625\,L_J, 1.625\,L_J$ and so on. There is a 0.75 probability that this is not exceeded.

Joint flexibility also reduces the bending associated with local variations in ground movement or pipe restraint (see section 3.2.7.) The distribution of bending associated with local restraint is $M_x/M_{max} = A(\lambda x)$ as for moment release. For local variations in ground movement the distribution is no worse than

$$M_x/M_{max} = B\left(\frac{\lambda x}{2} + \frac{\pi}{4}\right)$$

where the function $B(\lambda x) = \exp(-\lambda x)\sin(\lambda x)$. The following further general observations can be made:

(4) If $L_J\lambda \geqslant 3.0$ the maximum bending moment associated with system disturbance is not affected by joint flexibility if these joints happen to be in the worst position.

(5) If $L_J\lambda \geqslant 3.5$ the maximum bending moment associated with system disturbance is not affected by joint flexibility with a probability of 0.75 that this is not exceeded.

The following examples illustrate the significance of joint movement.

Example 3.10 Joint relaxation in small-diameter pipeline

The pipe and tunnel details are the same as in Examples 3.2, 3.4 and 3.8.

$$i = 2.6\,\text{m}, \quad w_{max} = 13.6\,\text{mm}, \quad d = 0.10\,\text{m}, \quad \lambda = 1.923\,\text{m}^{-1}.$$

For a joint spacing of $L_J = 5.49\,\text{m}$ (18 feet), $L_J\lambda = 10.6$.

For a joint spacing of $L_J = 3.66\,\text{m}$ (12 feet), $L_J\lambda = 7.0$.

From the general observations in section 3.2.8 and Figure 3.16, no allowance need be made for reduction in maximum bending strain associated with joint flexibility. The distributions of bending would, however, be affected up to a distance of $2/\lambda = 1.0\,\text{m}$ each side of the joints. Note that a joint spacing of less than $3/\lambda = 1.6\,\text{m}$ is required before significant reductions in longitudinal bending occur.

Example 3.11 Joint relaxation in large-diameter pipeline

The pipe and tunnel details are the same as in Examples 3.1, 3.3 and 3.7.

$$i = 2.6\,\text{m}, \quad w_{max} = 13.6\,\text{mm}, \quad d = 0.50\,\text{m}, \quad \lambda = 0.481\,\text{m}^{-1}.$$

$$\text{For } L_J = 5.49\,\text{m}\,(18\text{ feet}), \quad L_J\lambda = 2.64.$$

$$\text{For } L_J = 3.66\,\text{m}\,(12\text{ feet}), \quad L_J\lambda = 1.76.$$

From the general observations in section 3.2.8 and Figure 3.16, joint flexibility *may* cause significant reductions in the maximum strain compared with a rigidly-jointed pipeline. In practice the following calculations would only be carried out if significant bending had been shown in a rigidly-jointed pipeline. When considering joint flexibility it is convenient if the bending moment in the continuous rigidly-jointed pipeline has been calculated at intervals of 1, 2, 3 or 6 feet as these are common factors of the usual joint spacings of 12 or 18 feet.

(i) For the pipeline parallel to the tunnel line and $L_J = 5.49\,m$ (18 feet)

Where the pipeline is parallel to the tunnel it is only necessary to find the joint position relative to the tunnel face ($x = 0$) that gives the least reduction in bending moment. Since $L_J\lambda = 2.64$, moment release at a joint does not carry over to adjacent joints. The bending distribution for a continuous rigidly jointed pipeline is shown in Figure 3.15a. The maximum bending strain has been calculated in Example 3.3 as -91 microstrain at $x/i = 1.22$ ($x = 3.172\,m$). $L_J = 5.49\,m = 2.11i$, and so from inspection of Figure 3.15a the least reduction in bending occurs when $x_J = 0, 5.49\,m, 10.98\,m$ and so on. Assuming that the maximum still occurs at $x = 3.172\,m$, this maximum is calculated as follows.

At $x_J = 0, \varepsilon_{xp} = 0$, moment release at $x = 3.172\,m$ is zero.

At $x_J = 5.49\,m, \varepsilon_{xp} = -0.1175\dfrac{w_{max}}{i^2} \times \dfrac{d}{2} = -59$ microstrain.

At $x = 3.172\,m$, moment release from $x_J = 5.49\,m$ is

$$\text{release} = -M_J A(\lambda x) = -M_J A(\lambda[5.49 - 3.172]) = -M_J A_{1.115}$$
$$= +59 \times 0.439 = +26 \text{ microstrain.}$$

Maximum bending strain at $x = 3.172\,m, \varepsilon_{xp} = -91 + 26 = -65$ microstrain.

Figure 3.17a shows the bending moment envelope for incremental joint positions of 1.83 m. The maximum *is* associated with a joint at $x = 0$ and occurs at $x = 2.56\,m$, not at $x = 3.17\,m$. From Figure 3.17a

$$\varepsilon_{xp\,max} = -0.137\dfrac{w_{max}}{i^2}\dfrac{d}{2} = -69 \text{ microstrain.}$$

The relationship between moment release and rotation at the joint is

$$\theta_J = \dfrac{M_J}{E_p I_p \lambda}. \tag{3.24}$$

Thus, to go from the rigidly-jointed to the perfectly flexibly-jointed case, the differential rotation at $x = 5.49\,m$ is

$$\theta_J = 2 \times \dfrac{M_J}{E_p I_p \lambda} = \dfrac{2}{\lambda}\left(\dfrac{\partial^2 w_p}{\partial x^2}\right)_J = \dfrac{2}{\lambda}(-0.1175)\dfrac{w_{max}}{i^2}$$

$$= \dfrac{2 \times -0.1175 \times 13.6 \times 10^{-3}}{0.481 \times 2.6^2} = 9.8 \times 10^{-4}\,\text{rad} = 0.06°.$$

(Note that the moment is released at both sides of the joint, so the total differential rotation is twice the value given by equation (3.24).)

The maximum joint movement occurs when the maximum moment is

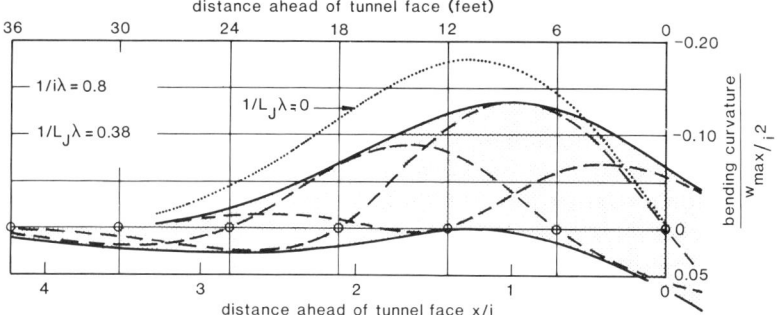

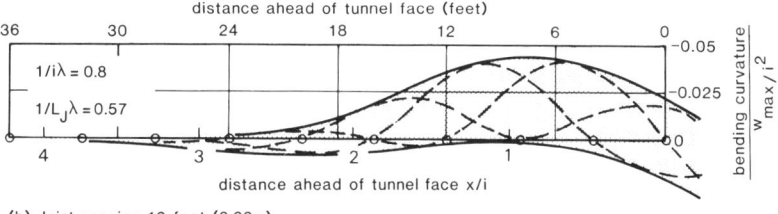

Figure 3.17 Bending in large-diameter pipe parallel to tunnel.

released at $x = 3.172$ m.

$$\theta_{J\max} = 2 \times \frac{M_J}{E_p I_p \lambda} = \frac{2 \times -0.181 \times 13.6 \times 10^{-3}}{0.481 \times 2.6^2} = 1.5 \times 10^{-3} \text{ rad.}$$

$$= 0.09°.$$

(Example 3.9 calculates a rotation of between 0.07° and 0.10° for perfectly rigid pipes.)

(ii) For the pipeline parallel to the tunnel line and $L_J = 3.66$ m (12 feet)

$L_J = 3.66$ m, $L_J \lambda = 3.66 \times 0.481 = 1.76$. For release of M_{J0} both sides of joint number (0) the carry-over to joint numbers (-1) and $(+1)$ is $M_{J0} A(\lambda L_J) = 0.137 M_{J0}$. The carry-over to joint numbers (-2) and $(+2)$ is $M_{J0} A(2\lambda L_J) = -0.038 M_{J0}$ and can be neglected for practical purposes. Consider the joints at $x = 0, \pm 12$ feet, ± 24 feet and so on. The 'fixity' state curvature (which is equivalent to moment) is shown in Figure 3.14a and has been calculated at 2 ft intervals. The joint relaxations required for perfectly hinged joints ($M_J = 0$) are found in Table 3.4.

Midway between joints (0) and (1) at $x = 6$ feet the bending is

$$\frac{\partial^2 w_p}{\partial x^2} \bigg/ \frac{w_{\max}}{i^2} \text{ for continuous pipeline} = -0.1437$$

Table 3.4 Relaxation table for iterative solution of deformations

1. x value of joint (feet)	48	36	24	12	0	− 12	− 24	− 36	− 48
2. Joint number	(4)	(3)	(2)	(1)	(0)	(− 1)	(− 2)	(− 3)	(− 4)
3. 'Fixity' state curvature × 10⁴ / w_{max}/i^2	+ 5	+ 44	− 453	− 1777	0	+ 1777	+ 453	− 44	− 5
4. Relaxation of − 1777 at joint (1) and + 1777 at joint (− 1)			+ 243	**+ 1777**	0	**− 1777**	− 243		
5. Relaxation at joints (2) and (− 2)		+ 29	**+ 210**	+ 29		− 29	**− 210**	− 29	
6. Relaxation at joints (3) and (− 3)	− 10	**− 73**	− 10				+ 10	**+ 73**	+ 10
7. Residuals	(− 5)	(0)	(− 10)	(+ 29)	0	(− 29)	(+ 10)	(0)	(+ 5)
8. Relaxation at joints (4) (2) (1)			− 4	**− 29**	0	**+ 29**	+ 4		
		+ 2	**+ 14**	+ 2		− 2	**− 14**	− 2	
	+ 5	+ 1						− 1	**− 5**
	0	**− 3**	0	**− 2**	0	**+ 2**	0	**+ 3**	0
9. Residuals	0	0	0	0	0	0	0	0	0
10. Sum relaxations	+ 5	− 76	+ 224	+ 1746	0	− 1746	− 224	+ 76	− 5
11. Check 'carry over'	− 10		− 10						
		+ 31	+ 239	+ 31	0				
12. Line (3) + (10) + (11) = residuals	0	− 1	0	0	0				
13. Corrected relaxations	+ 5	− 75	+ 224	+ 1746	0	− 1746	− 224	+ 75	− 5
14. Line (13) × 10^{-4}	0	− .007	+ .022	+ .175	0	− .175	− .022	+ .007	0

'Fixity' state $= -0.1437$

Relaxation from joint (1) $= +0.175 \times 0.584 = +0.1022$

Relaxation from joint (0) $= 0 \times 0.00 = 0$

Released curvature factor $= -0.041$

$$\text{hogging bending strain at } x = 6 \text{ feet, } \varepsilon_{xp} = \frac{0.041 \times 13.6 \times 10^{-3}}{2.6^2} \times \frac{0.50}{2}$$

$$= 21 \text{ microstrain.}$$

Figure 3.17(*b*) shows the bending moment envelope for incremental joint positions of 1.22 m. The maximum is associated with a joint at approximately

$x = 1.22$ m (2 feet) and occurs at $x = 2.26$ m. From Figure 3.17b, $\varepsilon_{xpmax} = 22$ microstrain. It should be noted that the bending moment now bears very little resemblance to the continuous rigidly jointed case or the so-called 'ground curvature' (Figure 3.15a) with $1/i\lambda \to 0$. The pipeline flexural rigidity has reduced the bending strain by 26 per cent and joint flexibility gives a further 56 per cent reduction, that is a total reduction of 82 per cent. This reduction is associated with differential joint rotation of only 0.08°.

(iii) For the pipeline transverse to the tunnel line

The bending moment envelopes for joint spacings of $L_J = 18$ and 12 feet are shown in Figure 3.18. The most favourable position for the joints is $|y| = 0, L_J, 2L_J$, and so on. The least favourable is $|y| = 0.5L_J, 1.5L_J$, and so on. Note that joint position y_J is the same as $L_J - y_J$, since the maximum bending moment merely occurs at position $-y$ instead of at y. In general it is not practicable to take account of the true joint position relative to the tunnel, and 75 per cent of the possibilities are included by taking $y_J = 0.375L_J$. In this case for joints at 18 feet centres the bending curvature factor for rigid joints (Example 3.1) is reduced from 0.60 to 0.475. A reduction of 40 per cent is associated with the pipe flexural rigidity and a further 12.5 per cent with joint flexibility. With joints at 12 feet centres the bending factor is further reduced to 0.185, a total reduction of 81.5 per cent from the so-called 'ground curvature'. For limitations on joint rotation, the design joint position is $y_J = 0.125L_J$. For

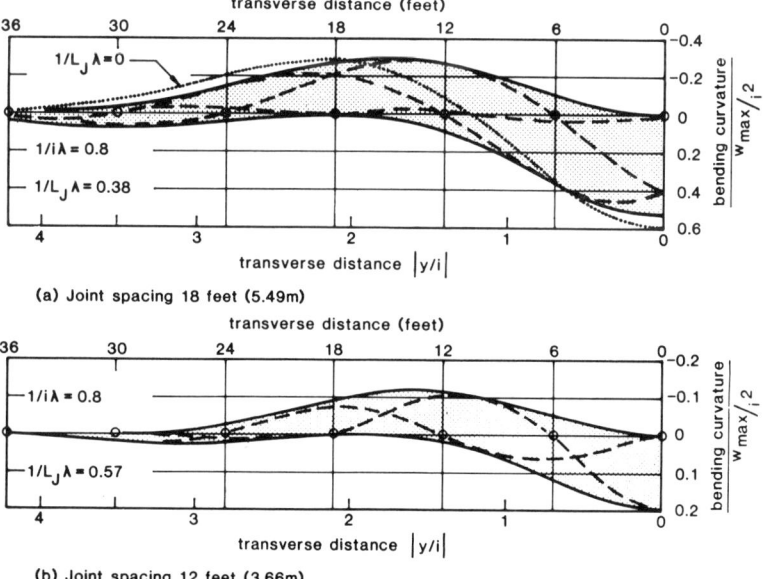

Figure 3.18 Bending in large-diameter pipe transverse to tunnel.

a joint spacing of 18 feet the maximum rotation in the joint is 0.26° compared with the value of 0.24° given by the simple calculation in Example 3.9. With joints at 12 feet centres the rotation is 0.29°.

The allowance for system disturbance to this pipeline has been calculated in Example 3.7 as $\varepsilon_{p\,max} = 59$ microstrain. The worst distribution for this effect is

$$\frac{\varepsilon_p}{\varepsilon_{p\,max}} = A(\lambda x),$$

associated with local restraint.

For joints at 18 feet centres ($L_J\lambda = 2.64$) and the design case of x_J or $y_J = 0.375 L_J$, the reduction in system disturbance effects is

$$\frac{\varepsilon_{p\,max} - \varepsilon_p}{\varepsilon_{p\,max}} = [A(0.375 L_J)]^2 + [A(0.625 L_J)]^2$$

$$= [A(0.99)]^2 + [A(1.65)]^2$$

$$= 0.515^2 + 0.176^2 = 0.30 = 30 \text{ per cent reduction.}$$

For the elastic continuum model (Figure 3.16 distribution), the reduction is only

$$\frac{\varepsilon_{pmax} - \varepsilon_p}{\varepsilon_{pmax}} = 0.395^2 + 0.12^2 = 0.17 = 17 \text{ per cent reduction.}$$

For joints at 12 feet centres ($L_J\lambda = 1.76$), the system disturbance effects are reduced by 73 per cent in the subgrade reaction model and only 53 per cent in the elastic continuum model.

3.2.9 Ultimate soil pressure

An estimate of the maximum restraint to differential soil–pipe movement can be obtained by equating the ultimate soil resistance with the bearing capacity of a strip footing. The general form for a strip footing given by Terzaghi and Peck (1967) is

$$\text{ultimate bearing capacity, } q_u = cN_c + \gamma z_p N_q + 0.5\gamma BN_\gamma.$$

For a pipe and differential soil–pipe movement:
 z_p is the depth from ground surface to pipe axis level,
 $B = d$, which is the pipe outside diameter,
 γ is the soil unit weight,
 $q_u = p_u = p_{1u} + p_{2u}$ (equations (3.15) and (3.16)), where $p_u d$ is the ultimate load on the pipeline/unit length,
 c is the soil cohesion,
 and N_c, N_q are bearing capacity factors with respect to cohesion and surcharge and N_γ accounts for the influence of the weight of the soil.
The bearing capacity factors are dimensionless and depend on the angle of

internal fraction, ϕ, and the failure mechanism. The failure mechanism depends on the depth/width ratio, z_p/d, on whether the soil is dense or loose (or stiff or soft), and on the direction of loading relative to the ground surface. The soil–pipe adhesion, c_a, also has a small influence on N_c. Where w_p is less than w, the break-out resistance will be greatly affected by the presence of a stiff road pavement, since for low values of z_p/d this will cause plastic flow around the pipe, similar to deep burial, rather than the passive wedge failure generally associated with shallow burial. For pipelines the bearing capacity factors N_q and N_γ can be merged to give the ultimate bearing pressure,

$$p_u = cN_c + \gamma z_p N_q. \tag{3.25}$$

For vertical ground movement where the pipe is pushing downwards into the soil, $w_p > w$, the usual bearing capacity factors for strip footings apply. Where the ground movement is horizontal and lateral to the pipe, the bearing capacity factors given by Hansen (1961) may be used. Model tests by Audibert and Nyman (1977) substantiate the use for pipes. For the case of vertical ground movement and $w_p < w$, soil failure occurs above the pipe. Tests on model pipes with granular soils have been reported by Matyas and Davis (1983) and a theoretical solution for the break-out resistance is given by Vesic (1971). Tests on model pipes and remoulded clay of shear strength 40 to 50 kN/m² have been reported by Roberts and Regan (1977). A full-scale test on a 100 mm diameter pipe at $z_p/d = 9.5$ was also carried out. British Gas Corporation Engineering Research Station have conducted vertical pull-out tests on short lengths of 4–8 inch diameter gas distribution mains. The tests represent the likely range of conditions that apply to a small-diameter distribution main at z_p/d in the range 3.0 to 12.0. A summary of the results is given by Casson (1985). All the data mentioned for cohesive soil relate to undrained conditions and the loading rates are far quicker than could be caused by tunnelling-induced movements. Bearing capacity factors N_c and N_q based on the theoretical analysis, model tests and full-scale tests are summarized in Figure 3.19 and these factors agree with the full-scale tests. For combined horizontal and vertical break-out movement, Nyman (1984) has suggested a design method based on an analogy to the behaviour of buried inclined anchor plates.

Very little information is available on the displacement required to develop the peak load. The full-scale tests reported by Casson (1985) required differential soil–pipe movement of 2.5–53 mm, with most of the larger movements occurring in granular soils. The majority of differential movements in cohesive soils were in the range 20–33 mm at peak load. Nyman (1984) and Audibert and Nyman (1977) found yield displacements of approximately $0.015z_p$ for dense and $0.025z_p$ for loose cohesionless soils and horizontal/vertical break-out of a pipe. The foregoing data relate to tests where the pipe is pulled through the soil. In the tests of Matyas and Davis (1983) the soil was allowed to settle relative to the pipe, simulating the loading

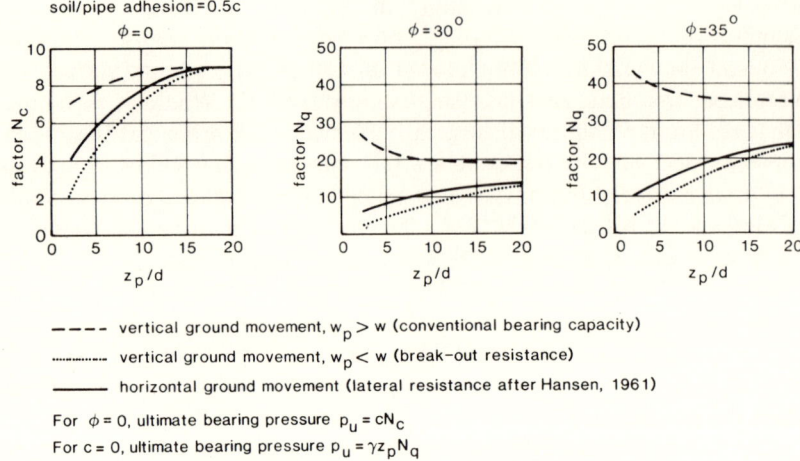

── ── vertical ground movement, $w_p > w$ (conventional bearing capacity)
········· vertical ground movement, $w_p < w$ (break-out resistance)
────── horizontal ground movement (lateral resistance after Hansen, 1961)

For $\phi = 0$, ultimate bearing pressure $p_u = cN_c$
For $c = 0$, ultimate bearing pressure $p_u = \gamma z_p N_q$

Figure 3.19 Bearing capacity factors for buried pipelines.

condition applicable to tunnelling-induced ground movement. For these tests the differential movement at peak load was approximately $0.007z_p$ for sand of $\phi = 30$–$36°$. However, the load for this displacement is only about a third to one-half the value indicated in Figure 3.19. A construction for p/p_u against $(w - w_p)/(w - w_p)_u$ is given by Audibert and Nyman. Linear design up to 50 per cent of the ultimate bearing pressure corresponds to a displacement of 0.13 $(w - w_p)_u$ at $p/p_u = 0.50$.

3.3 Ground movement parallel to a pipeline

Where ground movement is parallel to a pipeline, axial tensile and compressive forces are induced. The fixity conditions are crucial in determining whether tensile or compressive forces occur, as illustrated in Figure 3.20. In general, the compressive forces that may be induced by tunnelling movements should not be relied on to reduce tensile bending stress associated with transverse ground movement. Axial tensile forces are generally a secondary effect compared with longitudinal bending, but do combine with those positions where hogging bending occurs—ahead of the tunnel face in parallel pipelines and offset from the tunnel centre line in transverse pipelines. Metal pipelines are about 1000

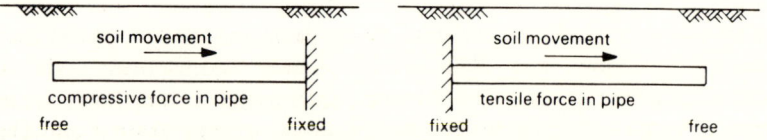

Figure 3.20 Effect of fixity on force in pipeline.

times more rigid than the soil they displace, and for a continuous rigidly-jointed pipeline the pipe movement is a fraction of the ground movement about five diameters from the pipe. Where a small-diameter relatively flexible branch joins the main pipeline, significant junction stresses may be created in the branch due to this restraint on movement. This aspect of the general problem is discussed in section 3.4.6, and the significance of junction stresses in pipeline networks subjected to earthquake movements has been considered by Singhal and Meng (1983).

The empirical relationships for horizontal ground movement are given in Chapter 2. For a pipeline transverse to a tunnel

$$v \simeq 1.65\left(\frac{y}{i}\right)v_{max}\exp\left[\frac{-y^2}{2i^2}\right] \qquad (3.26)$$

where the maximum

$$v_{max} \simeq -0.607\frac{n}{z_0-z}iw_{max} \qquad (3.27)$$

at $y/i = 1.0$. The negative value of v_{max} indicates ground movement towards the tunnel centre line, $y = 0$. Parallel to and directly above the tunnel drive, the horizontal ground movement is

$$u \simeq u_{max}\exp\left[\frac{-x^2}{2i^2}\right] \qquad (3.28)$$

where the maximum

$$u_{max} \simeq -0.399\frac{n}{z_0-z}iw_{max} \qquad (3.29)$$

at $y = 0, x = 0$. The negative value of u indicates that ground displacement from the *original* position is always in the direction of decreasing x, that is, in the present notation opposite to the direction of tunnel advance. From equation (3.28), the residual horizontal displacement in the x-direction is zero beyond $x = -3i$. This does not always represent the true situation, and if necessary a design movement beyond $x = 0$ of

$$u \simeq 0.5u_{max}\left(1+\exp\left[\frac{-x^2}{2i^2}\right]\right) \qquad (3.30)$$

may be taken. This gives a residual displacement of $0.5u_{max}$, and has the effect of slightly increasing the tensile stress in a parallel pipeline. To be consistent, similar non-recoverable deformation properties should be assumed for the soil in the axial soil–pipe interaction analysis. Geddes and Kennedy (1985) have outlined this approach to soil–structure interaction. From the above equations it can be seen that the maximum horizontal ground movement transverse to a tunnel is up to 50 per cent of the maximum settlement. Parallel to a tunnel the maximum is up to 30 per cent of the settlement. Since differential soil–pipe movement of approximately 10 mm will generate the full soil–pipe shear strength, soil–pipe slip may occur, particularly in a continuous rigidly-jointed pipeline transverse to a shallow tunnel.

For a pipeline parallel to but offset from the tunnel line in the plane $y = y_p$, the horizontal ground movement along the pipeline is

$$u_{y=y_p} = u_{max} \exp\left[\frac{-x^2}{2i^2}\right] \exp\left[\frac{-y_p^2}{2i^2}\right] \tag{3.31}$$

that is

$$u_{y=y_p} = u_{y=0} \times \text{(reduction factor)}.$$

Thus the effect on a pipeline offset from the tunnel drive is the same as on the centre line, except that the magnitude of the movement and the magnitude of the effect are reduced. The reduction factor depends on the geometry and the ground-movement parameter, i. Because the form of ground movement offset from the tunnel drive is identical with the form directly above the tunnel, the reduction factor applies equally to axial interaction effects in rigid or flexibly-jointed pipelines. The factors for offset pipelines are simply calculated from equation (3.31). The reductions for parallel ground movement are slightly more than for transverse movement (see Table 3.2).

The expressions for ground 'strain', more correctly *differential* ground movement, may be equated with the effect on a buried pipeline by assuming zero pipeline stiffness, but for all practical cases pipe stiffness and slip at the soil–pipe interface greatly reduces the strain in the pipeline. In jointed pipelines, the axial forces that can develop are severely limited by the very small forces that can be transferred at the pipe joints and these joints then accommodate most of the differential ground movement. The following analysis has been developed to assess the effect of ground movement on iron pipelines. The axial stiffness of these pipelines is such that the load in the pipeline may be calculated on the basis that the pipe is perfectly rigid. This can either be used as a slightly conservative solution or as a starting point for an iterative solution that allows for the relaxation associated with elastic extension of the pipe. Tensile forces in the pipeline are associated with differential movement on the ground movement profile. (Compare transverse ground movement where bending is associated with differential slope.) Generally it is necessary to limit the axial tensile stress; occasionally, limitations on pipeline joint extension may be important.

Theoretical approaches to this problem are similar to the methods available for piles subjected to axial loading or shrinking or swelling soil. The three broad categories of analysis are:

(1) Load-transfer methods, which use measured relationships between the soil–pipe shear stress at a point and the pipe movement at that point.

(2) Methods based on the theory of elasticity that employ the equations of Mindlin (1936) for subsurface loading within a semi-infinite mass.
(3) Numerical methods, in particular finite-element methods.

The load-transfer method is equivalent to the subgrade-reaction method for transverse ground movement. The movement of the pipe at any point is related only to the shear stress at that point and is independent of stresses elsewhere on the pipe. The use of this method for axial soil–pipe interaction has been outlined by Leach (1983, 1984). An advantage of the method is that a non-linear relationship between load and displacement can be accommodated quite easily in the iterative procedure as can the limitation on soil–pipe shear stress. A drawback is that it is not possible to extrapolate the load-transfer data. For grey iron distribution mains a design load-transfer curve is given in Figure 3.21. This is based on pipe-jacking tests on distribution mains of 4 to 24 inch diameter. The average soil–pipe shear strength for established mains is equal to the vane shear strength of the soil around the main. For a recently installed pipe the soil–pipe adhesive strength, τ_a, was only 50 per cent of the soil shear strength. The maximum shear stress mobilized was in the range 10–100 kN/m², where 10 kN/m² is for a very soft clay and 100 kN/m² for a stiff clay. The soil–pipe adhesive strength depends on the soil properties, pipe depth, pipe age and coating. In the absence of site measurements, $\tau_a = 50$ kN/m² is an average design value for established iron distribution mains.

Poulos and Davis (1980) have carried out calculations for a pile in shrinking or swelling soil. The results can be adapted to the case of a pipe

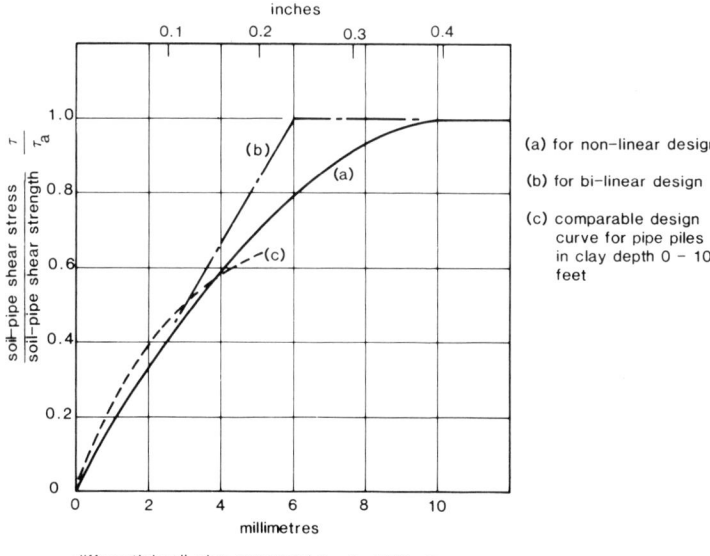

Figure 3.21 Design load-transfer curves for buried pipeline and parallel ground movement.

subjected to parallel soil movement. The analysis assumes Mindlin stress/deformation distributions in the soil, and is linear-elastic. The adaptation of the solutions for a pile is outlined in section 3.3.2.

The finite-element method has been used to model parallel soil movement, but the results do not cover a wide enough range of the variables to be generally applicable. For iron pipelines, typical soil properties and typical tunnel arrangements, the three methods of analysis give broadly similar results at least within the precision with which the form and magnitude of ground movement can be estimated. The main factors determining the behaviour of the pipeline are the magnitude and distribution of soil movement, the soil–pipe stiffness, the fixity conditions of the pipeline (including the effect of joints), and the shear strength at the soil – pipe interface. The precise nature of the elastic response of the soil is of secondary importance compared with determining the appropriate soil modulus and soil–pipe shear strength.

3.3.1 Elastic analysis

Poulos and Davis (1980) give the results of a linear-elastic analysis for a pile subjected to longitudinal soil movement (swelling soil). The assumptions are:

The soil is a homogeneous elastic half-space

The soil–pipe interface does not slip

The pile is incompressible and does not fail in tension or compression. The results for 'swelling' soil and axial load can be combined to give the maximum load in a pipeline caused by parallel ground movement associated with tunnelling. The load calculated on the basis of an incompressible pipe overestimates tensile forces.

3.3.2 Method of analysis

Since the load in a pipeline is caused by the differential ground movement, the movement profile parallel to the tunnel, as in Figure 3.22a, can be approximated by Figure 3.22b where the integral from $x = 0$ to $x \to \infty$ for a and b is the same. For typical pipe depth-to-diameter ratios, the free boundary above

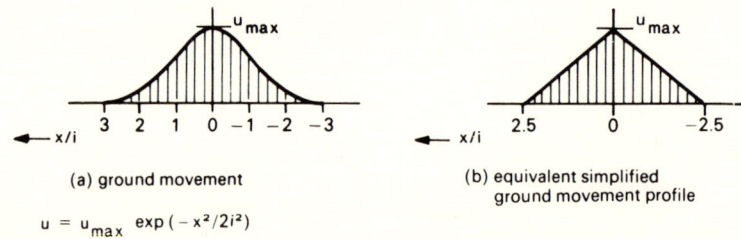

(a) ground movement

$u = u_{max} \exp(-x^2/2i^2)$

(b) equivalent simplified ground movement profile

Figure 3.22 Horizontal ground movement parallel to and directly above the tunnel drive.

the pipeline can be neglected. This slightly overestimates the load in the pipe. A pipeline parallel to the tunnel is assumed to extend a distance of $10i$ ahead of and behind the tunnel face position, that is, it is anchored $7.5i$ into undisturbed ground. On this basis the pipeline arrangement is similar to a pile in swelling soil where the interaction parameters are:

Notation after Poulos and Davis (1980)
d = pile diameter
L = pile length
z_s = zone of swelling soil
δ_g = maximum ground movement

Notation for buried pipelines
d = pipe diameter
$10i$ = pipe length
$2.5i$ = zone of ground movement
u_{max} = maximum ground movement

The common soil–pipe stiffness parameters are:
E_p = pipe elastic modulus E_g = soil elastic modulus
A_p = area of pipe wall = $\pi(d-t)t$

$$R_A = \text{pipe area ratio} = \frac{A_p}{\frac{1}{4}\pi d^2} = \frac{4(d-t)t}{d^2}$$

$$K^* = \text{soil–pipe stiffness factor} = \frac{E_p R_A}{E_g}$$

K^* is a measure of the relative compressibility of the pipe and the soil. The more compressible the pipe, the smaller the value of K^*. Piosson's ratio for the soil has a minor effect, and a value of 0.3 has been used throughout this section.

Poulos and Davis give the maximum pile load as $P_{max} = E_g d \delta_g \times I$, where the influence factor I depends on z_s/L and L/d. Thus for a buried pipeline the maximum tensile pipe strain depends on the soil–pipe stiffness factor K^*, the factor d/i and the ground movement factor u_{max}/i.

For a pipeline transverse to the tunnel, in the plane x = constant and $4i$ or more behind the tunnel face, the ground movement profile is shown in Figure 3.23a. The equivalent simplified movement is given in Figure 3.23b. The pipeline is assumed to extend $9i$ each side of the tunnel centre line, that is, it is anchored $6i$ into undisturbed ground. As for parallel ground movement due to the 'swelling' soil, the maximum tensile pipe strain depends on K^*, d/i and v_{max}/i. For this case $z_s = 2i$, $L = 8i$. An axial force is applied at $y = 0$ to make $v_p = 0$ at $y = 0$ as required by the symmetry of the ground-movement loading.

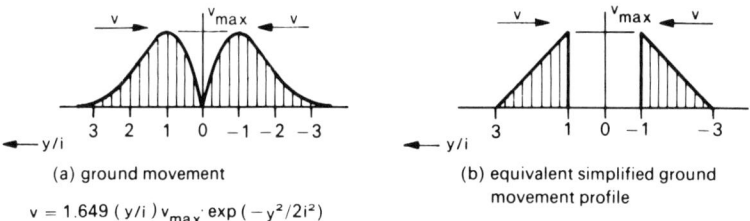

(a) ground movement
$v = 1.649\,(y/i)\,v_{max}\cdot \exp(-y^2/2i^2)$

(b) equivalent simplified ground movement profile

Figure 3.23 Horizontal ground movement transverse to the tunnel drive.

Assuming simple linear load distributions in the pipeline due to 'swelling' soil and axial load, the tensile strains due to ground movement and the compressive strains due to axial load can be superimposed to give the resulting maximum compressive and tensile strains. The method has been arranged so as to overestimate tensile strains and underestimate the relative magnitude of compressive strains.

3.3.3 *Solutions for a rigidly-jointed pipeline transverse to a tunnel drive*

The maximum soil movement parallel to a pipeline occurs when the pipe is transverse to the tunnel drive and the settlement trough is fully developed. For a pipeline with rigid (non-extendible) joints continuous across the tunnel line and anchored at least $6i$ into undisturbed ground, dimensionless curves for maximum pipe strain are shown in Figures 3.24 and 3.25. The parameter, i, is the settlement trough width parameter defined in section 2.4.2. The maximum tensile pipe strain has been calculated on the basis that the pipe is incompressible, that is, the differential soil–pipe movement is merely equal to the soil movement. This significantly overestimates the load for values of K^* less than 1000. In the design charts for pipe strain, the maximum pipe strains are inversely proportional to K^*. For a pipe with negligible stiffness, $K^* = 1$, the pipe strains are given by the maximum ground 'strains' in Chapter 2, that is,

maximum tensile (pipe $K^* = 1$), $\varepsilon_{y\,\mathrm{max\,tensile}} = 0.736 \dfrac{v_{\max}}{i}$ at $y = 1.73i$,

maximum compressive (pipe $K^* = 1$), $\varepsilon_{y\,\mathrm{max\,comp.}} = 1.65 \dfrac{v_{\max}}{i}$ at $y = 0$.

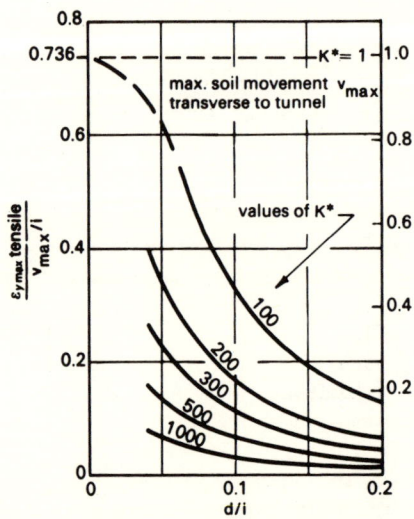

Figure 3.24 Maximum axial tensile pipe strain variation with diameter and stiffness factor K^*.

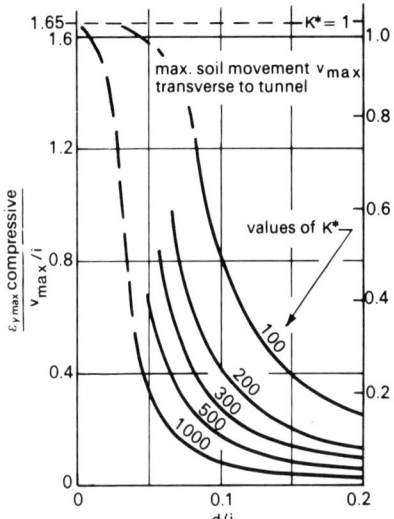

Figure 3.25 Maximum axial compressive pipe strain variation with diameter and stiffness factor K^*.

The maximum tensile pipe strain occurs at a transverse distance of between $1.73i$ and $3i$ (see Figures 2.10a and 2.10b), and for the purpose of calculating tensile bending plus axial strain is usually taken as coincident with the position of maximum hogging moment (see Figure 3.8). The maximum compressive pipe strain occurs at $y = 0$ and is usually ignored when calculating tensile bending plus axial strain. The reduction in pipe strain associated with pipe stiffness is greatest where the pipeline is transverse to the tunnel. This orientation has the narrowest zone of soil movement, and movement of the pipeline is prevented due to symmetry about the tunnel centre line (refer to Figure 3.20). The reduction factors compared with the strain for $K^* = 1$ are shown on the right-hand scales of Figures 3.24 and 3.25. Irrespective of the orientation of the pipeline, the reduction factor cannot be less than the values given by Figure 3.24 and 3.25.

Example 3.12 Calculation of axial strain in large-diameter transverse pipeline by elastic method

The pipe and tunnel details are the same as in Example 3.1.
Depth of pipeline axis to tunnel axis level, $z_0 - z_p = 3.5$ m.
Settlement trough width parameter, $i = 2.6$ m.
Maximum ground settlement, $w_{max} = 13.6$ mm.
Soil secant modulus (adjacent to pipe), $E_g = 10$ MN/m².
Pipe outside diameter, $d = 0.50$ m; pipe wall thickness, $t = 0.018$ m.
Pipe secant modulus, $E_p = 70$ GN/m² ($= 70 \times 10^9$ N/m²).

The maximum horizontal ground movement, at $y = 1.0i$, is given by equation (3.27) as

$$v_{max} = -0.607 \frac{n}{z_0 - z_p} i w_{max}.$$

For cohesive soil above the tunnel, $n = 1.0$. Thus,

$$v_{max} = -0.607 \times \frac{2.6}{3.5} \times 0.0136$$

$$= 0.0061 \text{ m } (6.1 \text{ mm}) \text{ towards the tunnel centre line.}$$

The soil–pipe stiffness factor, $K^* = \frac{E_p}{E_g} R_A$, where $R_A = \frac{4(d-t)t}{d^2}$

$$K^* = \frac{70 \times 10^9}{10 \times 10^6} \times \frac{4(0.5 - 0.018)0.018}{0.5^2} = 972.$$

The ratio, $\frac{d}{i} = \frac{0.5}{2.6} = 0.192$.

From Figure 3.24, the permanent tensile strain offset from the tunnel line and for $K^* = 100$ is

$$\varepsilon_{ypmax} = 0.135 \frac{v_{max}}{i} = 0.135 \times \frac{0.0061}{2.6}.$$

For $K^* = 972$, the strain is

$$\varepsilon_{ypmax} = \frac{100}{972} \times 0.135 \times \frac{0.0061}{2.6} = 3.26 \times 10^{-5}$$

$$= 33 \text{ microstrain tensile.}$$

This is additional to the tensile bending strain of 136 microstrain in Example 3.1. Total tensile strain = 169 microstrain.

From Figure 3.25, the compressive strain above the tunnel is

$$\varepsilon_{ypmax} = \frac{100}{972} \times 0.26 \times \frac{0.0061}{2.6} = 6.28 \times 10^{-5}$$

$$= 63 \text{ microstrain compressive.}$$

This implies a reduction in tensile strain in the bottom of the pipe to $282 - 63 = 219$ microstrain. In practice, this potential reduction in tensile strain would be ignored since axial soil–pipe slip can greatly reduce the effects of parallel ground movement without significantly affecting the bending associated with transverse ground movement.

If the pipeline is assumed to have perfectly telescopic joints at a transverse distance of $y_j = 2.6$ m, 7.8 m and so on (a spacing of 5.2 m = 17 feet), then from section 3.3.2 the maximum tensile strain in the pipe is

$$\varepsilon_{ypmax} = \frac{1.273 \, i \, v_{max}}{K^* \, d \, i} I_f$$

where the influence factor, I_f, depends on $z_s/L_J = 1.0$ and $L_J/d = 10.4$. The value of I_f is given by Poulos and Davis (1980) as 1.8. Thus, in this example,

$$\varepsilon_{ypmax} = \frac{1.273}{972} \times \frac{2.6}{0.5} \times \frac{0.0061}{2.6} \times 1.8 = 29 \text{ microstrain tensile.}$$

This value is virtually the same as the previously-calculated value for rigid joints of 33 microstrain. The release of the anchorage into undisturbed ground is offset by the release of the condition of no pipe movement at $y = 0$. For this position of flexible joints the tensile bending strain offset from the tunnel is also unaffected by joint relaxation (Figure 3.18a).

To allow for the effect of soil drainage, drained soil parameters can be substituted in the expression for K^*. Loading rates for transverse pipelines are such that a reduction in soil stiffness may occur during the build-up to maximum soil movement. For comparative purposes, $E'_g = \frac{2}{3}(1 + v'_g) E_g$, and so $E'_g/E_g = 0.87$. (Poisson's ratio has a minor effect, and $v_g = 0.3$ has been used throughout.) The reduction associated with soil drainage is 13 per cent.

Long-duration pile loading tests have indicated that at loads above about one-third of the ultimate, the settlement increases with time, long after consolidation should have finished. This time-dependent pile settlement at high loads is primarily the result of shear creep. In effect, this is equivalent to a decrease in E'_g with time, the decrease being greater as the shear stress approaches the ultimate value. For a pipeline transverse to the tunnel, shear creep is expected to give a significant reduction or relaxation in pipe strain with time when the soil-pipe shear stress is similarly more than one-third the ultimate value (see section 3.3.5).

Example 3.13 Calculation of axial strain in large-diameter transverse pipeline by 'load-transfer' method

The pipe and tunnel details are the same as in Example 3.12. The assumed load-transfer relationship for the calculation is the linear curve (b) shown in Figure 3.21. Since $v_{max} = 6.1$ mm, $v - v_p$ will be less than 6 mm and soil-pipe slip will not occur. The method is equivalent to the subgrade-reaction method and the relationship between soil-pipe shear stress and differential displacement is

$$\text{soil-pipe shear stress,} \quad \tau_y = \frac{K_y}{\pi_d}(v - v_p) \quad (3.32)$$

The modulus (dimensions FL^{-2}), $K_y = \dfrac{\pi d \tau_y}{(v - v_p)} = \dfrac{\pi d \tau_a}{6 \text{ mm}}$.

For comparison with Example 3.12, the soil-pipe shear strength, τ_a, may be taken as $E_g/200 = 50 \text{ kN/m}^2$. Thus, the modulus of the soil-pipe system for axial interaction is

$$K_y = \frac{\pi \times 0.50 \times 50 \times 10^3}{6 \times 10^{-3}} = 13.1 \times 10^6 \text{ N/m}^2.$$

On a pipe element length Δy, the strain is given by

$$\frac{\Delta \varepsilon_p}{\Delta_y} = \frac{\pi d \tau_y}{E_p A_p} \rightarrow \frac{d\varepsilon_p}{dy} = \frac{\pi d \tau_y}{E_p A_p} \qquad (3.33)$$

and

$$E_p A_p \frac{d^2 v_p}{dy^2} = K_y (v - v_p). \qquad (3.34)$$

Equation (3.34) is the equivalent of equation (3.5) for transverse ground movement. For a free length of pipe L, and end load P, the axial load along the pipe is given by

$$P_y/P = \cosh(\beta y) - \coth(\beta L)\sinh(\beta y) \qquad (3.35)$$

$$\text{where } \beta = \left[\frac{K_y}{E_p A_p}\right]^{1/2}$$

and is equivalent to the soil–pipe damping factor for transverse ground movement. β has dimensions of L^{-1}. For a semi-infinite pipe, equation (3.35) reduces to

$$P_y/P = \cosh(\beta y) - \sinh(\beta y) = \exp(-\beta y) \qquad (3.36)$$

Thus, for this linear analysis of a continuous rigidly-jointed pipeline the ground movement has an influence on distance of $\beta y = 3$ beyond the zone of ground movement. In this case a fully-anchored pipe must extend a distance of

$$y = \frac{3}{\beta} = 3 \left[\frac{E_p A_p}{K_y}\right]^{1/2}$$

$$= 3 \left[\frac{70 \times 10^9 \times \pi(0.50 - 0.018) \times 0.018}{13.1 \times 10^6}\right]^{1/2}$$

$$= 36.2 \text{ m} = 14i \text{ beyond } y = 3i.$$

In practice, due to increased soil stiffness for small movements, the true effective anchorage length will be much less than this and can be accommodated by 'fixing' the pipe against horizontal movement at not more than half this length beyond the zone of ground movement.

The iterative procedure to find the pipe movement, and hence soil – pipe shear stress and pipe strain, can be started at $v_p = 0$, that is, $v - v_p = v$. The soil–pipe shear stress is given by equation (3.32) and the differential pipe strain by equation (3.33). In the first iteration, $v_p = 0$ and so equation (3.36) can be used to integrate the pipe strain beyond $y = 3i$. The pipe movement at $y = 3i$ is

$$\varepsilon_{y=3i} \int_0^\infty \exp(-\beta_y) dy = -\varepsilon_{(y=3i)}/\beta = 12.1\varepsilon_{y=3i}.$$

This is combined with a numerical integration of equation (3.33) and the boundary condition $v_p = 0$ at $y = 0$. This simple first iteration gives $\varepsilon_{ypmax} = 25$ microstrain tensile and $\varepsilon_{ypmax} = 152$ microstrain compressive. The new pipe movement is used for the second iteration, and so on. The procedure closes

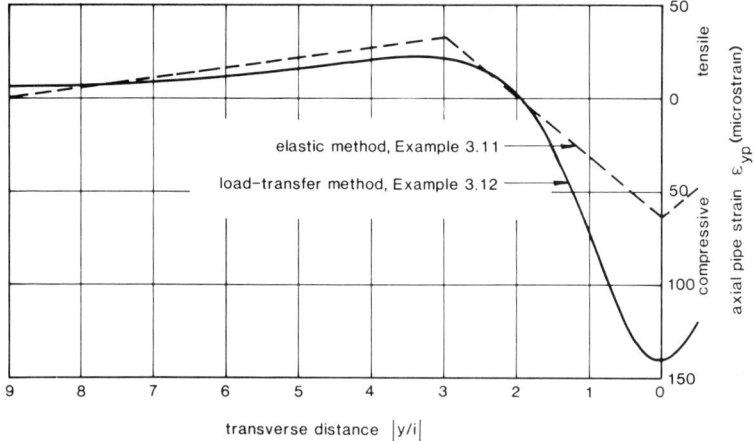

Figure 3.26 Axial strain in large-diameter pipe transverse to tunnel.

rapidly and the third iteration gives a maximum tensile strain of 22 microstrain and a maximum compressive strain of 140 microstrain. The maximum pipe movement is about 0.4 mm. A comparison between the results of this calculation and Example 3.12 is shown in Figure 3.26.

3.3.4 Solutions for a rigidly-jointed pipeline parallel to a tunnel drive

For a pipeline parallel to the tunnel, the maximum soil movement parallel to the pipeline occurs when the pipeline is directly above the tunnel drive. The reduction for offset parallel pipelines is given by equation (3.31). For a rigidly-jointed pipeline anchored by extending $7.5i$ into undisturbed ground, Figure 3.27 relates the maximum tensile pipe strain to the ratio d/i and soil–pipe stiffness factor, K^*. In Figure 3.27, the maximum pipe strain is inversely proportional to K^*. For a pipe with negligible stiffness, $K^* = 1$, the limiting value for strain is

$$\text{maximum tensile (pipe } K^* = 1\text{)}, \varepsilon_{x\,\text{max tensile}} = 0.607 \frac{u_{\max}}{i} \text{ at } x = i.$$

For the symmetrical form of ground movement illustrated in Figure 3.22, the strain is zero at $x = 0$ and the compressive strain at x/i behind the tunnel face is numerically equal to the tensile strain at x/i ahead of the face. The maximum pipe strain occurs between a distance of i and $2.5i$ ahead of the tunnel face and for the purpose of calculating tensile bending plus axial strain is usually taken as coincident with the maximum hogging moment (see Figure 3.12).

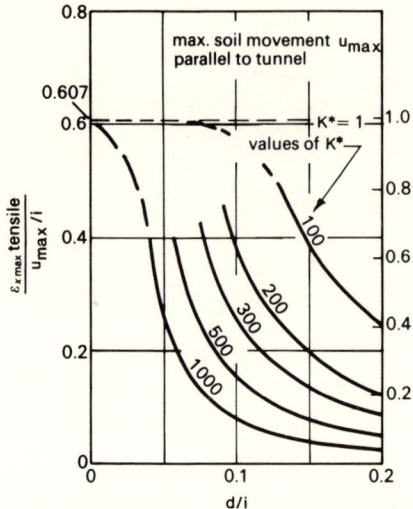

Figure 3.27 Maximum axial tensile pipe strain variation with diameter and stiffness factor K^*.

Example 3.14 Calculation of axial strain in parallel pipeline by elastic method

Large-diameter pipeline

The pipe and tunnel details are the same as in Example 3.3. The maximum horizontal ground movement, at $x = 0$, is given by equation (3.29) as

$$u_{max} = -0.399 \frac{n}{z_0 - z_p} i \, w_{max} = -0.399 \times \frac{2.6}{3.5} \times 0.0136$$

$= 0.0040$ (4.0 mm) in the opposite direction to that of tunnel advance.

As in Example 3.12, $K^* = 972$, $d/i = 0.192$.
From Figure 3.27, the maximum temporary tensile strain ahead of the tunnel face position is

$$\varepsilon_{xpmax} = 0.0273 \frac{u_{max}}{i} = 0.0273 \times \frac{0.0040}{2.6} = 4.2 \times 10^{-5}$$

$= 42$ microstrain tensile.

This is additional to the tensile bending strain of 91 microstrain in Example 3.3. Total tensile strain = 133 microstrain.

Small-diameter pipeline

The pipe and tunnel details are the same as in Example 3.4.
$E_g = 10 \, \text{MN/m}^2$, $\quad E_p = 70 \, \text{GN/m}^2$, $\quad d = 0.100 \, \text{m}$, $\quad t = 0.010 \, \text{m}$.

GROUND MOVEMENTS AND BURIED PIPELINES

Soil–pipe stiffness factor

$$K^* = \frac{E_p}{E_g} \frac{4(d-t)t}{d^2}$$

$$= \frac{70 \times 10^9}{10 \times 10^6} \times \frac{4(0.100 - 0.010)0.010}{0.1^2} = 2520$$

The ratio, $\dfrac{d}{i} = \dfrac{0.1}{2.6} = 0.0385$.

From Figure 3.27. the maximum temporary tensile strain ahead of the tunnel face position is

$$\varepsilon_{xpmax} = \frac{1000}{2520} \times 0.425 \times \frac{0.0040}{2.6} = 2.60 \times 10^{-4} = 260 \text{ microstrain tensile}.$$

This is additional to the tensile bending strain of 50 microstrain.
Total tensile strain = 310 microstrain.

Example 3.15 Calculation of axial strain in parallel pipeline by 'load-transfer' method

The pipe and tunnel details are the same as in Example 3.14 and the assumed load–transfer relationship is given in Example 3.13. The soil–pipe shear strength of 50 kN/m^2 is fully mobilized at a differential soil–pipe movement of 6 mm. From equation (3.33),

$$\frac{d\varepsilon_{xp}}{dx} = \frac{\pi d \tau_x}{E_p A_p} = \frac{\pi d \tau_a}{E_p A_p} \frac{\tau_x}{\tau_a}.$$

The linear load–transfer relationship is

$$\frac{\tau_x}{\tau_a} = \frac{(u - u_p)}{6 \text{ mm}} = \frac{(u - u_p)}{0.006}.$$

Since, by symmetry, $\varepsilon_{xp} = 0$ at $x = 0$ and assuming that everywhere $u_p = 0$, then

$$\varepsilon_{xpmax} = \frac{\pi d}{E_p A_p} \tau_a \frac{u_{max}}{0.006} \int_0^\infty \exp\left[\frac{-x^2}{2i^2}\right] dx$$

$$\varepsilon_{xpmax} = \frac{\pi d}{E_p A_p} \tau_a \frac{u_{max}}{0.006} \left[\frac{\sqrt{2\pi} i}{2}\right] \text{ tensile strain}.$$

The above is essentially the first step in an iterative procedure for pipe strain, starting with the shear stress based on zero pipe movement. Since there is at least a small elastic extension of the pipe, this first step must overestimate the pipe strain.

Large-diameter pipeline

Maximum tensile strain, $\varepsilon_{xpmax} < \dfrac{\pi \times 0.50 \times 50 \times 10^3 \times 0.004 \times \sqrt{2\pi} \times 2.6}{70 \times 10^9 \times \pi(0.05 - 0.018)0.018 \times 0.006 \times 2}$

$$= 8.9 \times 10^{-5}.$$

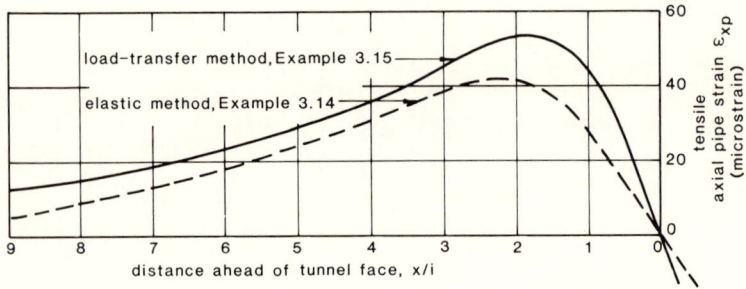

Figure 3.28 Axial strain in large-diameter pipe parallel to tunnel.

The maximum tensile strain is thus less than 89 microstrain. At the third iteration, the calculated strain reduces to a maximum of 53 microstrain at $x = 2i$. The results of this calculation and Example 3.14 are shown in Figure 3.28.

Small-diameter pipeline

Maximum tensile strain, $\varepsilon_{pmax} < \dfrac{\pi d}{E_p A_p} \tau_a \dfrac{u_{max}}{0.006} \left[\dfrac{\sqrt{2\pi} \, i}{2} \right]$

$$= \frac{\pi \times 0.10 \times 50 \times 10^3 \times 0.004 \times \sqrt{2\pi} \times 2.6}{70 \times 10^9 \times \pi(0.10 - 0.01)0.01 \times 0.006 \times 2}$$

$= 1.7 \times 10^{-4}$.

The maximum tensile strain is thus less than 170 microstrain. At the third iteration, the calculated strain reduces to a maximum of 97 microstrain at $x = 1.75i$.

3.3.5 Solutions for pipe displacement and soil–pipe shear stress

For the rigidly-jointed pipeline transverse to the tunnel drive, as described in section 3.3.3, the maximum horizontal pipe movement occurs where the axial strain is zero at a transverse distance between i and $3i$ depending on the soil–pipe stiffness and the pipe diameter. Figure 3.29a shows the variation of dimensionless maximum pipe movement, v_{pmax}/v_{max} with soil–pipe stiffness and diameter/trough width ratio. The pipeline movement, v_p, was calculated by integrating the pipe strain diagram starting at $v_p = 0$ at $y = 0$. For the pipeline parallel to the tunnel, as described in section 3.3.4, the maximum pipe movement is shown in Figure 3.29b. This maximum occurs where the axial strain is zero at $x = 0$. The integration was started at the free end of the pipeline ($x = 10i$) on the assumption that the movement of the free end is zero for a very flexible pipe ($K^* = 100$) and the movement is half the incompressible pipe movement for a pipe of stiffness $K^* = 1000$. Figure 3.29 generally overestimates the pipe movement, which will underestimate the bending induced in

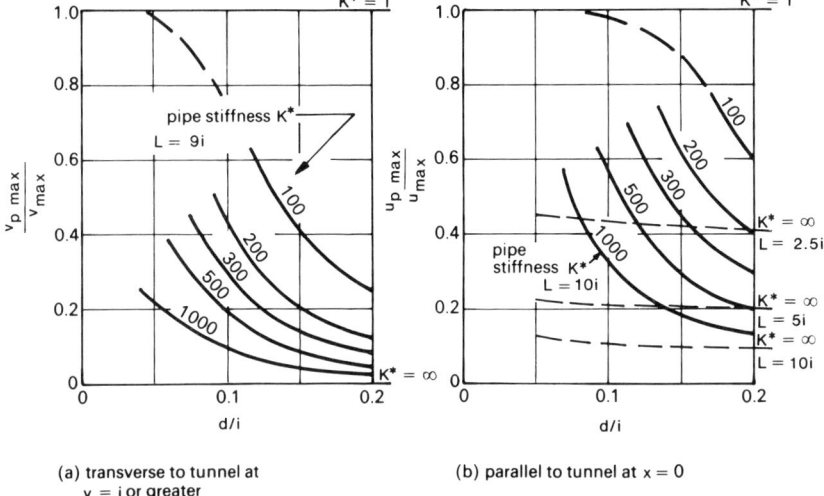

Figure 3.29 Maximum pipe movement.

flexible branches off the main pipeline. For this purpose, the main pipeline movement could be assumed to be not less than the movement of an incompressible pipe. This is zero for the transverse case and is given in Figure 3.29b for various anchorage conditions in the parallel cases.

The maximum soil–pipe shear stress occurs where the differential soil–pipe movement is a maximum. This is coincident with the point of maximum pipe movement and zero axial strain (that is, change from tensile to compressive strain). Substituting for the soil–pipe stiffness factor, K^*, in equation (3.33) and rearranging the variables, then the soil pipe shear stress, τ, is given by

$$\frac{\tau_y}{E_g} = \frac{K^* d}{6} \frac{\Delta \varepsilon_p}{\Delta y} = f\left(\frac{v_{max}}{i}, \frac{i}{d}\right)$$

and

$$\frac{\tau_x}{E_g} = \frac{K^* d}{4} \frac{\Delta \varepsilon_p}{\Delta x} = f\left(\frac{u_{max}}{i}, \frac{i}{d}\right).$$

The relationship between maximum soil–pipe shear stress and the pipe diameter/trough width ratio is shown in Figure 3.30. The results of a finite-element analysis are shown in Figure 3.30b and the simplified analysis used here overestimates the shear stress by up to 30 per cent.

3.3.6 Effect of pipe-joint extension

The longitudinal axial pulling effects on a pipeline have so far been restricted to a continuous rigidly-jointed pipeline. The axial strain induced by parallel

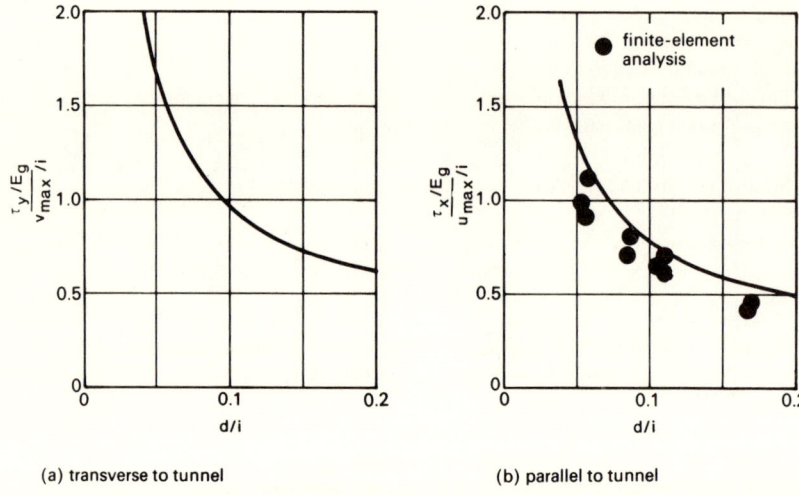

(a) transverse to tunnel   (b) parallel to tunnel

Figure 3.30 Maximum soil–pipe shear stress assuming full bond at soil–pipe interface.

ground movement is continuous across the joints and there is no differential movement at the joints. The axial pull-out resistance of the common joint types is given in section 3.4.3. Welded and flanged joints can be classified as non-extendible for the purpose of analysis. Rubber gasket push-in joints, bolted-gland mechanical joints and rubber gland mechanical couplings can be classified as perfectly flexible, that is, incapable of transferring significant axial load, and can be modelled as perfectly extendible. The lead–yarn joint can transfer only very small axial forces in the medium and long term, which is applicable to tunnelling-induced movement. This joint can be modelled as perfectly extendible, with a small correction for the limited load transfer.

For a preliminary assessment of axial pulling effects on a flexibly jointed pipeline, the joints can be taken as perfectly rigid since there are no practical cases where relaxation of regularly-spaced joints causes an increase in maximum axial tensile stress. Only if significant stresses are induced in a rigidly-jointed pipeline need consideration be given to possible reduction associated with joint movement. A quick check on possible joint extension can be obtained by assuming that the pipe between the joints is perfectly rigid and the joints then accommodate all the differential parallel ground movement. The possible joint extension is then the differential ground movement between midpoints of adjacent pipes.

Example 3.16 Calculation of potential pipe-joint pull-out

The pipeline is parallel to the tunnel drive and the pipe and tunnel details are the same as in Example 3.14. The joint spacing is 5.49 m (18 feet). For a parallel pipeline it is only necessary to consider the worst position for the joints, since as

the tunnel advances then the joints are bound to be in this position at some stage.
$u_{max} = 6.1$ mm, $i = 2.6$ m.
Joint spacing, $L_J = 5.49$ m; $L_J/i = 2.11$.
The worst position for the joints is at $x_J = 2.745$ m, 8.235 m, and so on.
At midpipe point $x = 0$, take $u_p = u = 1.000 u_{max}$.
At midpipe point $x = 5.49$ m, $u_p = u = 0.108 u_{max}$.
Maximum differential ground movement between adjacent pipes, $\Delta u = 0.892 u_{max}$.
Potential joint pull-out $= 0.892 u_{max}$
$= 0.892 \times 4.0 = 3.6$ mm.

Where the pipeline is transverse to the tunnel, as in Example 3.12, then the potential joint movement depends on the position of these joints relative to the tunnel line. The maximum joint *closure* occurs with $y_J = 0$, 5.49 m, 10.98 m and so on.
At midpipe point $y = 2.745$ m, $v = 0.998 v_{max}$.
At midpipe point $y = -2.745$ m, $v = -0.998 v_{max}$.
Maximum differential ground movement between adjacent pipes, $\Delta v = 1.996 v_{max}$.
Potential joint *closure* $= 1.996 v_{max} = 1.996 \times 6.1 = 12.2$ mm.
The maximum potential joint pull-out is given by:
at midpipe point $y = 2.60$, $v = 1.000 v_{max}$.
at midpipe point $y = 8.09$, $v = 0.041 v_{max}$.
Maximum differential ground movement between adjacent pipes, $\Delta v = 0.959 v_{max}$.
Potential joint pull-out $= 0.959 \times 6.1 = 5.8$ mm.

Reduction in longitudinal axial pulling effects due to joint movement. The alternative approaches to the solution of axial pulling in a jointed pipeline correspond with the 'force' and 'deformation' methods for bending in a jointed pipeline (see section 3.2.8). Here the simplest approach is the 'force' method.

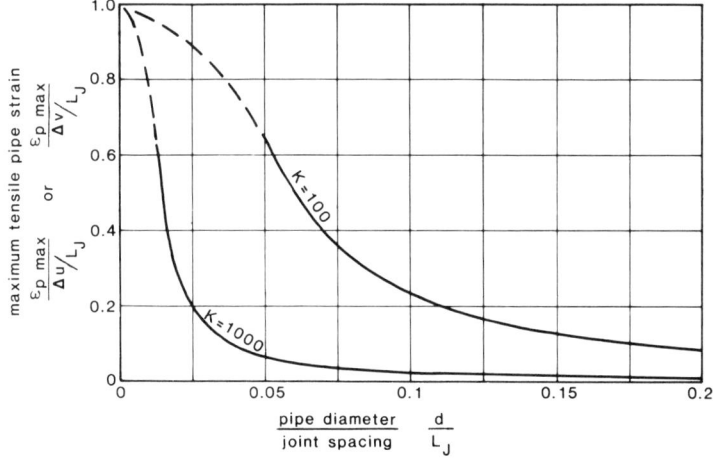

Figure 3.31 Maximum axial tensile pipe strain in flexibly jointed pipeline.

The pipeline is cut into its constituent pipe lengths and the horizontal ground movement is applied to this cut structure. If required, equal and opposite forces can be applied at the ends of the pipes to satisfy the load/displacement relationship of the joints.

Assuming that the axial force transferred at the pipe joints is zero then the force developed in a pipe length depends on the joint spacing, L_J, and the differential parallel soil movement, Δu or Δv, along the pipe length. The variation in maximum tensile pipe strain with differential soil movement and pipe-diameter/joint-spacing ratio is shown in Figure 3.31. As before, the pipe strain is inversely proportional to K^* with an upper limit of $\Delta u/L_J$ or $\Delta v/L_J$.

Example 3.17 Calculation of axial strain in transverse pipeline

The pipe and tunnel details are the same as in Example 3.12.

$i = 2.6\,\text{m}$, $\quad v_{max} = 6.1\,\text{mm}$, $\quad d = 0.50$, $\quad K^* = 972$.

For a joint spacing of $L_J = 18$ feet (5.49 m)

$L_J = 5.49\,\text{m} = 2.11i$; $\quad d/L_J = 0.5/5.49 = 0.091$.

Differential soil movement, Δv, is a maximum when $y_J = i$.
At $y_J = 2.6\,\text{m}$, $v = 1.000 v_{max}$.
At $y_J = 8.09\,\text{m}$, $v = 0.041\, v_{max}$.
$\Delta v = 0.959 v_{max}$.
From Figure 3.31, maximum tensile pipe strain is

$$\varepsilon_{ypmax} = \frac{100}{972} \times 0.27 \frac{\Delta v}{L_J} = \frac{100}{972} \times 0.27 \times \frac{0.959 \times 0.0061}{5.49}$$

$$= 2.96 \times 10^{-5} = 30 \text{ microstrain.}$$

For a joint spacing of $L_J = 12$ feet (3.66 m)

$L_J = 3.66\,\text{m} = 1.41i$; $d/L_J = 0.137$.
Δv is a maximum when $y_J = i\, \Delta v = 0.781 v_{max}$.

$$\varepsilon_{ypmax} = \frac{100}{972} \times 0.145 \times \frac{0.781 \times 0.0061}{3.66} = 19 \text{ microstrain.}$$

Example 3.18 Calculation of axial strain in parallel pipeline

The pipe and tunnel details are the same as in Example 3.14
$i = 2.6\,\text{m}, u_{max} = 4.0\,\text{mm}$, joint spacing $L_J = 3.66\,\text{m}$.

Large-diameter pipeline

$d = 0.50, K^* = 972, d/L_J = 0.137$.
At $x_J = 0$, $\quad u = 1.000 u_{max}$.

At $x_J = 3.66$ mm, $u = 0.371 u_{max}$.
$$\Delta u = 0.629 u_{max}.$$
From Figure 3.31, maximum tensile pipe strain is

$$\varepsilon_{xpmax} = \frac{100}{972} \times 0.145 \frac{\Delta u}{L_J} = \frac{100}{972} \times 0.145 \times \frac{0.629 \times 0.004}{3.66}$$

$$= 1.03 \times 10^{-5} = 10 \text{ microstrain.}$$

Small-diameter pipeline

$d = 0.10$, $K^* = 2520$, $d/L_J = 0.0273$, $\Delta u = 0.629 u_{max}$.
From Figure 3.31,

$$\varepsilon_{xpmax} = \frac{1000}{2520} \times 0.17 \times \frac{0.629 \times 0.004}{3.66} = 46 \text{ microstrain.}$$

3.3.7 Soil–pipe slip

With increasing magnitude of ground movement, the soil–pipe shear stress in the zone of ground movement reaches the limiting soil–pipe shear strength τ_a. Provided that the pipeline extends a minimum distance of $2.5i$ beyond the zone of ground movement, the shear stress in the anchorage length is less than τ_a. At full slip between the soil and the pipe, the shear force along the pipeline can be integrated to give the maximum load in the pipe. Taking τ_a as constant, then at full slip the maximum tensile pipe strains in rigidly-jointed pipelines are:
transverse to the tunnel

$$\varepsilon_{ypmax} = \frac{5 \; i \, \tau_a}{K^* d E_g} \tag{3.37}$$

parallel to the tunnel

$$\varepsilon_{xpmax} = \frac{10 \; i \, \tau_a}{K^* d E_g} \tag{3.38}$$

where K^* is the soil–pipe stiffness factor.
i/d is the ratio of settlement trough width to pipe diameter, and
τ_a/E_g is the ratio of soil–pipe shear strength to soil modulus.
Equations (3.37) and (3.38) give the maximum pipe strain when τ_a/E_g is less than τ/E_g from Figure 3.30. Comparing equation (3.38) with Example 3.15 shows that the pipe strain at *full* slip is about twice the value at initial local slip. For old grey iron pipelines, where surface corrosion in effect mechanically interlocks the soil and the pipe, E_g/τ_a is about 200. For a recently installed pipeline E_g/τ_a of about twice this value would be appropriate. Still lower values of soil–pipe shear strength are appropriate for plastic pipelines or pipes with protective sheathing or polyethylene sleeving.

3.4 Application of analysis to practical problems

The methods of analysis detailed in the foregoing sections, although based on linear-elastic theory, have been found to be quite adequate for practical application. The solutions for bending and axial stress have been used for the following purposes:

(1) The prior estimation of possible effects of proposed tunnels on adjacent buried pipelines, particularly grey iron distribution mains.

(2) The analysis of pipeline failures to establish if ground movement from adjacent tunnelling is either the main cause of failure or the likely final trigger for failure.

(3) The design of new pipelines and the operation of established pipelines as a contribution to an understanding of the causes of stress in pipelines and in identifying those features that are helpful in reducing stress levels.

Most of the low-pressure pipeline distribution network has been designed for internal pressure only, the pipe class being based on experience that the combination of diameter, wall thickness, material and depth should produce a pipeline adequate to carry the external loads. More recently, pipelines have been designed for internal pressure *and* external loads due to earth pressure, surcharge and traffic loading. It is still generally assumed that the principal effect of external loading is circumferential bending. In fact, for low-pressure metal pipelines higher stresses are generally associated with longitudinal bending and this is reflected in the relatively high incidence of transverse fractures in small-diameter brittle pipes. Also, for these pipelines it is easily shown by analysis or loading tests that the joint spacings commonly used in iron distribution mains make an insignificant contribution to the flexibility of a small-diameter pipeline. Flexible joints are frequently thought of as an aid to overall pipeline flexibility, whereas the contribution is only really significant for axial movements due to temperature changes or parallel ground movement. The design of sewer pipelines has recently incorporated the known effects of uneven pipe bedding and restraint to pipeline movement at the junction with a more rigid structure. Longitudinal bending moment resistance tests are now required on small-diameter pipes, and standard specifications require additional flexible joints close to the junction with a structure. The joint spacing at a structure is determined by the pipeline diameter (and pipe material) which is equivalent to the longitudinal flexural stiffness. These aspects have yet to be incorporated in most pipeline design, which frequently concentrates on circumferential bending strength, hoop tensile strength, flow capacity and route selection.

The analysis of pipelines for ground movement is statically indeterminate. The designer needs information on the load–deformation characteristics of the soil around the pipe, the pipe and the pipe joints. It is important to recognize that these three materials make up the pipeline itself. Representative values for these properties are given in the following sections. These are adequate for routine design for which specific testing is not practicable.

3.4.1 Soil properties

The application of the methods of analysis in sections 3.2 and 3.3 requires knowledge of appropriate values for the soil deformation parameters, E_g, E'_g, v_g and v'_g. It is recognized that if the analysis is linear-elastic then the appropriate elastic parameters are not simply a function of the soil but depend on the method of analysis, the means of evaluating the parameters and the type of construction process being modelled (Marsland, 1980). The elastic parameters are obtained from:

Laboratory or *in-situ* testing (see section 2.10)
Loading tests and back-analysis for the relevant parameters
Empirical correlation with other soil properties based on previous experience.

Young's modulus can be assessed from carefully conducted triaxial tests on high-quality samples, drained or undrained conditions being adopted as appropriate. The sample has to be reconsolidated to the estimated *in-situ* stress conditions before starting the modulus test. Plate loading tests can be used for *in-situ* determination of modulus and the modulus is calculated from the theoretical relationship between load and displacement (see section 3.2.1).

Loading tests and back-analysis are particularly useful as a means of establishing the soil parameters. The field conditions are accurately simulated, and some of the defects in the modelling may be removed if the same model is used for the back-analysis. Back-analysis of the data from six ground-movement/buried-pipeline cases suggests that empirical correlation with other soil properties is sufficiently accurate for routine design. In any event, the sensitivity of the design to changes in the assumed soil properties should be tested.

Typical values of elastic modulus and Poisson's ratio for various soil types are given in Table 3.5 for soil considered as a homogeneous elastic solid. The values suggested by Fry and Rumsey (1983) have been used in an empirical formula for predicting pipe bending stress resulting from adjacent trench excavation. The magnitudes of the ground movements adjacent to the pipe are similar to those caused by tunnelling. In their modelling the same soil stiffness is used for the trench wall and for the soil around the pipe. In this case the soil modulus values in Table 3.5 may not be directly applicable to the soil around the pipe although the model was fitted on this basis. The values suggested by Elson (1984) are for laterally-loaded piles where the method of installation clearly differs from that of a buried pipeline. Whilst the values in Table 3.5 may be conservative for predicting ground movements near trenches or movement of laterally loaded piles, the converse may apply to the effect of ground

Table 3.5 Typical soil deformation properties

Soil type (SPT N-value or c_u(kN/m^2) is given)	After Fry and Rumsey (1983) E_g(MN/m^2)	After Elson (1984)		
		E_g(MN/m^2)		v_g
Loose sand and gravel, $N = 4$ to 10	4 to 8	5 to 20		0.3 to 0.4
Medium dense sand and gravel, $N = 10$ to 13	8 to 24	16 to 20		0.2 to 0.35
Dense sand and gravel, $N = 30$ to $50+$	24 to $50+$	30 to 100		0.15 to 0.3
		undrained	drained	
Soft clay, $c_u = 20$ to 50	1 to 3	2 to 6	1 to 4	
Firm clay, $c_u = 50$ to 75	3 to 10	5 to 12	3 to 8	0.5 for undrained conditions. 0.1 to 0.3 for drained conditions.
Stiff clay $c_u = 75$ to 150	10 to 20	10 to 20	5 to 15	

movement on a pipeline. In the table the soil modulus to shear strength ratio is about 100. For ground movement and buried pipelines, $E_g/\tau \simeq 200$ is more appropriate. For the purpose of calculating relative values of pipe stress, the drained modulus for clay soils is often taken as $E'_g = \frac{2}{3}(1 + v'_g)E_g$, where v'_g is about 0.3.

Available information on soil stiffness around pipelines installed in trenches relates almost entirely to design for ring stiffness and buckling. In the majority of cases design is by a method that derives from the Marston–Spangler method. In this case the soil modulus is not a true Young's modulus (that depends on soil properties only) but includes some effects of pipe stiffness. However, back-analysis of cases of ground movement and the longitudinal bending of buried pipelines indicates that the soil modulus appropriate to the design for ring stiffness may also be used in the analysis of longitudinal bending stiffness. Values of soil modulus for the design of pipes are between 1 and 20 MN/m^2 depending on soil type and compaction. Typical values for *embedment* design are shown in Table 3.6. Iron and steel distribution mains are relatively thick-walled, that is, rigid to diametrical deformation, and at normal depths of cover only require a soil stiffness of 1 to 5 MN/m^2 for satisfactory performance in ring bending. Due to compaction of loose backfill with time, because of traffic loading and vibration, it is unlikely that an established pipeline will have a soil surround of stiffness less than 3 MN/m^2. Unless well-compacted, imported granular material has been used for the surround, it is unlikely that the soil modulus will be more than 15 MN/m^2. In the absence of any specific site data, 10 MN/m^2 is a reasonable assumption for the soil modulus around a distribution main in an urban area. Large-diameter

Table 3.6 Soil modulus for embedment materials (after Stanton and Staveley, 1979)

Embedment material	Soil modulus, E_g (MN/m^2) for Modified Proctor density				
	Untamped	80%Mp	85%Mp	90%Mp	95%Mp
Coarse-grained soils					
Gravel single size	5	7	7	10	14
Gravel graded	3	5	7	10	20
Sand and coarse-grained soil with less than 12% fines	1	3	5	7	14
Cohesive materials with low compressibility					
Coarse-grained soil with more than 12% fines; fine-grained soil with medium to no plasticity and containing more than 25% coarse-grained particles	–	1	3	5	10
Fine-grained soil with medium to no plasticity and containing less than 25% coarse-grained particles	–	–	1	3	7

flexible thin-walled pipes are constructed to a higher standard than are distribution mains. Compston *et al.* (1978) recommend that the soil stiffness for these pipelines should be determined by laboratory testing, but preliminary design should be based on $E_g = 10 \, \text{MN/m}^2$ on the assumption that a standard density of 95 per cent can be achieved economically with granular backfill. Nath (1977) analysed the performance of a 1.83 m diameter steel pipe surrounded with compacted sand in a trench in clay. From triaxial testing the sand had a secant elastic modulus of $20 \, \text{MN/m}^2$ for 2 per cent strain at a confining pressure of $40 \, \text{kN/m}^2$. For the clay in the trench wall a modulus of $7 \, \text{MN/m}^2$ was used in the modelling. Relatively higher values of soil modulus next to the pipeline may thus be applicable to cross-country thin-walled transmission pipelines where the construction standards are different from urban distribution mains.

3.4.2 *Pipeline properties*

The stress/strain properties of distribution pipe material in common use are summarized in Table 3.7. These properties are for short-term loading in direct tension. At normal operating temperatures in distribution mains creep is not significant in iron and steel, so these properties may also be used for long-term static loading. All plastics require detailed consideration of long-term properties, and an indication of the effect of creep is given in the footnotes to

Table 3.7 Typical pipe material properties for short-term static loading in direct tension (gradually applied, non-repeated loading without creep and at $20°C$)[1][3]

Material	Ultimate tensile stress (N/mm²)	Lower yield stress or proof stress (N/mm²)	Typical max. design stress for working loads (N/mm²)	Design stress/ultimate stress	Design stress/yield or proof stress	Secant elastic modulus to max. design stress (GN/m²)	Elastic strain equivalent to design stress (microstrain)
Grey iron							
Pit cast before year 1914	110 to 140	70 to 90 } 0.1% proof stress	27 to 35			67 to 87 } Ref[4]	400
Vertically cast to BS 78:1917	145 } ±20 N+/mm²	95	36	0.25	0.38	90	400
Vertically cast to BS 78:1965	155	100	38			93	400
Spun cast to BS 1211:1958							
over 16 in. dia. grade 12;	185	120	46			103	450
3 in. to 16 in. dia. grade 14	215	140	54			108	500
Ductile iron							
Spun cast to BS 4772		0.2% proof stress					
grade 420/12 material	420 min.	300 min.	155	0.37	0.52	165	940
Mild steel		yield or 0.2% proof stress		depending on duty			
BS 534:1966							
grade 320 material	320 min.	195 min.	generally 80–115 (occasionally up to 140)	0.25–0.36 (occasionally up to 0.44)	0.4–0.6 (occasionally up to 0.72)	210	380–550
grade 410 material	410 min.	235 min.	95–140				450–660

Plastic[2]							
UPVC to BS 3505	45 min	—	20	0.5	—	2.8	7000
MDPE	30 at 50 mm/min.	19 upper yield	7	0.4	—	0.7	10000
HDPE type 1	19 at strain rate 125 mm/min.	—	8	0.4	—	—	—
HDPE type 2	32 at strain rate 125 mm/min.	14 lower yield / 22 upper yield	8	0.25	0.57	0.9	9000

(1) For iron and steel pipes, creep is not significant at 20°C at the maximum design stress indicated. These stresses are therefore also used for long-term loading.
(2) After 50 years, plastic properties are UPVC at 10°C–UTS = 25 N/mm², max. design stress = 12 N/mm², creep modulus = 1.4 GN/m²
    MDPE at 20°C–U.T.S. = 8 N/mm², max. design stress = 5 N/mm², creep modulus at 3 N/mm² = 0.13 GN/m²
    HDPE type 1 at 20°C–U.T.S. = 6.5 N/mm², ,, ,, = 5 N/mm², ,, ,, = 0.13 GN/m²
    HDPE type 2 at 20°C–U.T.S. = 9.5 N/mm², ,, ,, = 5 N/mm², ,, ,, = 0.13 GN/m²
    HDPE type 1 at 10°C–U.T.S. = 9.5 N/mm²
    HDPE type 2 at 10°C–U.T.S. = 12.5 N/mm²
(3) For all the above materials the maximum design tensile bending stress is not less than the maximum design direct tensile stress.
(4) Secant elastic modulus for grey iron is for the elastic component of stress/strain curve, *not* stress/total strain curve.
(5) HDPE is high density polyethylene
    MDPE is medium density polyethylene

Table 3.8 Typical pipe dimensions

Nominal diameter		Vertically cast grey iron BS 78 class B mean values		Spun cast grey iron BS 1211 class B mean values		Spun cast ductile iron BS 4772 class K9 mean values		Mild steel BS 534: 1966 minimum values		UPVC BS 3505 class B mean values		HDPE and MDPE Minimum values			
													SDR = 41	SDR = 17	SDR = 11
mm	(inches)	d (mm)	t (mm)	d (mm)	t (mm)	d (mm)	t (mm)	d (mm)	t (mm)	d (mm)	t (mm)	d (mm)	t (mm)	t (mm)	t (mm)
75	3	96	9.7	96	7.4	98	6.0	89	3.2	89	3.1	90	—	5.1	8.2*
100	4	122	9.9	122	7.6	118	6.1	114	3.6	114	3.6	125	—	7.1	11.4*
150	6	177	10.9	177	8.4	170	6.3	168	3.6	168	4.7	180	4.4	10.2*	16.4
200	8	232	11.9	232	9.1	222	6.4	219	4.0	219	5.5	250	6.1	14.3*	22.8
250	10	286	13.2	286	9.9	274	6.8	273	4.0	273	6.8	280	6.9	15.9	25.5
300	12	334	14.5	334	10.9	326	7.2	324	4.5	324	8.1	315	7.7	18.0*	28.7
375	15	413	16.0	413	11.9	—	—	—	—	—	—	400	9.8	22.7	
450	18	492	17.5	492	13.2	480	8.6	457	6.3	457	11.4	450	11.0	25.5	
575	21	572	19.0	572	14.2	—	—	—	—	—	—	560	13.7	31.7	
600	24	650	20.3	650	15.2	635	9.9	610	6.3	610	15.3	630	15.4	35.7	
675	27	729	21.6	729	17.3	—	—	—	—	—	—	710	17.4	40.2	

*Dimensions of MDPE to British Gas Standard BGC/PS/PL2
HDPE is high density polyethylene, MDPE is medium density polyethylene.

the table. The typical maximum design stresses relate to tension over the full cross-section, and at every point along the section, as would be caused by internal pressure. For this type of statically determinate loading the ratio of yield stress to design stress is the load factor, that is, the value by which the load would have to be increased to cause structural failure. Some typical pipe dimensions are given in Table 3.8.

It is convenient to divide metals into two classes: ductile metals such as mild steel and spheroidal graphite (i.e. ductile) iron, and brittle metals such as the low-grade grey irons. An arbitrary dividing line is an ultimate elongation of 5 per cent. The usual failure criterion for ductile materials is *elastic* failure. In uniaxial tension this occurs at the yield stress indicated in the table. For combined stress the maximum shear stress theory is convenient for design, that is, elastic failure in the complex system occurs when the maximum shear stress equals $\frac{1}{2}\sigma_{yield}$. A brittle material cannot be considered to have definitely failed until it has broken, either through tensile fracture or shear fracture. In the elastic analysis of grey iron, an elastic failure criterion may be based on the 0.1 per cent proof stress, which is equivalent to about 0.65 times the ultimate tensile stress. For combined stress, elastic failure occurs when the maximum principal stress in the complex system reaches the value of the maximum stress at failure in uniaxial tension.

The foregoing relates to elastic failure of *material*. This may be localized in a *member* such as a pipeline, and may do no real damage if the material affected is so small or so located as to have a negligible influence on the strength of the pipeline as a whole. The significance of local overstressing (for example due to notches, holes or even accidental scratching of plastic pipes) depends on the mechanical properties and service conditions. For ductile metals under static loading it is usually of little importance. It may be assumed that no separate allowance is required for stress concentration, and the typical maximum design stresses are nominal stresses calculated as load divided by the net transverse cross-section. Fatigue properties and resistance to impact are much more likely to be affected than static strength.

For brittle materials, stress concentration is often a serious consideration. Irregularities of form such as holes, sharp shoulders, surface pitting or normal casting features may produce high localized stresses. The ratio of the true maximum stress to the stress calculated by the ordinary formulae of mechanics using the net section, but ignoring the changed distribution of stress, is the *factor of stress concentration* for the particular stress raiser. Roark and Young (1975) give data on factors of stress concentration for elastic stress. Even in relatively brittle materials, plastic yielding that occurs on overstressing greatly mitigates stress concentration and causes it to have much less influence on breaking strength than might otherwise be expected from consideration of elastic stress only. The *factor of stress concentration at rupture* represents the practical significance of stress concentration. This factor is the ratio of the computed stress at rupture for the plain specimen to the computed stress at rupture using the net section for the specimen containing the stress raiser. An alternative design approach is to use the *factor of strength reduction* which incorporates the stress concentration at rupture *and* the reduction in strength due to the reduced section at the stress raiser. Roark (1965) gives some data on factors of stress concentration at rupture for grey iron.

Pipelines of ductile material do not ordinarily fracture under static bending loading, but fail through excessive deformation. If the wall thickness is sufficient to preclude local buckling, the maximum bending moment that can be sustained is that which corresponds to plastic yielding throughout the section. Thus, in some cases higher tensile stresses are allowed in bending. For grade 420/12 ductile iron the maximum design tensile stress for ring bending due to earth and traffic loading is increased from $155N/mm^2$ (uniaxial tension) to $250N/mm^2$. A similar increase in allowable stress is appropriate for bending stress caused by the interaction of ground movements on a pipeline. For brittle materials the bending moment required to cause rupture depends on the material, the form factor, the extent of the material subjected to high stress, redistribution of bending moments in statically indeterminate systems and the absolute scale for conditions involving abrupt stress variation. All of the foregoing are taken into account by the rupture factor for the particular material, form, loading condition and size. The *rupture factor* is the ratio of the fictitious maximum tensile stress at failure, as calculated by the appropriate formula for elastic stress, to the ultimate tensile stress as determined by a conventional tensile test. By applying the appropriate rupture factor to the design tensile stress in axial tension, the load factor for different loading conditions and types of stress can be made similar. Roark (1965) gives further information on rupture factors.

A listing of the design codes and guides for buried pipelines is given by Edgell and Yarwood (1981). There is no detailed guidance on the design of grey iron (flake graphite) pipe, although this was the principal material in the construction of pipelines for the distribution of gases and liquids for more than 150 years. Extensive use of this material ceased in the 1970s, with the large-scale introduction of ductile iron (spheroidal graphite) pipe. Over 80 per cent of the UK gas and water distribution system is grey iron pipework. The following is a summary of the major innovations in the manufacture and use of grey iron in the UK (North American practice is similar):

19th century onwards—volume production of pit-cast pipes
1846—patent for vertical casting process
1850s—introduction of vertically cast pipes
1914—volume production of vertically cast pipes
1917—first British Standard for vertically cast pipes
1920s—introduction of centrifugal casting process
1930s—volume production of centrifugally-cast ('spun') pipes from 3 to 21 inches in diameter
1945—first British Standard for spun-cast pipes

Table 3.9 Typical material properties for grey iron pipes

Type of casting	Pit-cast, before 1914			Vertically cast, 1914–1970			Spun-cast, 1930–1970		
Diameter (inches)	$\geqslant 3$ $\leqslant 6$	> 6 $\leqslant 16$	> 16	$\geqslant 3$ $\leqslant 6$	> 6 $\leqslant 16$	> 16	$\geqslant 3$ $\leqslant 6$	> 6 $\leqslant 16$	> 16
Metal grade and phosphorus content[a]	less than grade 10 $1\frac{1}{2}\% < P < 2\%$			grade 10 $1\% < P < 1\frac{1}{2}\%$			grade 12 $\frac{1}{2}\% < P < 1\%$		
Tensile strength of casting (N/m² ± 20)[b]	140	130	120	180	160	140	250	220	190
Equivalent grade of casting									
BS 1452:1961 (tonf/in²)	less than grade 10			12	10	10	14	14	12
BS 1452:1977 (N/mm²)	less than grade 150			180	150	150	220	220	180
US (psi × 10³)	less than grade 20			20	20	20	30	30	20
Proof stress									
0.01%	← 0.28 × tensile strength →								
0.1%	← 0.65 × tensile strength →								
Strain at failure[c] (microstrain)									
Total ±1000	4000	4100	4200	3900	4200	4500	4800	4900	5100
Elastic ±200	1900	1800	1800	2100	2000	1900	2600	2400	2200
Non-elastic	2100	2300	2400	1800	2200	2600	2200	2500	2900
Design properties[d]									
Elastic modulus (GN/m²)									
tangent at zero stress	95	90	85	110	103	95	126	120	113
secant to 0.25UTS	90	86	81	104	98	90	119	113	107
secant to 0.4UTS	87	83	78	101	94	87	115	109	103
elastic strain equivalent to max. design stress (microstrain)									
axial tension = 0.25 UTS	390	380	370	430	410	390	525	485	445
bending tension = 0.4UTS	645	625	615	715	680	645	870	805	740
Poisson's ratio	← 0.26 →								
Coeff. of thermal expansion (per °C)	← 11.0 × 10⁻⁶ →								

[a] Different quality-control tests have been used for the different types of pipe at various times. Here these tests have all been reduced to the same basis. The metal grade is the tensile strength in tonf/in² of a 1.2 inch (30 mm) diameter cast bar machined to a gauge diameter of 0.798 inches (20 mm).

[b] For a given grade of metal the tensile strength of the casting depends on the cooling rate in the mould. This depends on the casting process and thickness and shape of the metal.

[c] The elastic strain depends on the graphite content. The non-elastic strain depends on the phosphorus content, the low-grade irons used for pipes typically having a high phosphorus content, as indicated in the first item of the table.

[d] The secant elastic moduli are for the elastic component of the stress/strain curve.

1948—discovery of ductile iron
1969—volume production of ductile iron pipes
1970s—extensive use of grey iron pipes ceases.

Since grey iron is no longer used for distribution mains, the relevant material properties and design data are not readily available in the design office. For static tensile loading the required data are given in Table 3.9. This offers sufficient information to construct the full tensile stress/strain diagram. An example is shown in Figure 3.32 for the lowest and highest strength irons in the Table. Angus (1976) discusses in detail the engineering properties of cast irons and the following is a summary, together with other data.

At no stress does grey iron show a truly elastic response. This is illustrated in Figure 3.32. In the design of grey iron pipelines for normal service conditions these effects can safely be ignored where design is in terms of working stresses. In any event, design based on the *elastic* secant modulus will overestimate stress and elastic strain. For fibre stresses up to 30 per cent of the tensile strength, the error in the stresses calculated by conventional elastic formulae is less than approximately 10 per cent. Low-strength irons are not notch-

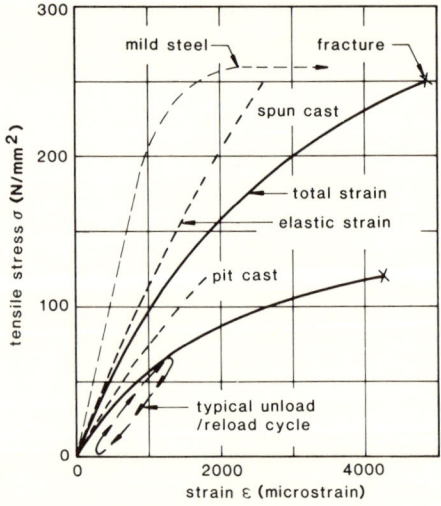

Figure 3.32 Typical stress/strain diagram for grey iron pipe.

Notes:

(1) Stress/strain diagrams are for static loading in direct tension.
(2) Spun cast is identified in Table 3.9 as 3-inch to 6-inch diameter. Pit cast is greater than 16-inch diameter. This represents the range of strengths for grey iron pipes.
(3) The stress/strain relationships are of the form

$$\sigma = \varepsilon/(a + b\varepsilon)$$

where $a = 1/E_0$, E_0 = tangent elastic modulus at zero stress, and b is found by substituting σ and $\varepsilon_{\text{elastic}}$ or $\varepsilon_{\text{total}}$ at failure as given in Table 3.9.

GROUND MOVEMENTS AND BURIED PIPELINES

sensitive, and may be considered as intrinsically strong material already weakened by the presence of graphite flakes which serve as notches or stress raisers. For static loading in direct tension, the maximum factor of strength reduction associated with normal casting features or *surface* pitting in pipes is 1.25. This allowance is included in the maximum design stress of 0.25 UTS. For the materials and diameters in Table 3.9, the rupture factor for longitudinal bending (third-point loading) or ring crushing is 1.6. It is thus customary to increase the design stress for *bending* tension to 1.6×0.25 UTS, that is 0.4 UTS. These design stresses for direct and bending tension include an allowance of 1.25 for stress concentration, but separate allowance must be made where appropriate for the reduction in strength caused by the holes drilled for service branches. Figure 3.33 *a* shows the factor of stress concentration for elastic stress and at rupture for a top-entry service hole. At rupture the stress concentration is up to 2.0, which implies a further allowance of up to 1.6 over and above the normal allowance of 1.25. In addition, the reduction in section has to be allowed for. The strength reduction for axial tension and bending is shown in Figure 3.33 *b*. For low-pressure distribution systems the maximum size of service connection that may be drilled into a main is one-quarter of the main diameter. For bending in the hogging mode the factor of strength reduction is 2.5. The allowable nominal stress is thus $1.25 \times 0.4 \text{ UTS}/2.5 = 0.2$ UTS. Similarly, for axial tension the allowable nominal stress is $1.25 \times 0.25 \text{ UTS}/2.15 = 0.15$ UTS. For medium-pressure mains the

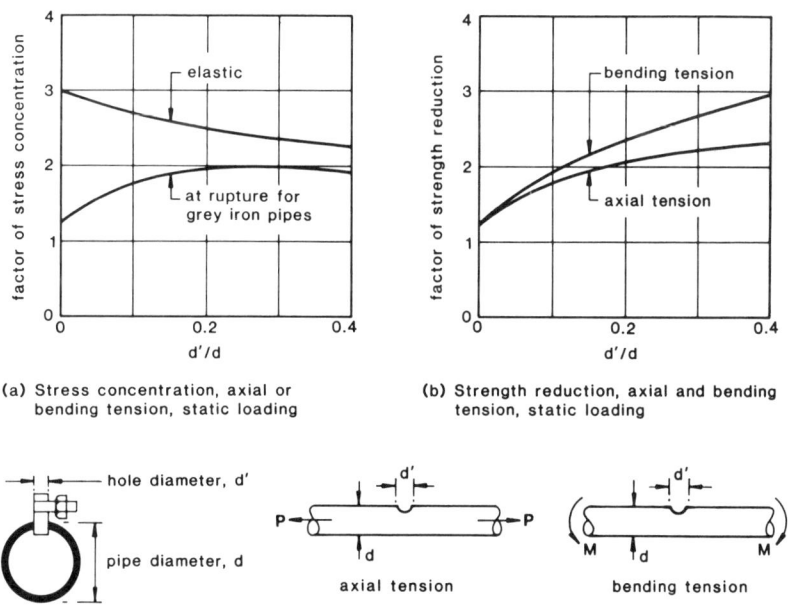

Figure 3.33 Stress concentration at service tappings in grey iron distribution mains.

size of service connection is limited to one-sixth of the main diameter. The equivalent nominal design stresses can be found by using Figure 3.33 b.

3.4.3 *Pipe-joint properties*

Joints in cast iron distribution mains are an important factor in the interaction of ground movement on buried pipelines. Joint movement or relaxation reduces the stresses induced by lateral and longitudinal ground movement. Excessive rotation or pull-out can cause leakage. Although the consequences of leakage at joints are not usually as severe as pipe fracture, a significant increase in leakage constitutes pipeline failure. The methods used to join cast iron pipes have developed through the years, and may be classified into three main types:

Rigid: e.g. turned and bored, flanged
Semi-rigid: e.g. lead–yarn
Flexible: e.g. Tyton rubber gasket push-in joint for liquids (US Pipe and Foundry Company), bolted-gland mechanical joint for gases

Joints on other types of pipeline can similarly be classified as rigid (e.g. welded, flanged, cement-caulked or mortar–yarn) or flexible (e.g. push-in rubber gasket, rubber gland mechanical couplings). Over 50 per cent of the UK gas and water distribution system has the semi-rigid lead–yarn joint. Flexible joints were introduced in the 1920s for both liquids and gases, with use becoming extensive by the 1930s. With the volume production of ductile iron pipes from 1969 onwards, the use of lead–yarn joints has ceased. North American usage of lead–yarn and flexible joints is similar. In the absence of detailed records of the joint type, or perhaps recent excavation to repair joint leakage, a preliminary assessment can be based on the recorded pipeline installation date, namely:

Up to 1934:   semi-rigid lead–yarn joint
1935–1969:   flexible rubber gasket push-in joint or bolted gland mechanical joint
1970 onwards: ductile iron pipes with flexible joints.

The primary advantage of flexible joints over lead–yarn joints was simple and more rapid jointing. However, it was recognized that pipeline flexibility and the ability to accommodate axial movement were important secondary advantages. Figure 3.34 illustrates the principal joints for iron mains. Typical dimensions are given in Table 3.10. A description of other joints less common in grey iron mains is contained in the report published by the Department of Energy (DoE 1977). The lead–yarn joint consisted of caulked lead backed by tightly-packed jute spun yarn. The yarn varied from untreated pure jute to yarns impregnated with tar; it served to centre the spigot in the socket and

GROUND MOVEMENTS AND BURIED PIPELINES

Nominal diameter (inches)	Load–yarn joint					Nominal diameter (mm)	Rubber gasket push-in joint			Bolted-gland mechanical joint		
	d_S (mm)	d_W (mm)	d_L (mm)	d_C (mm)	$\frac{d_C}{d_S}$ (deg)		d_S (mm)	d_W (mm)	θ_{Jmax} (deg)	d_S (mm)	d_W (mm)	θ_{Jmax} (deg)
3	89	27	44	9.7	6	75	82	38	5	—	—	—
4	89	27	44	9.7	6	100	82	38	5	76	35	4
6	89	27	44	9.7	6	150	82	38	5	82	45	4
8	102	32	51	9.7	5	200	89	38	5	88	50	4
10	102	32	51	9.7	5	250	96	38	5	91	50	4
12	102	32	51	9.7	5	300	104	38	5	93	50	4
15	114	35	57	9.7	5	—	—	—	—	—	—	—
18	114	35	57	11.2	5	450	107	38	4	100	55	4
21	114	35	57	11.2	5	—	—	—	—	—	—	—
24	127	38	63	11.2	5	600	125	38	4	108	65	4
27	127	38	63	11.2	5	—	—	—	—	—	—	—

d_S is depth of spigot, d_W is spigot withdrawal with joint in undeflected position, d_L is depth of caulked lead, d_C is thickness of caulked lead, d_C/d_S is rotation at spigot/socket metal binding, θ_{Jmax} is maximum rotation.

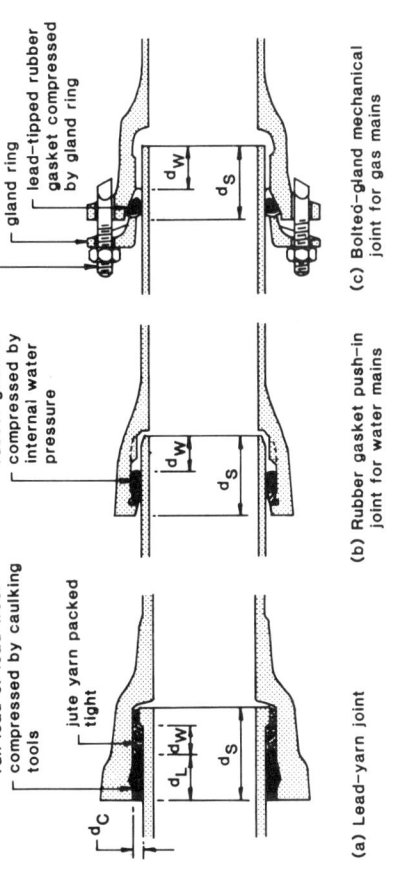

Figure 3.34 Principal types of joint for iron distribution mains.

(a) Lead–yarn joint
(b) Rubber gasket push-in joint for water mains
(c) Bolted-gland mechanical joint for gas mains

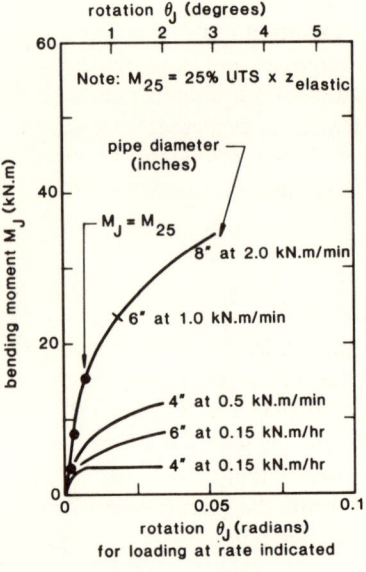

(a) Lead-yarn joints with hardened jute packing, gas mains – Reference (1)

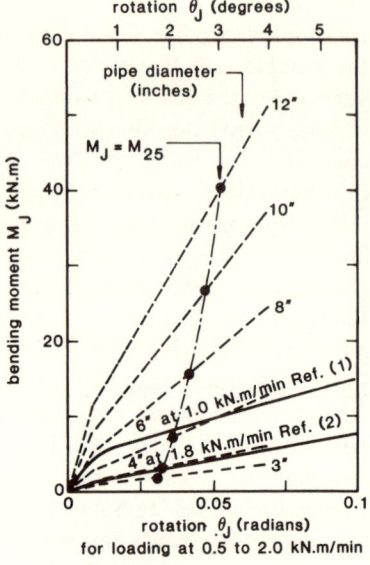

(b) Lead-yarn joints with soft jute packing, water mains

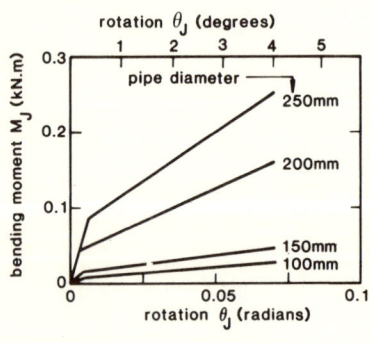

(c) Rubber gasket push-in joints, water mains – Reference (3)

Table of joint rotational stiffness values for short-term loading

Pipe diameter	load-rate kNm/min	stiffness α_1 kNm/rad	rotation deg	stiffness α_2 kNm/rad	rotation deg
Lead–yarn with hardened jute packing					
4 inch	0.5	1100	0–0.25	375	0.25–1
6 inch	1.0	2500	0–0.25	900	0.25–1
8 inch	2.0	2500	0–0.25	900	0.25–1
Lead–yarn with soft jute packing					
3 inch	0.5	98	0–0.5	49	0.5–4
4 inch	0.5	158	0–0.5	79	0.5–4
6 inch	1.0	332	0–0.5	166	0.5–4
8 inch	2.0	636	0–0.5	318	0.5–4
10 inch	2.0	966	0–0.5	483	0.5–4
12 inch	2.0	1318	0–0.5	659	0.5–4
Rubber gasket push-in type					
100 mm	–	1.52	0–0.27	0.33	0.27–4
150 mm	–	3.44	0–0.26	0.48	0.26–4
200 mm	–	12.21	0–0.21	1.76	0.21–4
250 mm	–	13.83	0–0.36	2.62	0.36–4

Figure 3.35 Rotational stiffness of joints in iron pipelines. References: (1) Harris and O'Rourke (1983); (2) Pocock et al. (1980); (3) Singhal and Benavides (1983); (4) Takagi et al. (1985). (1), (2), for lead–yarn joints; (3), (4) for rubber gasket joints.

prevent molten lead entering the pipe, and was also intended to provide some rotational flexibility. Lead wool joints were a variation of run lead, to avoid the trouble and hazard of molten lead in confined and wet trenches. A variation of the ordinary yarn and caulked lead wool was a double layer, that is, lead–yarn–lead. This arrangement was intended to protect the yarn from contamination with tars (from impure gas) and thus retain the rotational flexibility. Harris and O'Rourke (1983) have tested and examined in detail 21 joints from pit-cast grey iron gas mains installed in New York between 1878 and 1914. Two joints were cement-caulked, four were lead wool and the remainder were run lead. Only three of the joints had soft fibrous packing. In the remainder the jute was heavily impregnated with hydrocarbons and was hard and solid. This proportion of cement-caulked, lead wool and run lead joints is typical. It may also be assumed that the yarn in gas mains that have carried manufactured gas is hard and solid rather than soft and fibrous. The behaviour of the lead–yarn joint is thus different for gas and water mains, although the original construction may have been similar. Also, the change from manufactured to natural gas can dry and shrink the yarn. In water mains the wet yarn expands and may help to seal leakage paths.

Under short-term loading, lead is markedly non-elastic in stress–strain response. For other than very short-term loading, creep at ordinary temperatures is substantial. The test results for lead–yarn joints relate almost entirely to the short-term response. The loading rates are 100 to 1000 times more rapid than would be caused by tunnelling. This short-term response must therefore be regarded as a conservative upper limit for the very slowly applied loads induced by tunnelling movements. Also, for pipelines transverse to a tunnel or similar long-term loadings, even if there is significant joint restraint at first application of the load, this will substantially relax in the long term.

The resistance of lead–yarn joints to short-term bending is shown in Figure 3.35a and b. The former applies to gas and the latter to water mains. The joints deform at an initial stiffness, α_1, until substantial yield of the lead at about 0.25° for hardened jute and 0.5° for soft jute. Thereafter the lead is highly stressed, and deforms readily up to the maximum joint rotation which is limited by binding of the spigot and socket parts of the joint. Figure 3.35b is derived from the theoretical equation for joint slip given by O'Rourke and Trautmann (1980):

$$M_{J\,\text{slip}} = \tfrac{3}{8}\pi d^2 C_a d_L \tag{3.39}$$

where M_J is the bending moment at the joint, d is the pipe outside diameter, C_a is the pipe/lead adhesive (shear) strength, d_L is depth of caulked lead, θ_J is the rotation across the joint, and $\alpha = M_J/\theta_J$ is the joint rotational stiffness. Examination of data for this type of joint indicates a distinct change in joint performance at about 0.5°. For short-term loading the mean value of C_a is 1.75 N/mm². Thus with $\theta_J = 0.5°$, $C_a = 1.75\,\text{N/mm}^2$, d from Table 3.8 and d_L

from Table 3.10, the initial rotational stiffness can be calculated. From 0.5° to 4° the stiffness (α_2) is at least half the initial value. Figure 3.35b shows that the calculated stiffness from equation (3.39) with $\theta_{Jslip} = 0.5°$ agrees with the test data available for 4 inch and 6 inch diameter pipes with soft jute packing.

Rubber gasket joints have little resistance to bending moment. Figure 3.35c shows a stiffness 100 to 1000 times less than the short-term response of lead–yarn joints. Again, there is a distinct slippage at about 0.25°. Associated with slippage are stiffnesses α_1 and α_2. The performance of bolted-gland mechanical joints is probably similar to that of push-in joints. For both lead–yarn and rubber gasket joints the initial rotation may be considered linear-elastic. Beyond the slippage moment, the unloading curve may be taken as parallel to the initial stiffness, α_1. Reloading retraces the unloading curve to the original bending moment. Beyond this point the virgin moment/rotation curve is followed.

Leakage caused by rotation at lead–yarn joints in gas mains depends on the condition of the jute packing, the loading rate and the rotation. If the lead is well bonded to the joint surfaces, then no significant increase in leakage occurs up to about 4° of rotation, when the lead is being forced out of the joint socket. With soft jute packing, again no significant increase occurs until 4° of rotation because in this case the lead is always highly stressed and deforms readily to seal the joint. For short-term loading, hardened jute packing and inadequate bonding of the caulked lead, then leakage increases almost linearly with degrees of rotation up to about 0.5°. At this point the lead begins to deform significantly and leakage generally decreases with further rotation. For tunnelling-induced movements where the loading is very slowly applied, it is anticipated that the lead will gradually deform. In those cases where there is no *initial* leakage then there will be no leakage caused by the gradual movement of the joint due to short-term flexibility or long-term relaxation. Where there is some initial leakage this will probably increase even for very slowly applied loading. For this last case a tolerable rotation as a result of ground movement cannot be given, and the only sensible action is to repair the pipeline joints so that leakage is stopped and movement can be accommodated. The movement required to cause leakage at lead–yarn joints in water mains appears to be slightly higher than for gas mains. A suitable limit for tunnelling-induced movement is 1.5°.

Rubber gasket joints are able to accommodate rotation without leakage. The rotation required to cause metal binding is given in Table 3.10. Three-quarters of this available rotation could be taken up in laying the pipeline, if necessary. The initial rotation in the 21 caulked joints examined by Harris and O'Rourke (1983) was between 0.1° and 1.9°, with a mean deviation from straight of 1.0°. A suitable design case is to assume an initial rotation of up to 1.5° for both lead–yarn and rubber gasket joints. Although metal binding causes stress concentration in the joint (stress concentration factor at rupture up to 1.25) no specific allowance need be made in an elastic analysis for pipe

Table 3.11 Maximum axial pipe stress at joints in iron distribution mains

mm	Nominal diameter inches	$\frac{\text{Pipe stress}}{\text{Yield stress}} \times 100\%$		
		Grey iron with lead–yarn joints		Grey or ductile iron with rubber gasket joints
		Medium-term	Long-term	
100	4	5	3	
200	8	4	2	0.1 or less
300	12	3	2	
600	24	3	2	

bending stress. The allowable stresses include an allowance for stress concentration of up to 1.25 since this can also be caused by normal casting features or surface pitting. A summary of allowable joint rotations is given in Table 3.12.

The resistance of lead–yarn joints to axial pull-out is

$$P_{\text{Jslip}} = \pi d C_a d_L \tag{3.40}$$

where P_J is the axial force at the joint and the other notation is the same as for equation (3.39). O'Rourke and Trautmann (1980) have summarized some test data on this type of joint, and the mean value of C_a is 1.75N/mm^2. This compares with the value given by Andrews (1972) of up to 1.4N/mm^2. These values of C_a are for short-term loading, and cannot be induced by tunnelling movements because of creep during the relatively slow application of the load. For medium-term loading a suitable design value is 1.0N/mm^2. For long-term loading a design value of not more than 0.5N/mm^2 may be taken.

Rubber gasket joints have little resistance to axial pull-out. Singhal and Benavides (1983) have carried out load/displacement tests on 35 push-in type joints from 100 mm to 250 mm diameter. The mean value of P_{Jslip} was 0.2 kN at 100 mm diameter, increasing to 1.7 kN at 250 mm diameter. The equivalent values for lead–yarn joints and medium-term loading ($C_a = 1.0 \text{N/mm}^2$) are 16.9 kN and 45.8 kN. The displacement required to mobilize the full slippage force in rubber gasket joints was 2 mm at 100 mm diameter, increasing to 5 mm to 250 mm diameter. The contribution that axial joint stiffness can make to axial stress in a pipeline is summarized in Table 3.11. Clearly, only insignificant forces can be transferred by rubber gasket joints, and lead–yarn joints transfer only very small forces.

The maximum axial pull-out, d_W, for spigot-and-socket-type joints is given in Table 3.10. Rubber gasket joints can tolerate this maximum without leakage. Lead–yarn joints can accommodate $0.75 d_W = 20$ to 30 mm without a significant increase in leakage. If there is no bead on the spigot this is increased to $0.75(d_S - d_L) = 30$–50 mm. For design purposes these maximum values are reduced to take account of initial withdrawal, joint rotation and future

Table 3.12 Allowable joint rotation and pull-out in iron distribution mains for tunnelling-induced movement

Type of distribution main	Joint rotation from initial position (degrees)	Joint pull-out from initial position (mm)
Lead–yarn joints in gas main with initial leaks	zero	zero
Lead–yarn joints in gas main initially sound	1.0	10
Lead–yarn joints in water mains	1.5	15
Rubber gasket joints in gas or water mains	2.5	25

movement due to other causes. A summary of allowable joint pull-out is given in Table 3.12.

3.4.4 Transverse ground movement

The principal effect of tunnelling-induced ground movement on a buried pipeline is longitudinal bending of the pipeline associated with transverse soil movement. The manner in which the various parameters interact is summarized in Table 3.13, the ground-movement parameters being listed first in approximate order of importance. The tunnel depth-to-diameter ratio has the strongest influence. Secondly, the effects of the soil–pipe interaction parameters are given. Here the allowable bending strain is the most important variable. The resistance of a pipeline to system disturbance depends on the longitudinal flexural rigidity (including the flexural rigidity of the pipeline joints), the longitudinal bending strength (including the reduction in strength caused by corrosion or service holes) and the pipe diameter. By substituting the typical pipe properties and dimensions of Tables 3.7, 3.8 and 3.9 in equation (3.18) a measure of the sensitivity of a pipeline to *overstress* caused by system disturbance is obtained. This is given in Table 3.14, where for comparative purposes the soil elastic modulus is taken as $10\,MN/m^2$ and the ground movement as 100 mm. For the flexibly-jointed pipelines, joints are assumed to lie at distances $0.375L_J$ and $0.625L_J$ from the position of maximum stress (see section 3.2.8). The maximum design stress in longitudinal bending has been taken to be 1.6 times the design stress for direct tension associated with internal pressure (see section 3.4.2).

The values of pipe stress/maximum design stress in Table 3.14 give the relative resistance of the pipelines for a failure criterion based on yield. Grey iron fractures without substantial yield, so the values are also a measure of the relative risk of fracture associated with ground movement. In any particular case the true risk will largely depend on existing local stress levels due to previous disturbance, stress due to internal pressure, stress due to external

Table 3.13 Effect of tunnel, pipe and soil parameters on pipe bending stress and strain

Variation in parameter	Effect on pipe bending stress and strain		
	Smooth ground movement profile or irregular slope and $1/\lambda > 0.7i$ transverse or $1/\lambda > 1.0i$ parallel	Irregular slope to ground movement profile and $1/\lambda < 0.7i$ transverse or $1/\lambda < 1.0i$ parallel	Non-uniformity in soil-pipe system
Increasing tunnel depth-to-diameter ratio	Increases trough width } → decreases maximum settlement → decreases stress and strain Decreases ground loss }		
Increasing ground loss at tunnel face	Increases settlement trough volume → increases maximum settlement → increases stress and strain		
Increasing allowable bending deformation of pipe	Decreases stress and strain as a proportion of allowable stress and strain		
Increasing pipe diameter at d/t = constant or d/t increasing within limits for commercially available pipes		Practically no effect on stress and strain	Decreases stress and strain
Increasing pipe elastic modulus		Decreases strain and increases stress	
Increasing depth of pipe below ground surface	Increases effective soil modulus, increases maximum settlement at pipe level, decreases trough width } → Increases stress and strain		

Table 3.13 (*Contd.*)

Variation in parameter	Effect on pipe bending stress and strain		
	Smooth ground movement profile or irregular slope and $1/\lambda > 0.7i$ transverse or $1/\lambda > 1.0i$ parallel	Irregular slope to ground movement profile and $1/\lambda < 0.7i$ transverse or $1/\lambda < 1.0i$ parallel	Non-uniformity in soil-pipe system
Increasing soil modulus around pipe		Increases stress and strain	
Yielding of soil	Decreases stress and strain compared with linear-elastic analysis		
Yielding of pipe	Decreases stress and increases *total* strain compared with linear-elastic analysis		
Pipe position parallel to and offset from tunnel drive compared with parallel to and directly above tunnel drive	Decreases maximum transverse ground movement at pipe position →	decreases stress and strain in plain pipeline but may cause high stresses in pipeline networks	
Pipe position parallel to tunnel compared with transverse case	Decreases stress and strain		Increases the extent of system disturbance effects
Pipe position oblique to tunnel drive compared with transverse case	Decreases stress and strain		Increases the extent of system disturbance effects
Flexible joints or non-continuous compared with rigid joints and continuous	Decreases stress and strain in metal pipes greater than 150 mm (6″) diameter for joints at 3.66 m centres and in metal pipes greater than 250 mm (10″) diameter for joints at 5.49 m (18 feet) centres. Otherwise little effect on maximum stress and strain.	Decreases stress and strain	Decreases stress and strain in metal pipes greater than 200 mm (8″) diameter for joints at 3.66 m centres and in metal pipes greater than 300 mm (12″) diameter for joints at 5.49 m (18 feet) centres. Otherwise little effect on maximum stress and strain.

loading such as traffic, and seasonal effects associated with ground temperature and moisture changes. The initial slight advantage of grey spun iron over statically cast iron will be lost with time due to the relatively worse effect of corrosion on the thinner wall of spun-cast pipes. In general, stress levels increase with age due to corrosion, the accumulative effect of disturbances to the system and increasing traffic loading. To ensure an adequate service life it is necessary to limit the stress increases associated with ground movements. Setting the allowable stress increase at a low level also reduces the risk of pipe fracture. The possibility of pipe fracture being caused *solely* by ground movement due to tunnelling is confined to relatively brittle materials. Even for small-diameter (less than 300 mm) grey iron pipelines, which constitute the greatest risk and also the major part of the gas and water distribution system of the UK, the risk of fracture would be low *provided* that the pipeline was not already overstressed (stress greater than 0.25 UTS), excessively corroded (reduced to less than 0.75 of the original wall thickness) and provided that the ground movement did not exceed the order of 100 mm. Reference to section 3.4.8 shows that high stress levels can be 'locked-in' by poor construction standards, and these will be magnified by dynamic loads from traffic on damaged or uneven road surfaces. Corrosion can also have a severe weakening effect on metal pipelines unless positive measures have been taken

Table 3.14 Resistance to pipe overstress associated with ground movement

Nominal pipe diameter (mm)/inches	Pipe stress/maximum design stress						
	grey iron			mild steel grade 410	uPVC short-term	HDPE SDR = 17 short-term	ductile iron grade 420
	pit-cast	vert. cast	spun-cast				
100/4							
rigid joints	2.06	1.72	1.47	1.32	1.23	1.26	0.79
flexible at 18ft/(5.49 m)	2.06	1.72	1.47	1.32	1.23	—	0.79
flexible at 12ft/(3.66 m)	2.06	1.72	1.47	—	—	—	—
200/8							
rigid joints	1.40	1.20	1.06	0.87	0.72	0.63	0.55
flexible at 18ft/(5.49 m)	1.40	1.20	1.06	0.87	0.72	—	0.55
flexible at 12ft/(3.66 m)	1.30	1.10	0.99	—	—	—	—
300/12							
rigid joints	1.05	0.90	0.81	0.68	0.49	0.50	0.43
flexible at 18ft/(5.49 m)	1.00	0.85	0.77	0.66	0.49	—	0.41
flexible at 12ft/(3.66 m)	0.77	0.64	0.60	—	—	—	—
600/24							
rigid joints	0.67	0.60	0.55	0.42	0.26	0.25	0.26
flexible at 18ft/(5.49 m)	0.40	0.34	0.28	0.28	0.26	—	0.15
flexible at 12ft/(3.66 m)	0.09	0.08	0.07	—	—	—	—

at the installation stage to prevent corrosion. Although there are relatively few tunnels where the ground surface movement exceeds 100 mm, there are many old grey iron pipelines that are locally overstressed or excessively corroded. It is these pipelines that fracture as a result of seasonal ground movements associated with soil temperature and moisture changes, or ground movements associated with traffic loading. In general, consideration should be given to the effects of tunnelling on grey iron pipes when the movement is expected to exceed 10 mm. Detailed consideration should be given to the effects on flexible pipes when the movement is expected to exceed 50 mm. For grey iron pipelines, the depth from pipeline axis to tunnel crown should be greater than twice the tunnel diameter, that is, $(z_0 - z_p)/2R \leqslant 2.5$. For tunnels overlain by stiff clay this may be reduced to $(z_0 - z_p)/2R \geqslant 1.5$. Ground movement in this zone close to a tunnel cannot be adequately estimated.

The direction of the tunnel relative to that of a pipeline has an important influence. A tunnel parallel to a pipeline exposes every part of the pipeline to the maximum effect. Any point of weakness is found by the wave of bending caused by the tunnel. In addition, the reduced strength at service tappings of up to 50 per cent (see section 3.4.2) will coincide with the maximum bending tension. Conversely, only a small part of a transverse pipeline is exposed to the maximum bending. This maximum is in the sagging mode where the pipeline strength is not reduced by service tappings. Irrespective of the results of analysis, the preferred tunnel/pipeline orientation should be transverse crossing. Tunnelling parallel to and directly below a pipeline should be preferred to a small offset parallel. A small offset does not reduce the bending by very much but does introduce additional pulling effects on branches (see section 3.4.6, Examples 3.20 and 3.21).

3.4.5 Parallel ground movement

The effects of parallel ground movement on iron pipelines are significantly restricted by the limited ability of the joints to transfer axial load. Continuous pipelines in steel or plastics are able to accommodate relatively large axial strains without failure. For these pipelines the potential axial strain is restricted by the soil–pipe shear strength. Plastic pipelines and externally sleeved or coated steel pipelines can only generate relatively small soil–pipe shear stresses.

Damaging effects of parallel ground movement are likely to be limited to axial joint movement and bending induced in branches off the main pipeline. For a jointed pipeline, most of the parallel ground movement is absorbed as joint slip. If the joints are already leaking (as may be the case for lead–yarn joints) then any movement will increase the leakage rate. Otherwise, horizontal ground movement of 20–40 mm, depending on joint type, is required to cause leakage associated with joint pull-out. This implies a maximum settlement of at least 40–80 mm. At this level of movement bending

GROUND MOVEMENTS AND BURIED PIPELINES 209

effects will also be severe. Significant effects can be induced in branches off the main pipeline. These are considered in the following section.

3.4.6 Pipeline networks

In practice, urban distribution systems are laid as pipeline networks. Where a tunnel passes beneath a network of connected pipes, the restraining effect of small-diameter or very flexible branches off the main pipeline can be neglected when considering the effects of ground movement on the main. The restraint to movement of a small-diameter pipe because of the connection to a more rigid pipe or structure cannot generally be ignored. Although it may not be practicable to analyse the complete network in detail, the effects that occur at changes in stiffness should always be borne in mind. The following examples illustrate some network effects. The first example deals with a layout that may occur at a street junction.

Example 3.19 Vertical bending of branch at connection to main pipeline

The plan layout is shown in Figure 3.36. The problem can be approached by the 'force' method mentioned in section 3.2.8. The joint between the main and the branch is cut, but the presence of the cut branch does not significantly affect deflection of the main pipeline. The restraint to soil movement around the main can, in the first instance, be neglected and the cut branch assumed to settle as though the main pipeline is not present. The solution for the network is found by restoring geometrical continuity at the cut.

The tunnel details are the same as for Example 3.1. The main pipeline is 0.50 m diameter transverse to the tunnel, as in Example 3.1. The branch pipeline is 0.10 m diameter parallel to the tunnel, as in Example 3.4. After the settlement wave has passed the junction, the deflection of the 'cut' branch is $w_{pbranch} = w_{max} = 13.6$ mm. The slope of the 'cut' branch is zero. The deflection of the main pipeline is shown in Figure 3.14c and is 11.91 mm above the tunnel centre line (Example 3.6). The forces required to restore continuity at the cut are shear, Q_J, and moment, M_J. For a rigid joint between the branch and the main, and assuming the main is completely rigid to torsion, the slope of the branch at the joint is zero. By symmetry, the effect of Q_J and M_J on the semi-infinite branch is the same as a point load of $2Q_J$ on an infinite branch. The load–deformation equations are as follows.

For the branch (Hetenyi, 1946):

$$w'_{pbranch} = \frac{-2Q_J \lambda_{branch}}{2K_{branch}}.$$

$$w'_{pbranch} = \frac{-1.923}{11.1 \times 10^6} Q_J = -1.732 \times 10^{-7} Q_J.$$

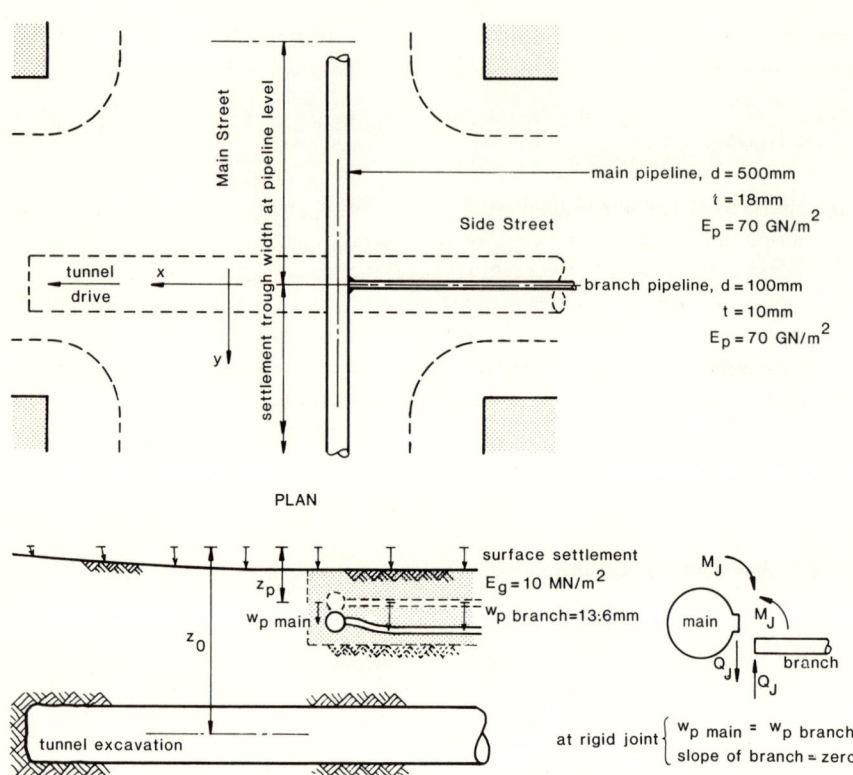

Figure 3.36 Example of vertical bending of branch at connection to main pipeline.

For the main (as above):

$$w'_{pmain} = \frac{Q_J \lambda_{main}}{2K_{main}} = \frac{0.481}{2 \times 11.9 \times 10^6} Q_J = 2.021 \times 10^{-8} Q_J.$$

For geometrical continuity:

$$(11.91 \times 10^{-3}) + w'_{pmain} = (13.6 \times 10^{-3}) + w'_{pbranch}$$

$$w'_{pmain} - w'_{pbranch} = 1.69 \times 10^{-3}.$$

Thus,

$$Q_J(2.021 \times 10^{-8} + 1.732 \times 10^{-7}) = 1.69 \times 10^{-3}$$

$$Q_J = 8738 \text{ N} = 8.74 \text{ kN}.$$

Substituting for Q_J:

$$w_{pmain} = (11.91 \times 10^{-3}) + (2.021 \times 10^{-8} \times 8.738 \times 10^3) = 12.1 \text{ mm}.$$

$$w_{pbranch} = (13.6 \times 10^{-3}) - (1.732 \times 10^{-7} \times 8.737 \times 10^3) = 12.1 \text{ mm}.$$

The additional sagging bending strain in the main pipeline at the branch is

$$\varepsilon_{yp} = \frac{Q_J}{4\lambda} \times \frac{d}{2E_p I_p} = \frac{8.738 \times 10^{-3} \times 0.500}{4 \times 0.481 \times 2 \times 7 \times 10^{10} \times 7.93 \times 10^{-4}} = 2.04 \times 10^{-5}$$

$= 20$ microstrain.

The hogging bending in the branch at the joint is

$$\varepsilon_{xp} = \frac{2Q_J}{4\lambda} \times \frac{d}{2E_p I_p} = \frac{2 \times 8.738 \times 10^{-3} \times 0.100}{4 \times 1.923 \times 2 \times 7 \times 10^{10} \times 2.90 \times 10^{-6}} = 5.60 \times 10^{-4}$$

$= 560$ microstrain.

Some joint rotation is inevitable where the branch joins the main. For a perfectly flexible joint ($M_J = 0$), the load–deformation equations are as follows:

For the branch (Appendix B):

$$w'_{pbranch} = -\frac{2Q_J \lambda_{branch}}{K_{branch}} = -3.465 \times 10^{-7} Q_J.$$

For the main (as before):

$$w'_{pmain} = 2.021 \times 10^{-8} Q_J.$$

For geometrical continuity:

$$Q_J = \frac{1.69 \times 10^{-3}}{3.667 \times 10^{-7}} = 4.61 \text{ kN}.$$

$w_{pmain} = w_{pbranch} = (11.91 \times 10^{-3}) + (2.021 \times 10^{-8} \times 4.609 \times 10^3) = 12.0$ mm.

The bending moment distribution in the branch is

$$M_x = \frac{Q_J}{\lambda_{branch}} B(\lambda x), \text{ where } B(\lambda x) = \exp(-\lambda x)\sin(\lambda x).$$

The maximum occurs at

$$\lambda x = \frac{\pi}{4}, \quad x = \frac{\pi}{1.923 \times 4} = 0.4 \text{ m from the joint.}$$

$$\varepsilon_{xpmax} = M_{xmax} \times \frac{d}{2E_p I_p} = 0.3224 \times \frac{4.609 \times 10^3}{1.923} \times \frac{0.100}{2 \times 7 \times 10^{10} \times 2.90 \times 10^{-6}}$$

$= 1.90 \times 10^{-4} = 190$ microstrain sagging.

The calculation has so far ignored any local restraint to soil movement around the main pipeline, as though the soil was able to flow around the pipe. For ground movement large enough to generate the ultimate soil pressure this is partly the case (see section 3.2.9). In this particular example, the soil pressures are very small and the soil–pipe bond will be intact (see Example 3.6). Ignoring the presence of the branch, then on the outside of the main at axis level the soil movement is $w = w_{pmain} = 11.91$ mm. The subgrade-reaction analysis gives no information about the soil movement on a horizontal plane through the main. A three-dimensional linear-elastic finite-element analysis will overestimate the restraint to local soil movement. A reasonable and slightly conservative design (based on measurements around friction piles) is to assume that the effect is restricted to five times the main pipe diameter from the external surface of the pipe. Thus, at a horizontal distance of 2.75 m from the centre line of the main pipeline the soil movement is 13.6 mm. A linear or 'normal probability curve' variation can be taken between this point and the main pipeline. This soil restraint is applied to the 'cut' branch by the method in section 3.2.3, or alternatively the solutions in section 3.2.4 and 3.2.6 may be used with the substitution $2.5i = 2.75$ m, and the condition that the slope of the branch is zero. The geometrical incompatibility remaining after applying the soil restraint is then removed as appropriate for a rigid or flexible joint. The solutions are:

For a rigid joint:
$$w_{\text{pmain}} = w_{\text{pbranch}} = 11.93 \text{ mm},$$

and for the branch
$$\varepsilon_{\text{xpmax}} = 116 \text{ microstrain hogging at joint.}$$

For a flexible joint:
$$w_{\text{pmain}} = w_{\text{pbranch}} = 11.92 \text{ mm},$$

and for the branch
$$\varepsilon_{\text{xpmax}} = 10 \text{ microstrain sagging at } 0.4 \text{ m from the joint.}$$

The additional bending induced in the main and the branch are well within the general system disturbance allowances of 59 microstrain in the main pipeline (Example 3.7) and 253 microstrain in the branch (Example 3.8).

Example 3.20: Horizontal bending of branch at connection to main pipeline

If the tunnel line in Example 3.19 is down the centre of the side street (see Figure 3.36), then after the tunnel has passed the junction the branch pipeline will be translated both vertically and horizontally towards the tunnel. At the connection to the main pipeline this will be resisted, since the main is relatively stiff and the movement is less than the unrestrained soil movement.

For this example the offset distance to the branch pipeline is $y_p = 2.0$ m (as in Example 3.5). Away from the influence of the connection to the main, the vertical and horizontal displacement of the branch are merely the same as the imposed ground movement. From section 3.2.5 the displacements are:

$$w_{\text{pbranch}} = w = \exp\left[\frac{-y_p^2}{2i^2}\right] w_{\text{max}} = \exp\left[\frac{-2.0^2}{2 \times 2.6^2}\right] \times 13.6 \text{ mm}$$
$$= 10.12 \text{ mm}.$$

$$v_{\text{pbranch}} = v = \frac{-n y_p}{z_0 - z_p} \exp\left[\frac{-y_p^2}{2i^2}\right] w_{\text{max}}$$

$$= \frac{-1.0 \times 2.0}{5.0 - 1.5} \exp\left[\frac{-2.0^2}{2 \times 2.6^2}\right] \times 13.6 \text{ mm}$$

$= 5.78$ mm towards the tunnel.

(Note: $\overline{w+v} = \sqrt[2]{10.12^2 + 5.78^2} = 11.65$ mm as Example 3.5.)

The deflection of the main pipeline is shown in Figure 3.14c and at $y = 2.0$ m ($y/i = 0.77$) is $w_{pmain} = 9.69$ mm. The differential settlement at a 'cut' at the joint between the branch and the main is $10.12 - 9.69 = 0.43$ mm, compared with 1.69 mm in Example 3.19. (Alternatively the differential movement can be read from Figure 3.14d as $w_p - w = w_{pmain} - w_{pbranch} = -0.031 w_{max} = 0.42$ mm.) The vertical bending induced in the branch and the main are just $0.42/1.69 = 25$ per cent of the values in Example 3.19.

The axial horizontal movement of the main pipeline depends largely on the pipeline joint flexibility and the position of these joints. For rigid joints, or with the joints approximately equally spaced each side of the tunnel and beyond the junction with the branch, v_{pmain} is very small. For a symmetrical arrangement, v_{pmain} is zero above the tunnel centre line and v_{pmain} at $y = 2.0$ m can be simply calculated by integrating the pipe strain diagram shown in Figure 3.26. Thus, for the elastic method, at $y = 2.0$ m,

$$y/i = \frac{2.0}{2.6} = 0.77,$$

$$v_{pmain} = \frac{0 - (63 + 38)}{2} \times 10^{-6} \times 2.0 \text{ m} = -1.01 \times 10^{-4} = -0.101 \text{ mm}.$$

For the load transfer method,

$$v_{pmain} = -0.244 \text{ mm}.$$

The forces required to restore continuity at a cut between the branch and the main are shear, Q_J, and horizontal moment, M_J, as indicated in Figure 3.37. The load–deformation equations are as follows:

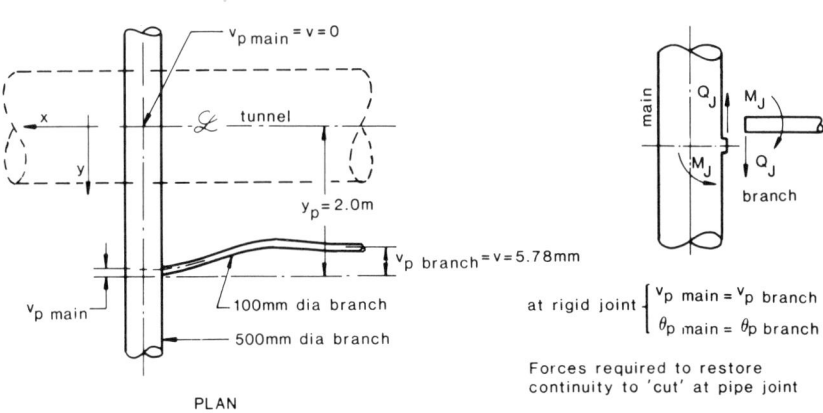

Figure 3.37 Example of horizontal bending of branch at connection to main pipeline.

For moment M_J (Appendix B):

$$v'_{pbranch} = \frac{-2M_J\lambda^2_{branch}}{K_{branch}} = \frac{-2 \times 1.923^2 M_J}{11.1 \times 10^6} = -6.663 \times 10^{-7} M_J.$$

$$\theta_{pbranch} = \frac{+4M_J\lambda^3_{branch}}{K_{branch}} = +2.563 \times 10^{-6} M_J.$$

$$v'_{pmain} = 0.$$

$$\theta_{pmain} = \frac{-M_J\lambda^3_{main}}{K_{main}} = \frac{-0.481^3 M_J}{11.9 \times 10^6} = -9.352 \times 10^{-9} M_J.$$

For shear Q_J (Appendix B):

$$v'_{pbranch} = \frac{+2Q_J\lambda_{branch}}{K_{branch}} = +3.465 \times 10^{-7} Q_J.$$

$$\theta_{pbranch} = \frac{-2Q_J\lambda^2_{branch}}{K_{branch}} = -6.663 \times 10^{-7} Q_J$$

$$\theta_{pmain} = 0.$$

For the elastic method (section 3.3.2),

$$v'_{pmain} = \frac{-Q_J}{2} \frac{I_f}{E_g d},$$

where I_f is an influence factor given by Poulos and Davis (1980) and is 0.043 for $v_g = 0.3$ and $L/d = 9i/d = 46.8$. Thus,

$$v'_{pmain} = \frac{-0.043}{2 \times 10 \times 10^6 \times 0.50} Q_J = -4.30 \times 10^{-9} Q_J.$$

For the load-transfer method (Example 3.13),

$$v'_{pmain} = \frac{-Q_J}{2} \frac{1}{E_p A_p \beta} \text{ where } \beta = \left[\frac{K_y}{E_p A_p}\right]^{1/2}$$

$$= \frac{-Q_J}{2} \left(\frac{1}{E_p A_p K_y}\right)^{1/2}$$

$$= \frac{-Q_J}{2} \left(\frac{1}{7 \times 10^{10} \times \pi \times 0.482 \times 0.018 \times 13.1 \times 10^6}\right)^{1/2}$$

$$v'_{pmain} = -8.37 \times 10^{-9} Q_J.$$

In the first instance neglecting the restraint to soil movement around the main, then for geometrical continuity and a rigid joint between the main and the branch the results are as follows.

For $v_{pmain} = v_{pbranch}$:

$$-(0.101 \times 10^{-3}) + v'_{pmain} = -(5.78 \times 10^{-3}) + v'_{pbranch}$$
$$5.679 \times 10^{-3} = -(6.663 \times 10^{-7})M_J + (3.465 \times 10^{-7})Q_J - (4.30 \times 10^{-9})Q_J.$$

For $\theta_{pmain} = \theta_{pbranch}$:

$$-(9.352 \times 10^{-9})M_J = +(2.563 \times 10^{-6})M_J - (6.663 \times 10^{-7})Q_J.$$

Thus,

$$Q_J = 3.1867 \times 10^4 \text{N}, \quad M_J = 8.2542 \times 10^3 \text{Nm}, \quad v_{pmain} = -0.238 \text{mm}.$$

The maximum bending strain in the branch occurs at the rigid joint:

$$\varepsilon_{xpmax} = M_J \frac{d}{2E_p I_p} = \frac{8.254 \times 10^3 \times 0.100}{2 \times 7 \times 10^{10} \times 2.90 \times 10^{-6}} = 2033 \text{ microstrain}.$$

For the main,

$$\varepsilon_{ypmax} = \frac{M_J}{2} \frac{d}{2E_p I_p} = \frac{8.254 \times 10^3 \times 0.500}{4 \times 7 \times 10^{10} \times 7.93 \times 10^{-4}} = 19 \text{ microstrain}.$$

Taking account of the restraint to soil movement around the main (as in Example 3.19) and imposing this restraint on the branch gives additional load–deformation equations:

$$v'_{pbranch} = +5.485 \times 10^{-3},$$

$$\theta_{pbranch} = -1.503 \times 10^{-3}.$$

Thus for the condition $v_{pmain} = v_{pbranch}$:

$$0.194 \times 10^{-3} = -(6.663 \times 10^{-7})M_J + (3.422 \times 10^{-7})Q_J.$$

and for a rigid joint, $\theta_{pmain} = \theta_{pbranch}$:

$$-(9.352 \times 10^{-9})M_J = +(2.563 \times 10^{-6})M_J - (6.663 \times 10^{-7})Q_J - 1.503 \times 10^{-3}.$$

Thus,

$$Q_J = 3.439 \times 10^3, \quad M_J = 1.475 \times 10^3, \quad v_{pmain} = v_{pbranch} = -0.116 \text{mm}.$$

The maximum bending strain in the branch at the joint is

$$\varepsilon_{xpmax} = 363 \text{ microstrain}.$$

For a flexible joint between the branch and the main, $M_J = 0$ and $\theta_{pmain} \neq \theta_{pbranch}$ in the above compatibility conditions. Thus $Q_J = 5.67 \times 10^2 \text{N}$, $v_{pmain} = v_{pbranch} = -0.103$ mm, and in the branch at 0.4 m from the joint $\varepsilon_{xpmax} = 23$ microstrain.

The geometrical continuity conditions may also be solved using the load–deformation relationship for the load-transfer model rather than the elastic model, that is, $v'_{pmain} = -8.37 \times 10^{-9} Q_J$. In this case, and allowing for the soil restraint around the main, the solutions are for a rigid joint, $\varepsilon_{xpmax} = 369$ microstrain; for a flexible joint, $\varepsilon_{xpmax} = 24$ microstrain.

Since some slight flexibility is inevitable at the joint between the branch and the main, the additional bending induced in the branch is within the general system disturbance allowance of 253 microstrain $\times$ 0.85 (reduction factor for $y_p = 2.0$ m) = 215 microstrain. The beneficial effect of a flexible joint at the change in stiffness from branch to main is clearly indicated by Examples 3.19 and 3.20. Because the ground movement does not generate the ultimate soil–pipe shear stress (no slip on the main), the local restraint to soil movement around the main reduces the bending in the branch by a factor of 5. Where horizontal ground movement *does* slip along the main pipeline then very high junction stresses may be caused in a branch that resists this movement.

The next example of a pipeline network is typical of tunnelling parallel to and offset from a pipeline with service branches. Here the pulling effect on the branch has to be considered.

Example 3.21 Axial pulling of branch off main pipeline parallel to and offset from tunnel

The plan layout is shown in Figure 3.38. The tunnel details are the same as in Example 3.1. The main pipeline is 0.50 m diameter parallel to the tunnel, as in Example 3.5. The offset distance is 2.0 m. The branch pipeline is 0.10 m diameter transverse to the tunnel as in Example 3.2.

As the tunnel advances past the branch pipeline then horizontal and vertical bending will occur in the branch near to the junction with the main. These effects are similar to those of Examples 3.19 and 3.20 but of reduced magnitude

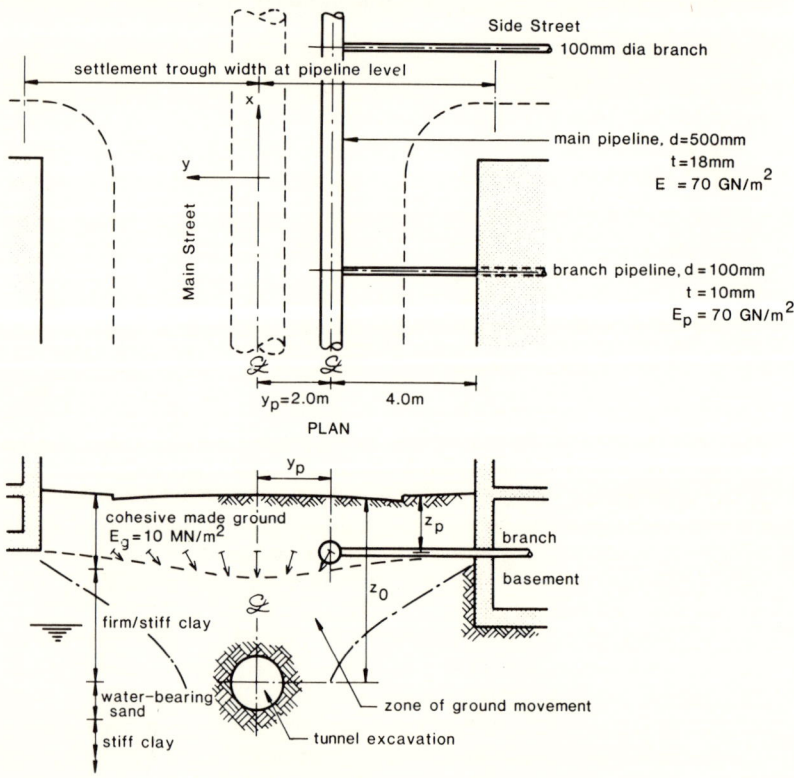

Figure 3.38 Example of axial pulling of branch off offset and parallel main pipeline.

due to the reduced level of differential soil–pipe movement. The maximum value of $w - w_{pmain}$ is only 0.29 mm compared with 0.42 mm in Example 3.20. The maximum value of $u - u_{pmain}$ is only 2.57 mm compared with 5.68 mm in Example 3.20. The theoretical residual displacement of the main pipeline in the direction parallel to the tunnel, u_{pmain}, is zero. For a continuous rigidly-jointed pipeline this will be the case in practice. Any horizontal bending induced permanently in the branch pipeline shown in Figure 3.38 will be more than covered by the general allowance for system disturbance. Where the main pipeline has flexible joints then there is more scope for residual displacement of the pipeline (this being balanced by residual displacement within the joints). The allowance for system disturbance in the transverse branch is equivalent to a differential displacement of 0.75×5 per cent w_{max}, which is 0.51 mm or 17 per cent of the maximum displacement for a jointed main. This should adequately cover any permanent horizontal bending.

In the fully developed settlement trough, the main pipeline is displaced horizontally 5.78 mm towards the tunnel. At a branch off the main this movement must either be accommodated as joint extension in the branch or as tensile strain in the branch (reduced by the horizontal bending in the main caused by this pull at the branch). The solution is found by cutting the joint between the main pipeline and the branch. This gives the extension of a flexible joint. A rigid connection is simulated by applying equal and opposite forces to the branch and the main to close the joint. The branch into the side street is considered first and the axial soil–pipe interaction model is the load-transfer method similar to that in Example 3.13 and 3.15.

The horizontal soil movement along the branch is similar to that in Example 3.12 and is of the form

y (m)	y/i	v (mm)	v/v_{max}
−2.0	−0.77	5.78	0.94
−2.6	−1.0	6.1	1.0
−7.8	−3.0	0	0

The pipe is free to move at $y = -2.0$ m, and in this sense is similar to a pipeline parallel to the tunnel. By the load-transfer method, the tensile strain and elastic extension of the branch are

y (m)	y/i	$\varepsilon_{ypbranch}$ (microstrain)	$v_{pbranch}$ (mm)
−2.0	−0.77	0	1.77
−2.6	−1.0	33	1.76
−3.9	−1.5	98	1.67
−5.2	−2.0	137	1.52
−6.5	−2.5	145	1.34
−7.8	−3.0	133	1.16
−33.8	−13.0	0	0

Note that this is similar to Example 3.15, where for 6.1 mm movement the pipe strain is $97 \times 6.1/4.0 = 148$ microstrain. If the joint between the branch and the main is thus allowed to extend freely, the joint opens a distance of $v_{\text{pmain}} - v_{\text{pbranch}} = 5.78 - 1.77 = 4.01$ mm. This gap can be closed by applying a lateral force of P_J to the main and a similar axial pull to the branch. The load–displacement equations are:

For the main (as in Example 3.19):

$$v'_{\text{pmain}} = -2.021 \times 10^{-8} P_J.$$

For the branch (refer to Example 3.13):

$$v'_{\text{pbranch}} = \frac{\varepsilon_{ypJ}}{\beta}, \quad \text{where} \quad \beta = \left[\frac{K_y}{E_p A_p}\right]^{1/2}$$

$$v'_{\text{pbranch}} = P_J \left(\frac{1}{E_p A_p K_y}\right)^{1/2}$$

$$K_y = \frac{\pi d \tau_a}{6 \text{ mm}} = \frac{\pi \times 0.10 \times 50 \times 10^3}{6 \times 10^{-3}} = 2.618 \times 10^6.$$

Thus,

$$v'_{\text{pbranch}} = P_J \left(\frac{1}{7 \times 10^{10} \times \pi \times 0.09 \times 0.01 \times 2.618 \times 10^6}\right)^{1/2} = 4.393 \times 10^{-8} P_J.$$

For geometrical continuity at a rigid non-extendible joint:

$$5.78 \times 10^{-3} + v'_{\text{pmain}} = 1.77 \times 10^{-3} + v'_{\text{pbranch}}.$$

$$4.01 \times 10^{-3} = (2.021 + 4.393) \times 10^{-8} P_J$$

$$P_J = 6.252 \times 10^4 = 62.5 \text{ kN}.$$

The maximum horizontal bending strain in the main (at the branch) is

$$\varepsilon_{yp\text{branch}} = \frac{P_J}{E_p A_p} \exp(-\beta y) = \text{maximum} \frac{6.25 \times 10^3}{7 \times 10^{10} \times \pi \times 0.09 \times 0.01}$$

$$= 146 \text{ microstrain}.$$

The axial tensile strain in the branch is

$$\varepsilon_{yp\text{branch}} = \frac{P_J}{E_p A_p} \exp(-\beta y) = \text{maximum} \frac{62.5 \times 10^3}{7 \times 10^{10} \times \pi \times 0.09 \times 0.01}$$

$$= \text{maximum } 316 \text{ microstrain}.$$

GROUND MOVEMENTS AND BURIED PIPELINES

The final distribution of tensile strain in the branch is

y (m)	$\dfrac{y}{i}$	$\varepsilon_{\text{ypbranch}}$ (ground movement)	$\varepsilon_{\text{ypbranch}}$ (axial pull)	$\varepsilon_{\text{ypbranch}}$ (total microstrain)
-2.0	-0.77	0	316	316
-2.6	-1.0	33	295	328
-3.9	-1.5	98	254	352
-5.2	-2.0	137	219	356
-6.5	-2.5	145	188	333
-7.8	-3.0	133	162	295

The final movement of the branch and the main at the joint is 4.52 mm. The above applies only where the branch is continuous. A flexible joint within 18 feet (5.49 m) of the junction would greatly reduce the effect. Consider now the branch pipeline shown in transverse section in Figure 3.38. With flexible joints at the connection to the building *and* the main, only small axial forces are induced in the branch. From section 3.3.6,

$$L_J = 4.0\,\text{m}, \quad d = 0.10\,\text{m}, \quad \Delta v = 6.10 - 1.63 = 4.47\,\text{mm}, \quad K^* = 2520.$$

From Figure 3.31,

$$\varepsilon_{\text{ypmax}} = \frac{1000}{2520} \times 0.195 \times \frac{4.47 \times 10^{-3}}{4.0} = 8.6 \times 10^{-5}$$

$$= 86\ \text{microstrain}.$$

(The load-transfer method gives smaller axial strain for the Figure 3.21 load–displacement relationship.)

If the branch is rigidly jointed to the building but free to move at the connection to the main, then the fixity conditions are similar to those in Figure 3.20 (right-hand side). The building foundation is located beyond the zone of ground movement and is therefore subject to insignificant movement. If the ground movement is sufficient to mobilize the full soil–pipe shear strength, then, using Equation (3.33), $\tau_a = 50\,\text{kN/m}^2$, $\varepsilon_{yp} = 0$ at $y = -2.0\,\text{m}$,

$$\frac{d\varepsilon_{yp}}{dy} = \frac{\pi d \tau_a}{E_p A_p} = \frac{\pi \times 0.100 \times 50 \times 10^3}{7 \times 10^{10} \times \pi \times 0.090 \times 0.01} = 79\ \text{microstrain/m}$$

maximum at $y = -6.0\,\text{m}$, $\varepsilon_{\text{ypbranch}} = 79 \times 4.0 = 316$ microstrain.

Allowing for the actual form and magnitude of ground movement, the displacement of the branch at $y = -2.0\,\text{m}$ is $v_{\text{pbranch}} = 0.51\,\text{mm}$ and $\varepsilon_{\text{ypbranch}} = 216$ microstrain (load-transfer method). Either the joint opens by $v_{\text{pmain}} - v_{\text{pbranch}} = 5.27\,\text{mm}$ or axial force P_J maintains geometrical continuity at a rigid joint. The load–displacement equation for the main is as before:

$$v'_{\text{pmain}} = -2.021 \times 10^{-8} P_J.$$

In the case of a branch, if the shear stress between the branch and the soil caused by pulling the branch at $y = -2.0$ m is ignored, then the axial restraining force at the connection to the building, P_{fix}, is also equal to P_J and the movement of the branch at the joint with the main is merely the elastic extension of the 4.0 m length under axial load P_J, that is,

$$v'_{\text{pbranch}} = 4.0 \times \frac{P_J}{E_p A_p}, \quad \text{since } v'_{\text{pbranch}} = 0 \quad \text{at } y = -6.0 \text{ m}.$$

$$v'_{\text{pbranch}} = \frac{4.0}{7 \times 10^{10} \times \pi \times 0.09 \times 0.010} P_J = 2.021 \times 10^{-8} P_J$$

Allowing for the restraint to movement of the branch due to the soil–pipe shear stress, then using equations (3.32), (3.33), (3.34) and (3.35) and the condition that $v'_{\text{pbranch}} = 0$ at $y = -6.0$ m,

$$v'_{\text{pbranch}} = 1.889 \times 10^{-8} P_J \quad \text{and} \quad P_{\text{fix}} = 0.9028 P_J$$

Thus

$$(5.78 - 0.51) \times 10^{-3} = (2.021 + 1.889) \times 10^{-8} P_J$$

$$P_J = 134.8 \text{ kN},$$

$$P_{\text{fix}} = 121.7 \text{ kN},$$

and

$$v_{\text{pbranch}} = v_{\text{pmain}} = (5.78 \times 10^{-3}) - (2.021 \times 10^{-8} \times 134.8 \times 10^3)$$

$$= 3.06 \text{ mm}.$$

The maximum axial strain in the branch at the connection to the building is

$$\varepsilon_{\text{ypbranch}} = 216 \mu\varepsilon \text{ (from ground movement)} + \frac{P_{\text{fix}}}{E_p A_p}$$

$$= 216 + \left(\frac{121.7 \times 10^3}{7 \times 10^{10} \times \pi \times 0.09 \times 0.01} \right) \times 10^6 = 831 \text{ microstrain}.$$

The maximum bending strain in the main is $\varepsilon_{\text{xpmain}} = 315$ microstrain.

From these sample calculations it is clear that high stress levels can be induced in branches from an offset parallel pipeline if the pipeline joints are rigid (that is, non-extendible) and particularly if the branch is anchored against movement at an adjacent structure. Pipe joints of the lead–yarn type can only transfer small axial forces. For the 100 mm diameter branch in the sample calculation, this is approximately 13.8 kN in the medium term and 6.9 kN in the long term. This limits the axial strain in the branch to 70 microstrain in the medium term at a joint, compared with 300–700 microstrain for a truly rigid joint.

In the last example, the main pipeline is restrained by $5.78 - 3.06 = 2.72$ mm. The horizontal break-out load (section 3.2.6) is $p = K_{\text{eff}} \Delta v_{\text{pmain}} / d$

GROUND MOVEMENTS AND BURIED PIPELINES 221

(Equation (3.15) plus (3.16)). The break-out load is

$$p = \frac{11.9 \times 10^6 \times 2.72 \times 10^{-3}}{0.50} = 6.47 \times 10^4 \text{N/m}^2 = 64.7 \text{kN/m}^2.$$

The horizontal break-out resistance (section 3.2.9) is $p_u = cN_c + \gamma z_p N_q$ (equation (3.25)). Taking $c = 50 \text{kN/m}^2$, and $\phi = 0$ then, from Figure 3.19,

$$p_u = 50 \times 4.6 = 230 \text{kN/m}^2, \text{ i.e. } \frac{p}{p_u} = 28 \text{ per cent only}.$$

From the above it is clear that very high stress levels could be generated in a rigidly-jointed pipeline before soil yield begins to limit the loading.

3.4.7 Performance of the soil–pipe interaction model

The solutions for longitudinal bending given in sections 3.2.4, 3.2.5 and 3.2.8 have been tested against the published data concerned with the effect of tunnelling on buried pipelines. Agreement between the analysis and the measurements or observations is good; the data on the effects of movement are sparse compared with the mass of data on the movement alone—see Howe et al. (1980), O'Rourke and Trautmann (1982), Howe and Hunter (1983), Hurrell (1983), Dorling (1984), Takagi et al. (1985), Hurrell (1985). More data are available on the effects of sewer trenches (typically up to 6 m deep) parallel to a pipeline. The magnitude of the ground movement is similar to that associated with tunnels in soil and the length of the wave of movement is similar to that from a shallow tunnel. Sources of ground movement/pipe strain data are given by Rumsey and Dorling (1985) and by Howe (1985a). The soil–pipe interaction modelling in section 3.3 produces results that are entirely consistent with the measurements. The importance of system disturbance effects, which depend on the magnitude of ground movement rather than on the overall form of movement, has been emphasized by Yeates (1985a). These effects, which may be caused by small local variations in ground movement or pipeline restraint, reconcile the measurements and the analysis. Leach (1984) has made a preliminary assessment of the model for soil–pipe interaction due to tunnelling movements. The assessment was based on estimated ground movement (similar to that in Chapter 2) rather than on measured movements; even so, the comparison between 'predicted' and measured longitudinal bending strain shows good agreement. The model for axial strain (load-transfer method) generally overestimated pipe strain and this could be attributed to soil–pipe slip or joint movement.

The allowance for general system disturbance effects, as in section 3.2.7, is virtually the same as an empirical formula that 'fits' all the reliable UK data on pipe strains caused by ground movement associated with trenching. The trenching formula is applicable to metal pipelines up to 300 mm diameter parallel to the source of ground movement. In this case joint flexibility does not significantly affect the resulting bending strain. Fry and Rumsey (1983) consider that the empirical formula has a possible range of error of ± 25 per cent. In general, it should be expected that the solutions in section 3.2 overestimate pipe bending moment since all the simplifications have been on the safe side where necessary. Given an accurate specification for the ground movements, it is expected that section 3.2 should predict pipe stress to within ± 25 per cent or better if the correct assumptions are made about joint stiffness and position. Conservative assumptions about joint flexibility can always be made. It should be remembered that it is a good prediction of ground movement if the maximum settlement can be estimated to within 50 per cent without having measurements relating to a similar tunnel constructed under similar conditions in similar ground.

There have been very few measurements of pipe strain induced by network effects similar to those in section 3.4.6. However, the fracture data for ground movement effects on distribution systems indicate that junction stresses can be highly significant. The results of analysis such as that in Example 3.21 are entirely consistent with the relatively high risk of damage when tunnelling at shallow depth parallel to and offset from a main pipeline with branches.

3.4.8 *Other considerations*

When assessing the effect of tunnelling on buried pipelines it is useful to have an indication of pipe stresses that arise from other causes. They may be summarized as:

1. Earth loads associated with main laying and trench reinstatement (permanent)
2. Earth loads associated with consolidation, frequently collapse settlement, of trench backfill (permanent)
3. Internal pressure (permanent and transient)
4. Vehicle wheel loading at surface (transient static)
5. Vehicle wheel loading at surface (dynamic)
6. Vehicle wheel loading at surface causing deformation of trench backfill and pipe embedment (permanent)
7. System disturbance caused by adjacent excavation due to proximity of other services (permanent and transient)
8. Restraint to free movement associated with temperature changes of pipe material (seasonal)
9. Soil movement associated with frost heave (seasonal and some permanent)
10. Soil movement associated with moisture changes in shrinkable clays (seasonal and some permanent)
11. Direct mechanical impact caused by adjacent excavation due to proximity of other services (permanent and transient)

12. Strength reduction and stress concentration due to corrosion and in particular localized corrosion pitting (permanent)
13. Strength reduction and stress concentration due to service branch tappings (permanent).

All of the above factors can cause significant stresses. Items 7, 10, 11, 12 and 13 are particularly associated with small-diameter pipelines; items 1, 2 and 3 apply particularly to large-diameter pipelines. Rarely can a pipeline failure be attributed to one main cause, but rather it is the last increment of loading or a clearly identifiable loading that is recorded as the cause of failure.

For gas *distribution* mains the effect of internal pressure on pipe stress levels is usually insignificant. In some water mains the internal water pressure, which causes uniform ring tension, may be large enough to contribute significantly to failure. Here the failure mode is a longitudinal split, or a blow-out, hole or perforation associated with long-term corrosion.

A summary of published data on measured *longitudinal* bending strains for 100 mm diameter spun grey iron pipelines is given in Table 3.15. The transient load attributable to traffic can be increased by a damaged or uneven road surface. The dynamic load is generally in the range 1.5 to 2.5 times the static load, but can reach 4.0 times the static load. Longitudinal bending stresses decrease with increasing pipe diameter (that is, increasing beam strength) and on the basis of section 3.2.7 are approximately proportional to the value of $d(1/E_p I_p)^{0.5}$. At d/t and E_p constant this implies stress approximately inversely proportional to diameter. For comparative purposes, Table 3.15 is for $d = 0.122$ m and $E_p I_p = 485$ kN m^2. A solution for longitudinal bending moment in a buried pipeline due to a point load at the ground surface is given by Pearson (1979).

Table 3.15 Typical longitudinal bending strain in 100 mm spun grey iron pipelines

	Pipe strain (microstrain) for different standards of pipeline construction		
Cause of pipe strain	*Very good* (granular bedding/ densely compacted backfill)	*Average* (trimmed trench bottom/ compacted backfill)	*Very poor* (uneven trench bottom/loose clay backfill)
'Locked-in' due to main laying and trench reinstatement	25	50	150+
'Locked-in' after consolidation of backfill due to traffic loading etc. (up to 1 year after installation)	25	50	150+
Total 'locked-in' due to installation	50	100	300+
Transient static load due to traffic on smooth road surface	50	100	250+

Seasonal thermal stress in a buried pipeline depends on the coefficient of thermal expansion of the pipe material, the bond between the soil and the pipe, restraint to pipeline movement and the spacing and pull-out characteristics of the pipeline joints. In the UK the typical seasonal temperature variation at mains depth is 15°C. For grey iron this implies an *unrestrained* thermal movement of ± 80 microstrain. Measured thermal changes in stress are equivalent to ± 50 microstrain, the remaining thermal movement being accommodated at the pipeline joints. It is known that a drop in temperature causes a significant increase in transverse fractures in gas and water mains. For this to be caused by the very small stresses induced by thermal restraint would indicate that a significant proportion of the distribution system is already highly stressed. The effect of frost loading on buried pipelines in North America has been investigated by Smith (1976) and confirms work by Monie and Clarke (1974). In the UK there is also a great increase in the number of fractures during prolonged periods of sub-zero temperatures. Owen (1985) has reported some preliminary work on the effects of soil moisture changes in shrinkable clay. Rumsey *et al.* (1981) have summarized other causes of longitudinal bending stress.

3.5 Design to accommodate ground movement

Ground movement is an inevitable consequence of any excavation. A requirement for the successful design of a tunnel is that construction should not excessively damage adjacent or overlying buildings, streets or utilities. Damage to a pipeline may be immediate (fracture or leakage), or the imposed stress due to disturbance may reduce the service life of the pipeline. The life expectancy of a buried pipeline is heavily dependent on the stress level experienced, together with in many cases the debilitating effect of corrosion. In the particular case where the owner of a pipeline perceives a risk of damage, it may be inadequate compensation that damage actually sustained will be repaired. Property damage or injury to persons may be a possible consequence of leakage from a gas pipe, particularly the relatively high leakage rate that would occur at a fracture. Leakage from a water main may also cause consequential damage to other shallow services, buildings, roads and sewers due to wash-out of fine soil into sewers and drains.

If immediately noticeable damage is caused to a buried pipeline, this is essentially equivalent to 'severe' building damage in the UK Department of the Environment classification (DoE, 1981); that is, repair work involving replacement is required. For a pipeline the cost of repair is usually far higher than the cost of making good building damage, unless this is in the 'very severe' category. The UK gas distribution industry's damage control policy has been outlined by Howe (1985a). An essential feature, as far as proposed tunnelling works are concerned, is that grey iron mains must be protected from undue increases in the severity of their operating environment. Ground movement causing only 'slight' (DoE, 1981) visible damage to buildings may cause significant stress in rigid pipelines. It is possible that if the tunnel route coincides with an important and perhaps already highly-stressed section of the gas distribution network then the cost of replacement pipelines in advance of the tunnelling works may exceed the cost of constructing the tunnel alone. In any event, pipeline owners need to assess the risk of damage from ground movements and the consequences. Tunnel promoters also have a responsibility for avoidance of damage. Designers of tunnels have to be satisfied that the cost of any diversion or replacement of pipelines in advance of the tunnelling works is reasonable in the particular circumstances. In the event of a pipeline failure in the vicinity of recent construction work, it may be necessary to identify the possible causes.

A preliminary assessment of the possible effect of tunnelling-induced movement may be based on Table 3.16. This is drawn from fracture data, measurements and theoretical analysis. The majority of tunnel construction is for sewers of less than 2.5 m diameter. Except for shallow tunnels or poor ground conditions, these tunnels can usually be constructed with less than 25 mm settlement. A great many tunnels produce movement in the range 10 to 25 mm settlement and detailed consideration should be given to the effect of this movement on brittle pipelines. Risk of pipeline failure has to be set in the context of overall pipeline reliability from all causes. For steel transmission pipelines the average failure rate is about 1 failure per 1000 km years. Grey

Table 3.16 Preliminary assessment of the effect of ground movement on a buried pipeline

Maximum surface settlement (mm)	Brittle materials (grey iron, asbestos cement, clayware)	Ductile materials (steel, ductile iron, uPVC, polyethylene)
$w_{max} \leqslant 10$	Pipe stress increase is not significant compared with other causes of stress such as installation, traffic load, seasonal movement	
$w_{max} > 10$	The effects of movement should be assessed in detail	—
$w_{max} > 25$	Significant stress increase virtually certain; possible failure of small-diameter pipes	—
$w_{max} > 50$	Possible failure of large-diameter pipes	Significant stress increase likely; the effects of movement should be assessed in detail

Note: Pipeline failure is defined here as an incident that leads to a significant leak or otherwise requires immediate repair. For brittle materials, depth from pipeline axis to tunnel crown has to be greater than twice the tunnel diameter except for tunnels overlain by stiff clay, where this may be reduced to a minimum clearance of one diameter. Minimum tunnel depth to axis level irrespective of diameter is 3.5 m

iron distribution mains have a failure rate between 10 and 100 per 1000 km years depending on location, diameter and the corrosive nature of the soil. Direct mechanical impact through adjacent excavation causes a failure rate in distribution mains of about 5 per 1000 km years in ductile iron compared with up to 40 failures per 1000 km years in non-ferrous pipelines.

For UK gas distribution mains (approximately 220 000 km), failures involving consequential explosion have an incidence of 0.1 per 1000 km years, with about 35 per cent of these explosions causing severe damage and 10 per cent causing fatalities (DoE, 1977). Because of the relatively high failure rate of distribution mains from all causes and the relatively low incidence of failure caused *solely* by tunnelling movements, together with the many variables such as pipe material, age, size, soil type, external factors and so on, it has not been possible to show that tunnels in general have a *statistically* significant effect on distribution pipeline failure rates. There is, however, ample evidence to show that high stress levels can be induced and failure may occur (Howe, 1985*b*).

3.5.1 *Consultative procedures*

The primary aim of consultation between the promoter of tunnelling works and the owner of pipelines likely to be affected is to avoid possible problems by the adoption of a procedure for early consultation at the preliminary design stage of a tunnel. Since the most important variables are the pipeline material and tunnel depth-to-diameter ratio (see Table 3.13), preliminary assessment can be made from Table 3.16 based on the first stage (that is, desk study) in a properly planned site investigation (see section 2.10). The essential features of a model consultative procedure are:

> Exchange of information
> Identification of problems and their resolution
> Agreement of cost sharing
> Agreement of a disputes procedure to cover both technical and financial differences.

Where a proposed tunnel is likely to affect a pipeline, the promoter of the tunnelling works should consult the owner as to any requirements that are reasonably necessary to safeguard the pipeline. In the UK this consultation is usually a statutory requirement falling within the provisions of the Public Utilities Street Works Act, 1950. This statutory provision has some of the essential features of a model consultative procedure and is outlined first since any other agreements are supplementary rather than in place of the statutory provision. The essential rules of the relevant section of this statute are:

(1) A prescribed procedure, to be followed by public utilities executing works in streets, for notifying other utilities, whose apparatus is likely to be

affected, of their intention to execute the works (minimum three days' clear notice).

(2) A duty on the proposer of the works to comply with certain requirements including requirements of the pipeline owner that are reasonably necessary for the protection of the pipeline (and is reasonable having regard to the time when the requirement is made).

(3) Without prejudice to the above, a duty to provide proper temporary support and a permanent foundation to the pipeline affected by the works.

(4) Provision for compensation equal to the expense incurred by the pipeline owner in making good damage caused to the pipeline by the construction of the tunnel works.

Notable features *absent* from the statutory rules are:

(5) Since the tunnel promoter has a duty to provide at all times proper 'support' to a pipeline, a pipeline owner cannot insist on replacement in advance of the works since if the third item above is complied with then this cannot be 'reasonably necessary' under item (2), above.

(6) A tunnel promoter cannot insist on replacement or diversion of a pipeline in advance of the works; he must if necessary design and so construct the works so as to comply with item (3).

(7) There is no provision for arbitration (this being specifically excluded from item (4)), since it is clear from the period of notice required that this is not consultation in the ordinary sense but notice of what is going to be done; this does not, however, exclude arbitration by agreement between the parties on any issue.

In addition to the provisions of statute, recourse can be made to common law. In general this imposes a duty (unless modified by statute) on a tunnel promoter to exercise all reasonable diligence and care to prevent the construction works from causing damage to others. In this case a pipeline owner may seek a remedy by means of civil action for an injunction and/or damages. A claim might succeed if it could be shown that the probability of damage was high and it would be in the public interest to deal with the matter in advance by precautionary expenditure rather than await the occurrence of actual damage and its possible consequences leading to injury to persons or damage to property. Whether a claim would succeed would depend entirely on the evidence. To avoid the expense and delay of litigation, it has been necessary in the UK to *supplement* the framework of statute and common law governing the activities of the water industry (which is responsible for the majority of tunnelling) and the gas industry (which is responsible for the majority of pipelines where failure is an immediate hazard).

For excavation involving trenches in soil greater than 1.5 m deep constructed in proximity to grey iron gas mains, a nationally agreed model consultative procedure (MCP) is available (WAA/BGC, 1984). Some of the background work to the procedure is given by Rumsey and Dorling (1985) and by Howe (1985*a*). A similar nationally agreed procedure does not as yet exist for tunnelling works, and consultation and negotiation takes place on an *ad hoc* local basis. Problems arise when too little information is supplied by the tunnel designer for a reasonable assessment of the possible effect of the tunnel. In other cases the route may be virtually decided before contact is made with affected parties. Since pipeline owners are responsible for the safe operation of the system, they are bound to assess the possible effects of excavation work in the vicinity of the plant. In the absence of the detailed information and experience that is available through the tunnel designer, this assessment may have to be very conservative. The arrangements for exchange of information, cost-sharing and disputes procedure are adequately covered by the MCP for trench works. Where the inital exchange of information shows that a detailed assessment *is* required then the ground movement estimate has to be provided by the tunnel designer. The 'risk assessment' is the responsibility of the pipeline operator. In no case should the tunnel designer try to restrain the pipeline operator from carrying out renewal or diversion of pipelines to the extent that he sees as necessary. Disputes should only be about the extent of costs that are reasonably necessary. Whether in fact a pipeline needs renewal to continue to operate safely does not depend on who is paying for any necessary renewal. The tunnel designer must be able to justify the estimated ground movement. Thorough site investigation, coupled with extensive case history data, is required for reasonable estimates to be made. A continuing involvement of the designer through the construction stage is essential if theory and practice are to come closer together.

The following summarizes technical details of a tunnel/pipeline assessment for grey iron gas distribution mains.

(1) The tunnel designer supplies scheme drawings indicating the tunnel dimensions, line and level, a copy of the site investigation report and information on the proposed construction techniques to the pipeline owner.

(2) The form and magnitude of the *probable* ground movement is estimated by the tunnel designer and is usually given to the pipeline owner as ground-loss factor, V_t or $V_s\%$ (see section 2.4.1) and settlement trough width parameter, i (see section 2.4.2). This information is based on case history data for the particular tunnelling conditions envisaged (see Chapter 2). This does not preclude the tunnel designer from including the possibility of forms of ground movement involving local variations from the recognized overall form where this may have a more severe effect.

(3) The assumed geometry, material properties and ground movement data are agreed prior to risk assessment.

(4) The pipeline owner estimates the increase in tensile strain in the pipework due to the proposed tunnelling using techniques that reasonably model the transient settlement wave, the permanent settlement trough and the influence of soil–pipe interaction.

(5) The calculated pipe strain is compared with an allowable strain increase criterion for grey iron gas mains:

GROUND MOVEMENTS AND BURIED PIPELINES 229

(a) Changes in total tensile strain of less than 100 microstrain will not usually be considered to put a main at risk
(b) Tensile strain increases of 100–200 microstrain may be considered to put a main at risk according to factors such as pipe diameter, age, location, consequences of failure, leakage history, pipework details, depth of cover, traffic and previous known disturbance
(c) Tensile strain increases exceeding 200 microstrain will be considered to indicate an unacceptable risk to gas mains.

Table 3.17 is a guidance table. Identification of a main at risk will normally result in renewal ahead of the works.

(6) In situations where pipelines remain in place, the tunnel designer will arrange for ground movement to be monitored during construction. The recorded movements will be released to the pipeline owner within two days of measurement. These may be used, at the discretion of the pipeline owner, in a reassessment of the risk to gas mains.

The allowable strain increases are in a sense arbitrary, and are related to the method of analysis used in the assessment. A case can be made for different allowable levels of strain for transverse or parallel pipelines. There should also be a minimum tunnel depth-to-diameter ratio, a minimum tunnel depth irrespective of diameter and a maximum allowable settlement irrespective of tunnel depth.

Where pipeline renewal in advance of a tunnel is required, the appropriate renewal length may be based on Figure 3.39. Since renewal is more likely to arise where the tunnel depth-to-diameter ratio is small, the tunnel diameter has been included in the renewal length. For these shallow tunnels the settlement trough width is affected by the tunnel diameter. The renewal lengths in Figure 3.39 are very similar to the MCP for trench works with trench depth replaced by tunnel depth-to-axis and trench width replaced by tunnel diameter.

For a particular pipeline material and allowable strain increase, assessment charts can be constructed on the basis of section 3.2 giving allowable surface settlement for combinations of pipe diameter and soil type above

Table 3.17 Risk assessment

Calculated maximum tensile strain (microstrain)	Diameter of grey iron distribution main	Conclusion
Less than 100	All	Significant stress increase unlikely
100–150	Less than 300 mm (12")	Significant stress increase possible
100–200	300 mm (12") or more	
150 or greater	Less than 300 mm (12")	Significant stress increase probable
200 or greater	300 mm (12") or more	

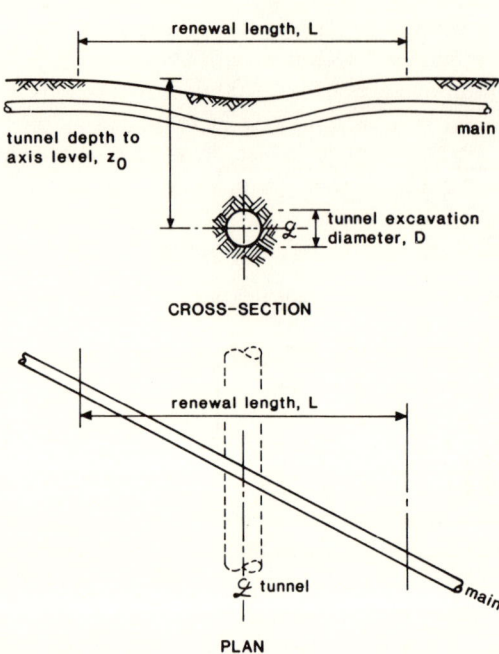

Figure 3.39 Renewal length for main crossing line of tunnel excavation. For cohesive soil above tunnel, $L = 2.0 z_0 + D$. For non-cohesive soil above tunnel, $L = 1.5 z_0 + D$. Site circumstances may require these lengths to be modified.

the tunnel. An example is shown in Figure 3.40, and for this case other variables such as pipe-joint flexibility and pipe-embedment stiffness have a small influence. Over a small range of diameters, for example 4″ to 12″ (100–300 mm), the pipe diameter does not have a major influence, and for simplification the charts for different diameters can be merged. A chart plotting surface settlement against tunnel depth for different combinations of ground loss factor V_t or $V_s \%$, and tunnel depth-to-diameter ratios can be overlain on the plot for allowable movement. It is found approximately that for each combination of tunnel depth-to-diameter ratio and pipeline orientation (that is, the geometry) there is one value of allowable ground loss for each type of settlement trough (that is, the ground conditions). So the major features influencing the performance of a pipeline are tunnel depth-to-diameter ratio (geometry), ground-loss factor and relationship between tunnel depth and settlement trough width (the ground conditions), together with the tolerance of the pipeline to strain increase. Again, this is similar to the MCP for trench works where the ground conditions and geometry are the factors that determine the 'risk' in the assessment procedure. Although there are other factors for trenches (Yeates, 1985b) these are generally less important than the two main variables.

The following example illustrates some of the difficulties that may arise

GROUND MOVEMENTS AND BURIED PIPELINES 231

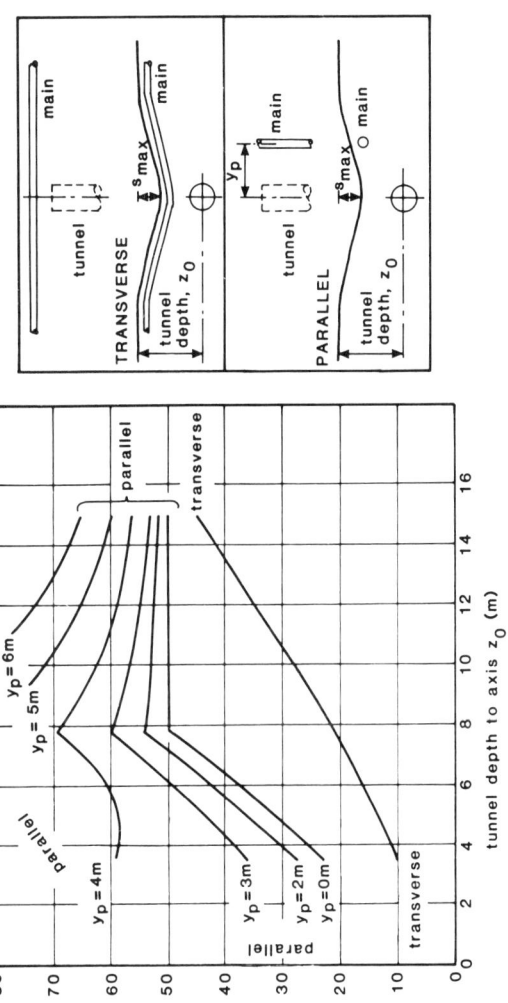

Procedure:

(1) Identify relevant line on chart according to position of main (see above)
(2) Using the tunnel depth to axis (z_0) provided by the tunnel proposer, read off relevant value for maximum surface settlement s_{max}
(3) Compare the estimated maximum surface settlement (w_{max}) provided by the tunnel proposer with value of s_{max} from the chart
  (a) If $w_{max} \geqslant s_{max}$, main is in Zone A
  (b) If $s_{max} > w_{max} \geqslant 0.5\, s_{max}$, main is in Zone B
  (c) If $w_{max} < 0.5\, s_{max}$, main is in Zone C.
(4) The worked examples illustrate the procedure.

Figure 3.40 Typical pipeline assessment chart.

between tunnel promoters and pipeline owners. It is not in any sense typical, but the unusual features are cautionary.

Example 3.22 Delay to tunnelling and cost associated with pipeline renewal in advance of construction

Details of proposed tunnel

Sewer tunnel length 420 m, excavated diameter 2.44 m. 75 per cent of length in stiff to very stiff boulder clay, $z_0 = 10$–14 m. 15 per cent of length in clay with water-bearing sand and gravel stratum in face, water head approximately 10 m, $z_0 = 13$–14 m. 10 per cent of length in old man-filled valley, generally cohesive, $z_0 = 9$ m minimum. Tunnel length (including break-out from existing shaft) in water-bearing sand and gravel is to be driven in compressed air with provision for pressure up to 1.25 bars.

Site investigation

The site investigation was comprehensive, involving desk study and two stages of site boring and testing, and is reported by Norgrove *et al.* (1979). Figures 2.35 to 2.37 are taken from this scheme and Figure 4.34 is an extract from the site investigation report.

Estimated cost

The estimated cost of constructing the tunnel was £0.6 M. The tunnel passes beneath congested narrow streets in a city, the streets being flanked on both sides by old tall buildings most of which have cellars extending under the footway (see Figures 2.35 and 2.36). The estimated cost for superficial building damage was £10 000.

Extent of utilities

Numerous utility pipes and cables run above the route of the tunnel. In particular, directly above the tunnel and running parallel are 24" (600 mm) and 12" (300 mm) grey iron gas distribution mains together with an 18" (450 mm) main offset by about 5 m. The sequence of events from 16 months before the intended start of construction was as follows.

Date	Progress in construction of scheme
Dec. 1978	Preliminary notice on utilities with plan and longitudinal section of proposed tunnel. Construction programmed to commence in April 1980.

Intervening period of discussion.

Sept. 1979	Assessment of possible tunnelling movements given to Gas Corporation (see Norgrove *et al.*, 1979). Ground-movement measurement

GROUND MOVEMENTS AND BURIED PIPELINES 233

instrumentation installed (including inclinometer access tubes and settlement rings), settlement studs fixed in road pavement, tilt meters fixed in building basement.

Intervening period for Gas Corporation assessment of effect of proposed works.

Jan. 1980   Decision by Gas Corporation that tunnel as proposed required the renewal of 12″, 18″ and 24″ gas mains along the route, at a cost of £1.0 M and requiring one year to complete following receipt of an order to carry out the works.

Intervening period of discussion, threat of civil proceedings for injunction to prevent work, re-assessment of probable movement and possible effects on gas mains.

Oct. 1981   Agreement reached on alternative route for part of length at additional construction cost of £0.1M, risk of moderate damage to buildings (including cathedral shown in Figure 4.34), plus £0.25M as share of £0.35 M gas main renewal cost over 70 m length of tunnel in filled valley. Construction now programmed to start in September 1982 to allow time for pipeline renewals.

Dec. 1981   Serious gas explosion resulting from fracture of 24″ main on originally proposed tunnel route. 24″ main to be renewed over whole length (except 70 m already ordered by Water Authority).

Apr. 1982   Original tunnel route adopted except for small change in line to put tunnel directly beneath new 24″ gas main.

July 1982   Gas leak due to fracture of 18″ main on route of proposed tunnel. 18″ main to be renewed over whole length (except 70 m already ordered by Water Authority).

Sept. 1982 Contract for construction of tunnel commenced and tunnel successfully constructed, December 1982 to April 1983. Ground surface movement was barely detectable. Pipe strains were negligible and have been reported by Hurrell (1983). Building damage confined to basement floor in a building due to compressed-air blow-out. Tunnelling difficulties confined to high compressed-air loss over a short length due, it is postulated, to the large number of boreholes and inclinometer tubes in the soil. Some were too close to the tunnel due to amendments to the original tunnel route for which the instrumentation programme was designed.

Two years' delay in the construction of this sewer tunnel was caused by equating the estimated ground 'strain' with the structural strain. Reference to Figure 3.20 shows that tensile ground 'strain', that is, differential ground movement, may cause either compressive or tensile structural strain. In general, there is no simple correspondence. If ground strain and structural strain are equal then no load can be induced in the structure. For the length of tunnel (70 m) where pipeline renewal was carried out because of the possible effect of the tunnel movements, the cost per metre was four times the tunnel construction cost over this length. It was fortunate for the tunnel promoter that the fractures in the 24″ and 18″ mains occurred before the tunnelling commenced. If these fractures had occurred during or after completion of the tunnel, even the most comprehensive set of measurements showing that the tunnel had apparently insignificant effect would have probably made no difference to the outcome of an enquiry into the likely cause. Fractures in large-diameter gas pipelines are very rare. The occurrence of two fractures in one street in the space of seven months is extremely rare, but is not bound to be associated with adjacent tunnelling. Even though measurement of negligible ground settlement would not have disproved cause and effect, it would be successful in showing that the tunnel promoter was acting diligently and with due regard to preventing damage to the property of others as far as possible.

3.5.2 *Measures to protect old pipe distribution systems*

When tunnelling beneath established pipeline networks, concern is likely to focus on grey iron pipes—particularly small-diameter pipes which are known to be sensitive to disturbance. High stress levels *can* be induced in less brittle materials, and it could be that this might eventually contribute significantly to failure. Reference to Table 3.14 shows that steel pipelines will reach the yield stress almost as easily as grey iron (at least at the minimum wall thickness given in Table 3.8). It is interesting to note that the performance of small-diameter steel distribution pipework, when subjected to earthquake movements, is no better than that of grey iron (see, for example, Mikaoka, 1978).

There are many choices available to the tunnel designer in reducing the effect of tunnelling on pipelines. Usually increasing the tunnel depth is beneficial, particularly in the case of a proposal involving shallow tunnelling. Occasionally amendments to the tunnel line may theoretically reduce the influence of the works, but it should not be expected that small amendments will make a significant difference. To be avoided are those tunnel/pipeline layouts that cause high stresses in networks (section 3.4.6). Controlling the ground loss at the tunnel is clearly beneficial. Here the emphasis should in the first place be on thorough preconstruction site investigation, a phased investigation providing better-quality data for the needs of the project. Only with adequate site investigation can there be confidence that the most appropriate tunnelling method is used. Where costly service diversions would otherwise be required, then more expensive tunnelling techniques may be a viable alternative. Ground-treatment processes to modify the geotechnical properties of the soil may also be an attractive alternative, particularly since only local treatment in the vicinity of a vulnerable pipeline may be required. The choices available in tunnelling methods and ground treatment are discussed in section 2.2.

Ground-surface movements can be measured quite easily, and besides providing a database for future tunnelling operations, measurements can provide advance warning of possible problems on critical sections. A large number of simple measurements of surface settlement will be more useful than a few very precise measurements. Measurements that show the actual effect of the tunnel may be very helpful in mitigating claims for damage. Measurements demonstrate that the tunnel promoter is interested and concerned about the effect of the tunnel works on adjacent pipelines, which can help in allaying fears of damage. Where the tunnel designer has had to make a pre-estimate of the form and magnitude of ground movement it is essential that there is feedback on the actual movements. Only in this way can experience and confidence be gained in estimating ground movements.

Pipe-joint leakage can be monitored before and during tunnelling and repairs made as necessary. Where there is a risk of leakage at joints in gas pipelines then a pre-tunnelling survey is essential. As with ground surface settlement measurements, the results of survey work should be freely available both to the pipeline owner and to the tunnel promoter. Pipe strains can be measured, but the technique requires considerable experience to obtain reliable results. The methods are outlined by Osborne (1985).

Generally, it is necessary to make an assessment of the effect of the tunnelling works *before* the work has started so that relaying or polyethylene sliplining of critical pipelines can be carried out before tunnelling begins. Renovation of old pipelines by thermoplastic pipe insertion has been extensively used in the gas industry and 'size-for-size' replacement of cast iron with polyethylene is now available for 75 mm and 100 mm diameter pipes. The method involves a machine being drawn through the iron main, expanding and fragmenting the pipe as it progresses. A PVC replacement sleeve is simultaneously drawn through. It supports the ground and protects the new polyethylene main. The polyethylene main is installed inside the sleeve by conventional means. A description of 'size-for-size' replacement with polyethylene is given by Freeman (1983). Conventional polyethylene sliplining is described by Reed (1978) and by Whipp and Glennie (1982). It is interesting to note that percussive moling can cause bending in adjacent pipelines of similar magnitude to tunnelling-induced movement (Dorling, 1984)!

3.5.3 *Design of new pipelines to accommodate ground movement*

When subjected to the moderate ground movements associated with tunnelling, all modern pipeline materials perform well when compared with grey iron. Where significant movement has to be accommodated, pipeline flexibility reduces the imposed load. Flexibility is provided either by the material itself or by joints within the pipeline. Up to the yield stress ferrous pipelines are of similar longitudinal flexibility. The advantage of ductile iron or steel over grey iron is the ability to yield without fracture. In the case of small-diameter pipelines, very close centres are required for flexible joints to significantly increase the overall longitudinal bending flexibility. In general, the most effective way of achieving pipeline flexibility is by using a relatively flexible material, that is plastic. It is emphasized that flexible pipe materials require at least as good bedding and backfilling as was used when the grey iron distribution system was installed. In particular, an uneven machine-dug trench bottom is not adequate if the pipeline has to carry significant external loads. Flexible pipes also rely on adequate compaction of backfill around the pipes. Where ground movement is transverse to a pipeline, ductile iron and high- or medium-density polyethylene at the diameter-to-wall thickness commonly used are known to perform well. Steel pipelines, if required to accommodate ground movement, can be sized accordingly. Flexible joints are beneficial, but if high security against joint pull-out is required then welded or anchored flexible joints should be used. Where there is a large change in pipeline flexural stiffness, a pinned joint should be used if possible since this reduces bending strains by about 70 per cent. This applies where a pipeline is connected to a structure and differential movement is expected, or where a small branch is attached to a relatively stiff main pipeline. Apart from the preceding case, flexible joints on small diameter pipelines do not have much beneficial effect at the spacings commonly used, that is, 3.66 m or more.

Where ground movement is horizontal and longitudinal to a pipeline the effect is reduced if the pipeline can accommodate axial movement at the joints. Smooth-surfaced pipelines or loose polyethylene sleeving (applied as a corrosion protection measure), or pipelines installed within sleeves, promote soil–pipe slip which reduces the pulling forces associated with horizontal ground movement.

4 Structural response to tunnelling settlement

When a tunnel is driven through soil the resulting free-ground ground-loss settlement profile can be approximated by functions of the normal distribution curve as described in Chapter 2. As the tunnel is advanced, a trough of settlement with a moving front develops (see Figure 2.2). Behind this front remains the settlement trough, with a constant cross-section of surface settlement defined, from equation (2.22), as

$$w = \frac{V_s}{\sqrt{2\pi} \, i} \exp(-y^2/2i^2). \tag{4.1}$$

Accompanying transverse displacements are defined, at the surface, from equation (2.23), by

$$v = -\frac{nyw}{z_0} \tag{4.2}$$

where for practical purposes n may be taken as 1.0.

In addition, ground surface slope is defined as

$$\frac{\partial w}{\partial y} = -\frac{yV_s}{\sqrt{2\pi} \, i^3} \exp(-y^2/2i^2). \tag{4.3}$$

When a structure lies within the zone of influence of such a trough, it will be subject to deformations; but conversely, the presence of the structure will modify the trough profile. No attempt was made to accommodate this effect in the preliminary examples in Chapter 2.

Solutions to this soil–structure interaction are now described at three levels of analysis, all of which still involve common simplifying assumptions. The first assumption is that the structure be considered essentially two-dimensional with respect to the settlement profile, so that the temporary disturbance of the advancing trough front is ignored. The second major assumption is that the condition causing most severe stress in the soil–structure system can be modelled by a short-term elastic analysis. This assumption is reasonable for overconsolidated clays, but incorrect for normally consolidated cohesive soils. An elastic model for sands is also reasonable, although it would be preferable to use a Gibson type of stiffness distribution.

In the case of brickwork, the elastic interactive analysis should be undertaken to assess the probability of cracking, and if this probability is high then an estimate of crack width should be made by manual superposition of rigid-body deformations of a 'hinged' structure on to a free-ground displacement profile.

Similarly, a steel frame should be analysed, interactively, to calculate stresses due to ground deformations and dead loads; if yield is induced in the frame, then plastic hinges can be expected to rotate substantially in order to allow the frame to follow the free-ground displacement profile. Reinforced concrete structures, however, can be expected to maintain integrity, despite minor cracking, except when subjected to severe ground movements.

4.1 Winkler ground model: a manual technique

In the simplest form of analysis, the Winkler ground model, in which the soil is considered as a series of discrete linear-elastic vertical springs, is used (see also Chapter 3). The structure is assumed to act as a simple beam in bending, which limits the application of this analysis to plain walls, rafts, and shallow service pipes.

Now, if the free ground displacement is defined by the vector **w**, then the loading to be applied to the soil–structure system is derived as follows.

Consider the Winkler ground model to provide vertical spring stiffness, s_1 (units kN/m) at each of several discrete points (see Figure 4.1). The displacements of the ground at these points, $\mathbf{d}_g$, under a vector $\mathbf{P}_g$ of upward loads, are simply defined by the matrix equation

$$\mathbf{P}_g = \mathbf{K}_{gd}\mathbf{d}_g$$

where $\mathbf{K}_{gd}$ is a diagonal matrix

$$\begin{bmatrix} s_1 & 0 & 0 \\ 0 & s_2 & 0 \\ 0 & 0 & s_3 \\ & & \text{etc.} \end{bmatrix}.$$

Also, if a matrix equation is written to define displacements $\mathbf{d}_s$ in the beam structure at the same interface nodes, then

$$\mathbf{P}_s = \mathbf{K}_s\mathbf{d}_s$$

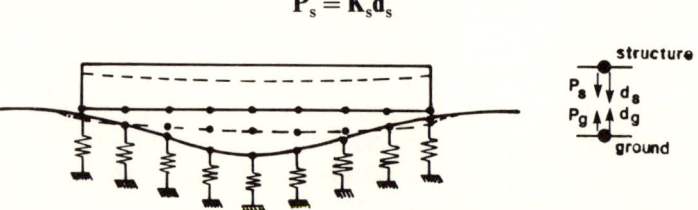

Figure 4.1 Interactive response of ground and structure to ground movements.

where $\mathbf{P}_s$ is the vector of downward forces on the structure, and $\mathbf{K}_s$ is a stiffness matrix for the beam, with respect to interface degrees of freedom only.

From equilibrium,

$$\mathbf{P}_s = \mathbf{P}_g,$$

and, for compatibility of displacement (assuming no separation),

$$\mathbf{d}_s + \mathbf{d}_g = \mathbf{w}.$$

By substitution,

$$\mathbf{K}_s \mathbf{d}_s = \mathbf{K}_{gd} \mathbf{d}_g$$
$$= \mathbf{K}_{gd}(\mathbf{w} - \mathbf{d}_s)$$

or

$$(\mathbf{K}_s + \mathbf{K}_{gd})\mathbf{d}_s = \mathbf{K}_{gd}\mathbf{w}. \tag{4.4}$$

This is to say that the soil–structure system response to the free-ground displacement, $\mathbf{w}$, is defined by the stiffness equation (4.4), which is equivalent to imposing a load $\mathbf{P}_e$, of

$$\mathbf{P}_e = \mathbf{K}_{gd}\mathbf{w}$$

on to the combined soil–structure system for calculation of $\mathbf{d}_s$, the structure displacement vector. Knowledge of $\mathbf{d}_s$ allows computation of structural stresses. The equation can be solved by numerical methods, but an alternative approximate method, suitable for manual computation, is available for deriving the sag moments induced by a tunnelling trough running below the mid-length of a structure.

In an extensive study of beams on elastic foundations, Hetenyi (1946) gives a closed-form solution for an infinite beam carrying a central point load. The expression for displacement of the beam is, in the notation of this text,

$$d_s = \frac{P\lambda}{2K}\exp(-\lambda y)(\cos \lambda y + \sin \lambda y) \tag{4.5}$$

where

$$\lambda = \sqrt[4]{\frac{K}{4EI}}$$

with K being the modulus of subgrade reaction.

By differentiating two and three times, the expressions for bending moment and shear are given, respectively, by

$$M = \frac{P}{4\lambda}\exp(-\lambda y)(\cos \lambda y - \sin \lambda y) \tag{4.6}$$

and

$$Q = -\frac{P}{2}\exp(-\lambda y)\cos \lambda y. \tag{4.7}$$

This allows the calculation of moments and shears in a beam due to the discrete forces in the equivalent load vector $\mathbf{P}_e$, by superposition of effects due to loads at several points along the infinite beam.

Example 4.1 Calculation of stresses in a long wall of uncracked reinforced concrete

Consider a wall 4 m high, 0.34 m thick, and composite with a 2.3 m wide base 0.5 m deep (see Figure 4.2). The elastic modulus for concrete, E_s, is 30×10^3 MN/m². The sandy soil beneath has an elastic modulus of $E_g = 10$ MN/m², Poisson's ratio $v_g = 0.3$ and a free-ground settlement profile defined by $w_{max} = 49$ mm, with $i = 1.02$ m. The free-ground displacement vector $\mathbf{w}$ is

49.0 mm at $y = 0$ m

30.3 mm at $y = \pm 1.0$ m

7.2 mm at $y = \pm 2.0$ m

0.65 mm at $y = \pm 3.0$ m

0.02 mm at $y = \pm 4.0$ m

The modulus of subgrade reaction, K, may be evaluated from the equation due to Vesic (1961) and quoted in Chapter 3, as

$$K = 0.65 \sqrt[12]{\frac{E_g d^4}{E_s I}} \frac{E_g}{(1 - v_g^2)}$$

The second moment of area of the wall and base, I, is calculated as 5.0 m⁴, so that

$$K = 0.65 \sqrt[12]{\frac{10 \times 2.3^4}{30000 \times 5.0}} \frac{10}{(1 - 0.3^2)}$$

$$= 4.22 \text{ MN/m}^2.$$

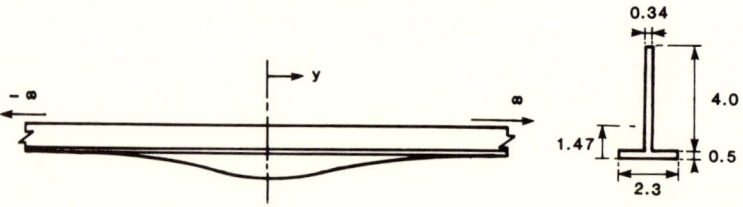

Figure 4.2 Beam on elastic foundation example.

Hence the damping factor λ is

$$\lambda = \sqrt[4]{\frac{K}{4E_sI}}$$

$$= \sqrt[4]{\frac{4.22}{4 \times 150000}}$$

$$= 0.0515 \, \text{m}^{-1}.$$

The equivalent load vector for the interface nodes of the soil–structure system is

$$10^6 \begin{bmatrix} 4.22 & & & \\ 0 & 4.22 & & \\ 0 & 0 & 4.22 & \\ & & & \text{etc} \end{bmatrix} \begin{bmatrix} 0.00002 \\ 0.00065 \\ 0.0072 \\ 0.0303 \\ 0.049 \\ — \\ — \\ 0.00002 \end{bmatrix} \text{(units Nm)}$$

$$\text{or} \begin{bmatrix} 94 \\ 2740 \\ 30000 \\ 127000 \\ 207000 \\ 127000 \\ — \\ — \\ 94 \end{bmatrix} \text{(units N).}$$

Hetenyi's expression for bending moment due to a central point load supported by an elastic foundation is equation (4.6). Hence the moment M_0 at the beam centre due to a central point load of 207 000 N is

$$M_0 = \frac{207\,000}{4 \times 0.0515} \exp(0)(\cos 0 - \sin 0)$$

$$= 1.01 \times 10^6 \, \text{Nm}.$$

By reciprocity, the moment M_1 at the centre due to a force of 127 000 N at $y = 1$ m is the same as the moment at $y = 1$ m due to a central load, that is

$$M_1 = \frac{127\,000}{4 \times 0.0515} \exp(-0.0515)(\cos 0.0515 - \sin 0.0515)$$

$$= 0.56 \times 10^6 \, \text{Nm}.$$

A similar central moment is generated by force of 127 000 N at $y = -1$ m.

Loads at $y = \pm 2$ m of 30000 N each cause a central M_2 of

$$M_2 = \frac{30000}{4 \times 0.0515} \exp(-0.103)(\cos 0.103 - \sin 0.103)$$

$$= 0.117 \times 10^6 \text{ Nm}.$$

For $y = \pm 3$ m,

$$M_3 = 0.009 \times 10^6 \text{ Nm}.$$

and loads at $y = \pm 4$ m cause negligible central moment. Hence the total central moment is 2.38×10^6 Nm, which induces bending stresses of 0.70 N/mm² tensile and 1.44 N/mm² compressive.

A calculation of central deflection $d_{s(y=0)}$ from equation (4.5) gives a value of 3.0 mm.

A modified calculation could be performed to give the maximum hog moment, at $y = \sqrt{3}i$, either for a long wall, or, with modified equations, for a half-length wall.

The results of this example calculation can be compared directly with later results from a finite-element analysis on a wall 6 m long. It is shown that the above tensile stress is a mere 3 per cent larger than the values from the finite-element analysis, while the deflections are much smaller.

The close agreement between the tensile stresses is partly coincidental, since the comparison is made between the infinitely long beam in Hetenyi's analysis and a discrete wall 6 m long in the finite-element analysis. In addition, the former incorporates neither transverse ground movement due to tunnelling nor the restraint of transverse soil stiffness.

In conclusion, then, the simple hand analysis gives a useful first estimate of the probable sensitivity of a wall or raft to tunnelling-induced ground settlements, but has some limitations which may lead to an inaccurate estimate of stresses, and will certainly lead to an underestimate of displacements.

4.2 Structure matrix stiffness plus Winkler ground model

At an improved level of modelling, the method developed in Couldery (1983) and reported in Attewell and Yeates (1984) uses a matrix stiffness formulation for a structural frame which is combined with a foundation stiffness model evaluated from a Winkler model. Consider the frame in Figure 4.3.

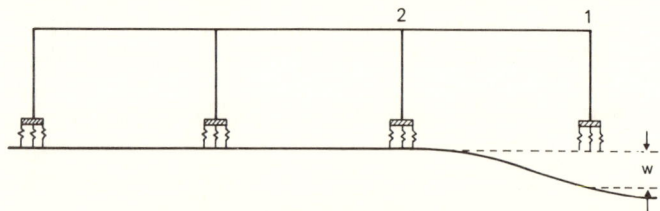

Figure 4.3 Settlement beneath a single footing of a multi-bay structure.

Under the influence of tunnelling-induced ground settlements each footing may settle differentially, thereby inducing bending deformations in the structural members. Footing settlement d_s (usually less than w, the free-ground displacement) creates a stress in the soil of $k(w - d_s)$ where k is the coefficient of subgrade reaction. The force P_s upon the structure necessary to deform the structure (primarily by bending of member 1–2, of length L) is very approximately

$$P_s = 12 \frac{EI}{L^3} d_s$$

or $P_s = k_s d_s$, where k_s is the structural stiffness.

For equilibrium, this equals the force in the subgrade, of $k(w - d_s)a$, where a is the area of the footing.

Thus, $(k_s + k)d_s = kaw$, where kaw is an equivalent force to be applied to the footing for calculation of footing displacement d_s.

In the general case of differential settlements at all footings, a load of $ka_i w_i$ is applied to each footing, where w_i is the predicted unloaded ground settlement at footing i having soil base contact a_i. This is analogous to the equivalent force vector $\mathbf{K}_{gd}$, of equation (4.4).

Analysis of a multibay multistorey framed structure involves the definition of structural, foundation (or interface), and ground nodes (Figure 4.4), and solution of a matrix stiffness equation

$$\mathbf{Kd} = \mathbf{P}.$$

In expanded form, and with subscript s referring to structure, f to foundation or interface, and g to ground nodes, the equation can be written as

$$\begin{bmatrix} \mathbf{K}_{ss} & \mathbf{K}_{sf} & 0 \\ \mathbf{K}_{fs} & \mathbf{K}_{ff} + \mathbf{N}_{ff} & \mathbf{N}_{fg} \\ 0 & \mathbf{N}_{gf} & \mathbf{N}_{gg} \end{bmatrix} \begin{bmatrix} \mathbf{d}_s \\ \mathbf{d}_f \\ \mathbf{d}_g \end{bmatrix} = \begin{bmatrix} \mathbf{P}_s \\ \mathbf{P}_f \\ \mathbf{P}_g \end{bmatrix} \quad (4.8)$$

where d, as before, represents displacement.

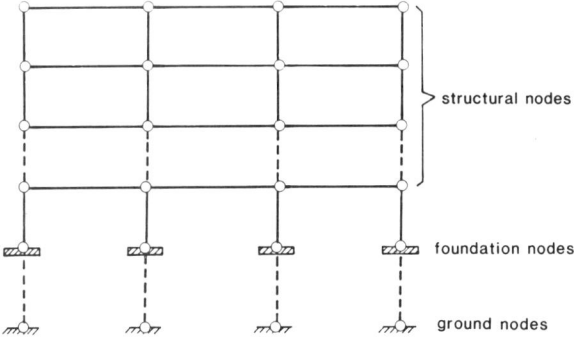

Figure 4.4 Designation of structural foundation and ground nodes.

The submatrices **K** are simply the structural stiffnesses of the structural components, and the submatrices **N** are stiffnesses of the ground elements (see Figure 4.4). The components of **K** are evaluated by standard matrix structural analysis, for example, Todd (1974), and the terms of $\mathbf{N}_{ff}$ are evaluated as $N_{ii} = ka_{ii}$ for vertical displacement terms, as $N_{ii} = \int ky^2 da$, for moment/rotation and $N_{ij} = 0$ for $i \neq j$.

Two possible solution techniques are available. The full stiffness equation (4.8) could be evaluated and solved for some imposed ground deformations, $\mathbf{d}_g$, caused by tunnelling. The preferred approach, however, is to evaluate an equivalent force vector, $\mathbf{P}_f$, to be applied to the foundation or interface nodes in the deflated matrix stiffness equation

$$\begin{bmatrix} \mathbf{K}_{ss} & \mathbf{K}_{sf} \\ \mathbf{K}_{fs} & \mathbf{K}_{ff} + \mathbf{N}_{ff} \end{bmatrix} = \begin{bmatrix} \mathbf{d}_s \\ \mathbf{d}_f \end{bmatrix} = \begin{bmatrix} \mathbf{P}_s \\ \mathbf{P}_f \end{bmatrix}. \tag{4.9}$$

This involves smaller matrices and fewer equations, and indeed offers a fully interactive analysis for solution of $\mathbf{d}_f$, the vector of foundation displacements, and $\mathbf{d}_s$. From this solution, structure member forces are evaluated.

One of the difficulties in the use of this method is in the choice of an appropriate effective area of soil associated with each footing which is to be used when evaluating the foundation–ground stiffness matrix. Additional shortcomings of the method are the difficulty of applying the analysis to walls and other continuous structural forms, and the lack of consideration of transverse displacements which can be shown to have a significant effect upon the displacements and stresses induced in a structure.

4.3 Finite-element analysis of ground and structure

In an attempt to produce further refinement in the prediction of response of a soil–structure system to tunnelling-induced ground movements it is proposed that a modified form of equation (4.4) be used:

$$(\mathbf{K}_s + \mathbf{K}_g)\mathbf{d}_s = \mathbf{K}_g \mathbf{w} \tag{4.10}$$

where $\mathbf{K}_g$ is now a fully-populated matrix, and the matrices and vectors are written with respect to both vertical and transverse displacements of the interface nodes. In the case of frames, nodal rotations must also be considered.

The equation can be formulated initially in terms of a soil substructure, $\mathbf{K}_g$, and a structure substructure, $\mathbf{K}_s$, defined with respect to interface nodal degrees of freedom only. These matrices are entirely different from the total matrices for the structure or the ground, and must be derived in a rather special way. A finite-element model of the ground alone is constructed, and then a unit displacement is prescribed at the first interface node while the remaining interface nodes are fully restrained. The calculated reactions at each of the interface nodes give one column of the soil substructure matrix, $\mathbf{K}_g$. The finite-element model is then 'loaded' by unit prescribed displacement at a second interface node, and so on to produce the full $\mathbf{K}_g$. A similar approach may be used to derive $\mathbf{K}_s$.

One possible method of solution of equation (4.10) would be a direct simultaneous equation approach, resulting in computed displacements of the interface nodes. The displacement vector could then be used as data for a numerical stress analysis of the structure. In many situations, however, this solution is inappropriate because of numerical difficulty. For example, in a typical situation of a wall 2 m high on a medium stiff clay, the terms in $\mathbf{K}_s$, the structure substructure matrix, are some thousand times greater than the terms in $\mathbf{K}_g$, the ground substructure matrix. $\mathbf{K}_s$ is, of course, singular since the only boundary restraints to the structure are through the soil. Hence, in equation (4.10) the matrix $(\mathbf{K}_g + \mathbf{K}_s)$ is ill-conditioned, and solution becomes unreliable (or unsuccessful with sophisticated self-checking routines).

A second approach, which is less sensitive to numerical problems, is to evaluate the matrix $\mathbf{K}_g$ and hence the vector $(\mathbf{K}_g\mathbf{w})$, still in terms of interface nodal degrees of freedom only, which is an equivalent load vector to be applied to the interface node degrees of freedom only. This load vector can then be applied to a full finite-element model of ground and structure, thus reducing the ill-conditioning problems of the discretized solution previously outlined. An additional advantage of the second approach is that $\mathbf{K}_s$, the structure substructure matrix, need not be evaluated. On the debit side, however, is the need to solve the whole ground–structure system, which involves a larger number of equations, although this problem is now less severe thanks to the power of modern computers.

A short computer program can be written to evaluate sets of $\mathbf{w}$ vectors for several trial tunnel lines, and to evaluate load vectors $\mathbf{K}_g\mathbf{w}$, formatted for direct input to a PAFEC[†] program.

The above method is found to perform well in the analysis of frames, but is slightly less satisfactory in the analysis of walls, where interface nodes are concentrated. A third approach is available for this situation, derived by consideration of equation (4.10). Noting that the right-hand side product $\mathbf{K}_g\mathbf{w}$ is unrelated to structure stiffness, we can write the equation for the case of a 'structure' of zero stiffness as

$$\mathbf{K}_g\mathbf{d}_s = \mathbf{K}_g\mathbf{w} = \mathbf{P}_e.$$

Now, in treating $\mathbf{K}_g\mathbf{w}$ as an equivalent interface node force vector, the desired free-ground displacement vector $\mathbf{w}$ can be imposed as prescribed displacements in an analysis of the ground alone, and the computed reactions from this analysis give the terms in $\mathbf{P}_e$, the equivalent force vector. The same $\mathbf{P}_e$ vector can then be applied as loading to a ground–structure analysis.

The computational effort for this third approach is similar to that in the

---

[†] PAFEC is a comprehensive package of finite-element analyses. For further details contact PAFEC Ltd., Strelley Hall, Main Street, Strelley, Nottingham, UK

previous technique, although it becomes more advantageous as the number of degrees of freedom at the interface nodes rises.

The choice of finite-element type for ground modelling is dictated primarily by economics. A three-dimensional rectangular prism element is desirable where possible, in order to provide an accurate assessment of the $\mathbf{K}_g \mathbf{w}$ vector, and also for use together with structural elements to give a representative ground-structure finite element model. This can be achieved satisfactorily where only a small number of interface nodes is required. An element mesh, $14 \times 6 \times 4$, is a manageable proposition. However, for more complex structures a three-dimensional solution becomes so large as to be unrealistic, and an alternative two-dimensional plane-strain or plane-stress approach must be used.

When two-dimensional analysis is used to model the ground below a structure, a situation which is essentially three-dimensional, then a rational evaluation of appropriate thickness of the ground 'slice' must be made. The evaluation must be a balance between correct diagonal stiffness terms (similar to Winkler stiffnesses), and correct off-diagonal terms giving cross-coupling between vertical and horizontal (and rotational) degrees of freedom. Such evaluations for granular and cohesive soils are tabulated in Table 4.1.

These values are proposed as reasonable for the examples undertaken here, but should not be used in general practice without confirmation.

Table 4.1 Ratios of two-dimensional soil width/structural foundation width

	Plane strain	Plane stress
Granular soils	1.0	1.1
Cohesive soils	1.0	3.0

The method outlined above is now used to analyse a variety of wall and frame structures, and the results are presented in the hope that engineers may find them of assistance when assessing the likely sensitivity of structures to tunnelling-induced ground movements. Ideally, a structure subjected to tunnelling-induced moments should be analysed, with soil–structure interaction, in order to include the specific features of the structure, of the ground and of the tunnel. However, it may be of some help in a preliminary assessment to use the results of the examples presented in sections 4.5, 4.6 and 4.8, but modified to suit particular site conditions as discussed in section 4.9.

In the majority of the examples the tunnel diameter is set at 2 m, and the tunnel axis is taken at 4 m, 5 m, 6 m, and 7 m depths, in order to encompass typical tunnelling situations. The surface settlement volume V_s is taken, typically and non-conservatively, as 1.5 per cent of the tunnel cross-section in cohesive soils, and 4 per cent in granular soils. The surface trough width parameter i is chosen, following the re-appraisals of O'Reilly and New (1982), as $(0.43 z_0 + 1.1)$ m for cohesive soils and $(0.28 z_0 - 0.1)$ m for granular soils

Table 4.2 Elastic properties of materials in the present examples

	$E(MN/m^2)$	v
Sands	10	0.3
Clays	6	0.49
Concrete	30×10^3	0.2
Brickwork	10×10^3	0.2
Steel	2.1×10^5	0.3

(see equations (2.31) and (2.32)). Assumed values for the elastic properties of the various materials are shown in Table 4.2.

For information on a more realistic assessment of soil mechanical properties reference should be made to specialist literature such as Scott *et al.* (1982). Consideration should be given to the variability of soil *in-situ* stiffness, the possible difference between this stiffness and that measured in laboratory samples and also the Gibson (1967) variation of stiffness with depth in granular soils.

The example structural components to be analysed are brickwork walls, both plain and with openings, on concrete foundations, on clay soil, or on sand, concrete cantilever retaining walls on sand, single and two-storey four-bay steel frames on clay and on sand, and a single example of a brick-infilled steel frame which was a case study of a building damaged by tunnelling.

In all the following examples the PAFEC finite-element package was used to evaluate $K_g w$ and to generate the soil–structure model.

In all the examples, two alternative tunnel lines are used to investigate structural strains. These are designated Case A and Case B, as in Figure 4.5.

Before the examples are presented, however, a brief assessment is made of the stresses likely to cause damage.

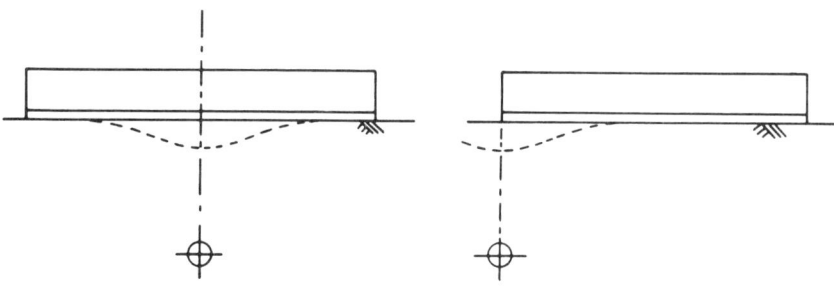

Case A: Tunnel below centre of structure   Case B: Tunnel below one end of structure

Figure 4.5 Tunnel lines: Case A and Case B.

4.4 A stress-based criterion for onset of damage

The criterion for onset of damage used in the examples which follow is that of maximum tensile elastic stress (which is directly convertible to strain), and also maximum elastic shear stress in the case of brickwork. This approach differs from the damage criterion of differential displacement divided by building length used by some workers, for the fundamental reason that this latter proposal, based upon surveys of damaged buildings, relates to the deflected form of the building in which cracks of various widths exist at the time of measurement. The elastic analysis outlined in section 4.3, however, models the uncracked structure, and conditions are evaluated with respect to the likelihood of the start of a crack; the deflection curvature of the example uncracked walls is extremely small, but as soon as a crack starts and runs through a wall then that wall may degenerate effectively into two separate components connected at one hinge point only, with a consequent substantial change in curvature (see Figure 4.16 and section 4.7). A second major difference is that tunnelling is a very rapid process, with the result that creep of structural members is less able to accommodate imposed deformations; in consequence any damage criteria should be made more severe than for more normal constructions which impose slowly increased loadings from a 'growing' structure.

A criterion for the onset of cracking in brickwork must be related to the age and quality of the building to be assessed. Old brickwork is likely to contain cracks, but the mortar is softer (either through age, or being lime-based) and is more tolerant of movement. For young brickwork in sound condition any of several criteria might be applied. British Standard Code of Practice CP 111: 1970 recommends an allowable bending tensile stress resisted by mortar adhesion only of 0.07 N/mm^2 for a 1:1:6 cement:lime:sand mortar, and an allowable in-plane shear stress of 0.1 N/mm^2 for 1:1:6 mortar increasing to 0.14 N/mm^2 for a 1:0.25:3 mix. British Standard BS 5280:1978 gives characteristic flexural strengths ranging from 0.7 N/mm^2 to 0.25 N/mm^2 depending upon mortar type and clay brick absorption; whilst characteristic shear strength is given as $(0.35 + 0.6 g_A)$ or $(0.15 + 0.6 g_A) \text{ N/mm}^2$, dependent upon mortar type, where g_A is imposed direct stress. Tests by Abu-el-Magd and MacLeod (1980) showed in-plane cracking at tensile stresses of between 0.8 to 1.7 N/mm^2 (a function of brick tensile strength and of shear bond failure). Samarasinghe *et al.* (1981) demonstrated shear strengths of about 0.7 N/mm^2. Other proposals, based upon strain, range from 500×10^{-6} to 5000×10^{-6} (for example Polshin and Tokar, 1957; Burland and Wroth with discussions, 1974; and Alexander and Lawson, 1981), which might suggest, for a linear stress–strain relationship, a stress range of 2.5 to 50 N/mm^2! This latter figure is clearly unrealistic, and might have arisen due to definition of strain by crack widening divided by a short gauge length. In summary, the crack criteria herein recommended for sound brickwork are the stresses from British Standard BS 5280 for characteristic values, and values for mean stresses should be derived by reference to the work of Abu-el-Magd and MacLeod (1980). If analysis shows that cracking is likely, then crack widths can be estimated from rigid body deformations, and severity of damage can be assessed in the accepted way by reference to crack widths and to Δ/L ratios, as in section 4.13 to which reference should be made.

Unreinforced concrete structures are rare, although unreinforced concrete is common in some strip foundations, concrete roads, and domestic floors. A crack, once initiated, will run through such components, and tensile strength will be lost entirely. Reinforced concrete, used extensively in frames, slabs and walls, is often cracked under normal working loads (or from thermal contraction or differential shrinkage), with reinforcement designed to carry tensile forces. The capacity to tolerate ground movements is rather different. If the member cross-section is heavily reinforced, then the cracked section may be sufficiently stiff and strong to carry the additional load effects without reinforcement yield, so that deflections and crack widths will be well controlled. If the cross-section is uncracked but lightly reinforced then the tensile strength of the concrete may be adequate to carry the additional loads without cracking. However, if the lightly reinforced section is precracked, or if it cracks because of ground movement, then a crack may be widened severely with yielding of the reinforcement, a highly undesirable situation of effective failure of the section. Concrete walls generally fall into the second category of lightly reinforced members, while long-span beams are typical of heavily reinforced members with inherent ductility. Normal cracking without reinforcement yield, that is crack widths of less than 0.5 mm, is now thought to have little effect upon durability with respect to progress of carbonation or to corrosion of reinforcement through depassivity (Beeby, 1983); such cracking is therefore tolerable except in the case of water-retaining structures where a crack width limit of 0.2 mm is required (British Standard BS 5337: 1976). Onset of cracking may be assessed with respect to a mean tensile strength of concrete which is often taken as one-tenth of the compressive strength of a cube, f_{cu} (see, for example, Neville, 1981), although a value proportional to the square root of compressive stress is also used (see, for example, British Standard Code of Practice CP 110: 1972). CP 110 limits shear in a lightly-reinforced member to 0.35 N/mm^2, a characteristic value. Little information is available on mean shear strength of concrete, so a criterion for tensile stress is preferable. The recommended approach, therefore, is to conduct an elastic interactive analysis to estimate maximum tensile stress, and from this to deduce the probability of cracking. If the section is lightly reinforced, then no cracking is acceptable because yield of reinforcement would result. In the case of heavily reinforced sections, an additional moment should be assessed, and added to the working load moments, for a check on serviceability. If the section is cracked prior to tunnelling, then an uncracked analysis could be used to estimate a moment due to ground movements which must be added to working moments for testing the serviceability of the cracked section.

Prestressed concrete members should be assessed for their moment capacity for working loads plus ground movements. Normally, no tensile stress is permitted.

Yield of steel in a steel frame can be predicted by reference to the British Standards BS 4360: 1979 and BS 449: 1969 or its successor BS 5950.[†] If a plastic hinge forms, due to ground movement plus imposed load, then large deformations will result, in a broadly similar way to those of cracked brick walls.

The following Table 4.3 of critical mean stresses is tentatively proposed for assessment of the onset of cracking or yield, with characteristic values in parentheses.

Table 4.3 Stresses likely to cause cracking in brickwork and concrete or yield in steel

	Tensile stress (MN/m^2)	Shear stress (MN/m^2)
Brickwork 1:1:6 (in sound condition) 1:0.25:3	0.5 (0.35)	0.3 (0.15)
	1.4 (0.5)	0.7 (0.35)
Plain concrete	$0.1 f_{cu}$ typically ≈ 3 MN/m^2	
Structural steel, grades 43	280 (240) ⎫ but dependent	
50	395 (345) ⎭ upon thickness	

4.5 Brick walls on clay soil

The finite-element interactive method described in section 4.3 is now used to analyse four examples of brick walls.

Example 4.2 1 m-high brick wall on clay soil

The first example is a solid, plain, brick wall, 6 m long, 1 m high and 0.22 m thick, on a concrete base 0.6 m wide by 0.3 m deep. A plane-strain finite-element analysis is used to evaluate $K_g w$ in terms of vertical and transverse forces at seven interface nodes. The $K_g w$ vector is then applied to a finite-element analysis of ground and wall, with the wall and base modelled by plane-stress elements, as shown in Figure 4.6. Results are presented in Figures 4.7a, b, which show maximum wall settlements for tunnels in Cases A and B, at varying axis depth, and maximum principal tensile and shear stresses induced in the brickwork. The levels of tensile and shear stresses are so low as to be unlikely to cause any cracking in sound brickwork, and it is proposed that this example wall on clay can tolerate the

---

[†] Users of American codes are directed to ASTM C34–74 (1974), ACI 318–83 (198?) and AISC (1978).

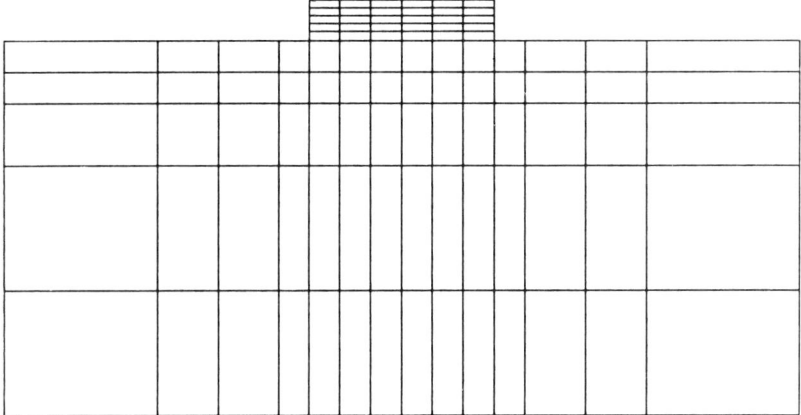

Figure 4.6 Mesh for 6 m × 1 m brick wall on clay.

tunnelling cases investigated without damage, and will show only limited settlement of less than 7 mm.

Example 4.3 2 m-high brick wall on clay soil

The next example is a brick wall 6 m long by 2 m high by 0.22 m thick on a 1 m × 0.3 m concrete footing. Results are plotted in Figure 4.8, showing tensile and shear stresses limited to 0.02 N/mm² and 0.04 N/mm², respectively. These are similar to the values found for the 1 m high wall and, again, damage is unlikely. The displacements in all four examples 4.2 to 4.5 are almost identical.

Example 4.4 2 m-high brick wall with opening on clay soil

When a similar 2 m-high wall with doorway opening is subjected to the tunnelling examples, the stresses are increased, as shown in Figure 4.9, with tensile stress of up to 0.07 N/mm² in the foundation and brickwork below the doorway from Case A, and tensile stress of up to 0.02 N/mm² in the lintel area from Case B. Because of the restraining effect of the concrete footing, it is again improbable that damage to the wall would be caused.

Example 4.5 House-front brick wall on clay soil

Last in this group of examples is a typical house-front brick wall, with a doorway and three window openings. The finite-element mesh that is used is far too coarse to predict stress concentrations at corners, but the stresses predicted with nodal averaging, shown in Figure 4.10, suggest that damage is unlikely to occur.

From these results it seems that, at first sight, tunnelling-induced ground movements cause little danger to brick construction on clay soil, primarily

252 SOIL MOVEMENTS

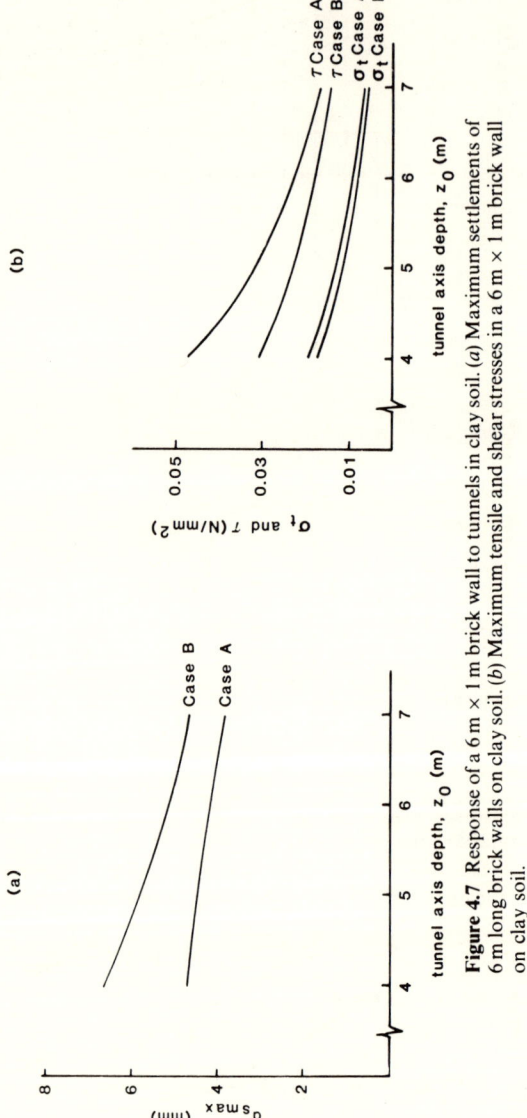

Figure 4.7 Response of a 6 m × 1 m brick wall to tunnels in clay soil. (*a*) Maximum settlements of 6 m long brick walls on clay soil. (*b*) Maximum tensile and shear stresses in a 6 m × 1 m brick wall on clay soil.

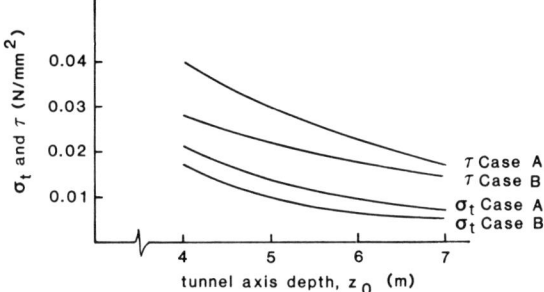

Figure 4.8 Maximum tensile and shear stresses in a 6 m × 2 m brick wall on clay soil.

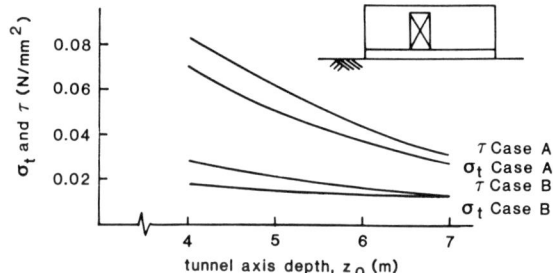

Figure 4.9 Maximum tensile and shear stresses in a 6 m × 2 m brick wall, with opening, on clay soil.

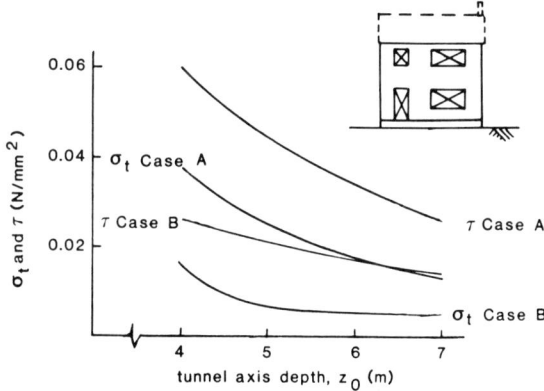

Figure 4.10 Maximum tensile and shear stresses in a brick house wall on clay soil.

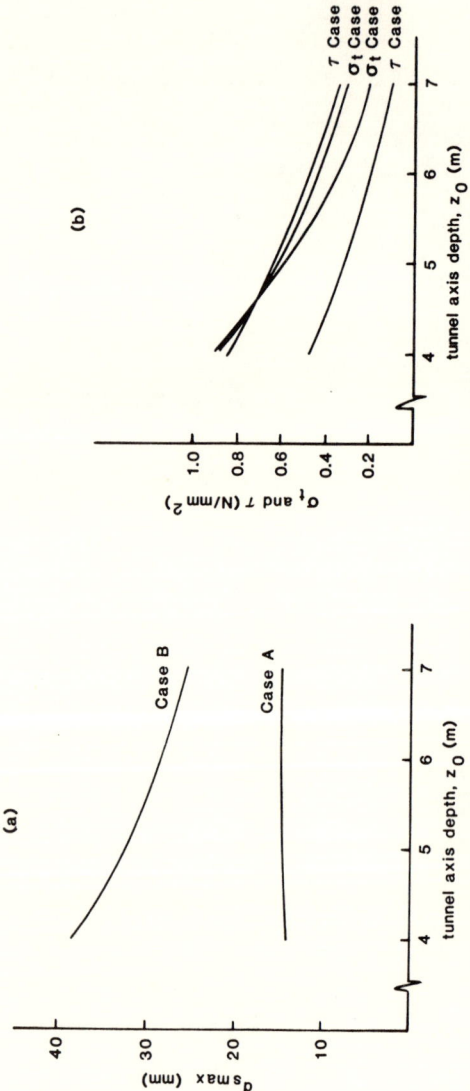

Figure 4.11 Response of 6 m × 1 m brick wall to tunnels in sand. (*a*) Maximum settlements of 6 m long brick walls on sand. (*b*) Maximum tensile and shear stresses in 6 m × 1 m brick wall on sand.

because tunnelling in clay can be conducted with small ground loss. However, additional consolidation settlement may greatly increase the long-term ground movements (see section 2.8), and the effect of this is an increase in **w** and hence $K_g w$, with corresponding increases in predicted wall stresses by a factor of the order of 2 to 4. Stress or strain levels to cause cracking in brickwork are difficult to define, both because of the variable quality of bricklaying and in particular the low tensile strength of mortar in the perpends, and also because of the deterioration of the mortar in the joints caused by weathering. Consequently, a conservative approach must be adopted in application of these results to real structures of similar proportions.

Driving a tunnel at greater depth produces only a slight reduction in peak tensile stresses in the walls.

4.6 Brick walls on sand

The same brick walls with concrete foundations are next analysed for response to tunnelling-induced settlements in sandy soils. This situation is generally more onerous on the structure because of the much larger (and often unpredictable) volume ground loss when tunnelling in sands, and in addition the surface settlement trough width parameter i is smaller, so that a narrow deep surface settlement trough tends to be formed.

Example 4.6 1 m-high brick wall on sand

Results for the 1 m-high wall in Figure 4.11 show substantial settlement of up to 38 mm, and tensile and shear stresses of 0.9 N/mm². These conditions are likely to cause cracking in low-to-medium strength brickwork.

Example 4.7 2 m-high brick wall on sand

The 2 m high wall suffers settlements similar to those of the 1 m wall in the above example, but with reduced stresses (see Figure 4.12), the maximum tensile stress being 0.7 N/mm². Cracks would therefore be less likely to occur than in the 1 m high wall.

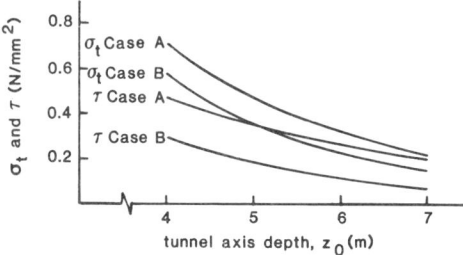

Figure 4.12 Maximum tensile and shear stresses in a 6 m × 2 m brick wall on sand.

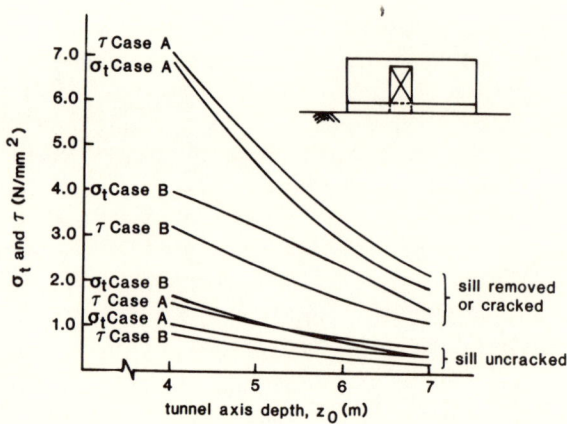

Figure 4.13 Maximum tensile and shear stresses in a 2 m brick wall, with opening, on sand.

Example 4.8 2 m-high brick wall with opening on sand

The 2 m wall with doorway was analysed first with a sound homogenous foundation below the opening, and then with the foundation removed below the doorway. The foundation variation has little effect upon calculated elastic-displacements, but the effect upon peak tensile stresses is dramatic; the standard form shows tensile and shear stresses of up to 1.6 N/mm², which are increased theoretically to 7.0 N/mm² by doorway foundation removal (see Figure 4.13). Both conditions might be expected to show cracking around the doorway lintel, but removal (or tensile failure) of the doorsill leads to a very severe level of cracking. It is interesting to compare this condition with the tunnelling-induced structural damage to the perimeter wall of a factory at Willington Quay, northeast England (see Figure 4.14 and Example 4.17) founded upon silty alluvial clay soil.

Example 4.9 House-front brick wall on sand

The brick wall of a house on sand is analysed with no foundation imperfection, the results being as shown in Figure 4.15. A tensile stress of 0.7 N/mm² may be sufficient to initiate cracking, but again, if the brickwork/concrete footing below the door were cracked, then a substantial increase in stress should be expected, together with a large increase in differential movements (both vertical and transverse).

In conclusion, a brick wall founded on granular soil subjected to tunnelling-induced movements is likely to suffer cracking, essentially because tunnelling through such soils tends to produce a deep narrow surface settlement trough which is potentially damaging to most structures. In addition, a wall with apertures experiences higher elastic stresses, and if there is a possibility of tensile or shear cracking causing effective segmenting of the foundation then substantial opening of cracks in the wall may result.

Figure 4.14 Damage to perimeter wall, Willington Quay, Tyneside, England.

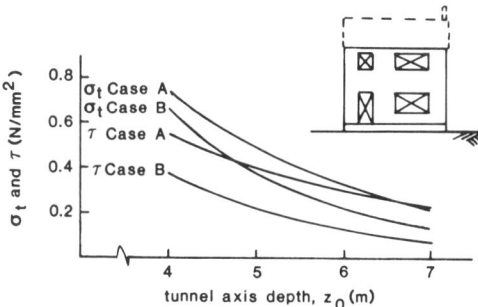

Figure 4.15 Maximum tensile and shear stresses in brick house-wall on sand.

An increase in tunnel axis depth from 4 m to 7 m typically reduces tensile stresses to about 30 per cent of the values due to a tunnel 4 m deep (to axis).

4.7 Crack widths in brick walls

If an interactive analysis of a wall on soil with a tunnel beneath it shows that a crack is likely to be initiated and propagated, then the structural stiffness of the wall will be modified drastically, and the elastic interactive analysis becomes quite inappropriate (see Figure 4.16).

The interactive analysis could be repeated with the structure modified to include a tension crack running through the wall. However such sophistication is perhaps misplaced, at the present stage of analytical development, in the light of the several imponderables including the tensile strength of brickwork and the crack direction.

Consequently, a simplistic analysis of cracking seems preferable, based upon either characteristic or mean tensile strengths as appropriate to the structure under consideration, and upon experience of typical crack paths as observed in settlement-damaged structures (see, for example, Thorburn, 1984) and in laboratory tests (see Abu-el-Magd and MacLeod, 1980).

First, the interactive analysis should be undertaken. If cracking is probable, then it is proposed that a particular cracking mechanism be selected which might split the structure into two discrete uncracked sections. If rigid-body displacements are then imposed on the two segments in order to follow closely the free-ground displacement profile, differential slopes between segments might be estimated.

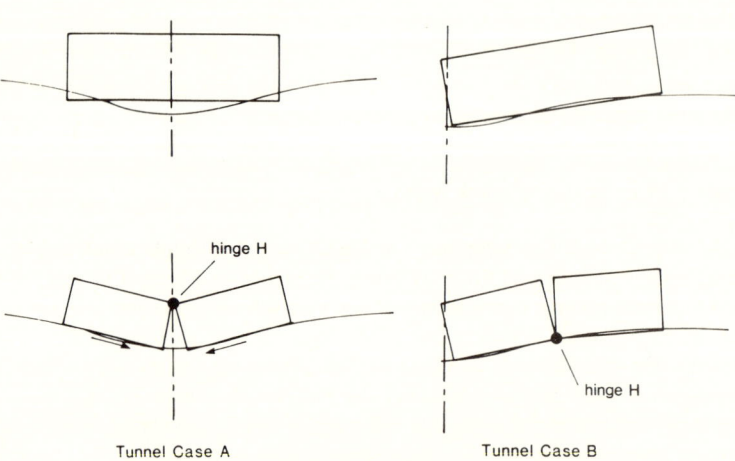

Figure 4.16 Uncracked and cracked deformations.

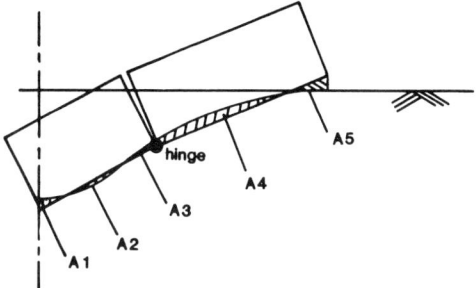

Figure 4.17 Cracked wall equilibrium.

In assessing probable rigid-body displacements of two (or more) uncracked segments of a cracked brickwork structure some simple assumptions are suggested. If a wall or pierced wall with continuous foundations is cracked in hog mode (curvature convex upwards), then a tensile or tensile/shear crack will run from the top course downwards, separating the wall into two segments which will be effectively pinned together at their foundation level. Three rules then govern the situation of a hinged rigid beam responding to a free-ground surface settlement, w (see Figure 4.17):

The total area between structure and line w is zero, to satisfy overall vertical equilibrium

The moment of the areas A1, A2 and A3 about the hinge H is zero; the moment of areas A4 and A5 about hinge H is also zero

The moment of all areas A1 to A5 about one end of the beam is zero.

A rather different set of conditions applies if segments of a wall with openings are supported by discrete footings. In this situation, after cracking is established, the rigid-body displacements are interrelated only by secondary ground response. In effect, the two segments may be treated as being independent, and may be assumed to follow the free-ground displacement profile with separate equilibrium of each segment.

Example 4.10 Estimates of crack width

Consider a brick wall 6 m long and 1 m high on sand through which a 2 m diameter tunnel is driven at 4 m axis depth. Reference to Figure 4.11 and to Table 4.3 shows a realistic probability that cracking will occur in some classes of brickwork, in sag mode A and in hog mode B.

Consider first the hog mode, B, where the free-ground maximum settlement is calculated as 49 mm, for $i = 1.02$ m. If the settlement trough is drawn to exaggerated vertical scale, and two rigid segments X and Y are superimposed on the profile, hinged together at a bottom corner, H, then an estimate of the slopes of X and of Y can be made, either graphically or by relative displacement values on the free-ground settlement, of about 0.023 (1 in 43) and 0.004 (1 in 250), respectively. The difference in slopes represents the rate of widening (or taper) of the crack, from the hinge, H, of about 0.019. When this slope difference is applied to a crack 1 m high, a maximum crack width of 19 mm is estimated, on the assumption of a vertical crack which might occur for low brick tensile strength and high shear bond strength in the brick–mortar interfaces. For the reverse case of strong bricks and weak bond, a diagonal crack might develop instead, probably of about the same width. This should be considered as an upper bound, since the wall may retain some integrity, or several narrower cracks may form.

A similar process might be applied to the sag condition, Case A, where a side slope of 0.020 might develop in each segment, leading to a crack taper of 0.04 and a crack width of 40 mm. However, the crack width would be substantially reduced by resistance from the soil to lateral outward movement of the foundation segments, and a realistic estimate of crack width cannot easily be made.

The ground restraint mechanism is primarily responsible for the statement that hogging curvature is observed to be more damaging than sag (although cracking below ground and out of view is less likely to attract attention than is cracking around windows and doors).

Consideration of walls, with openings, on continuous foundations might be treated in a similar way, although the preliminary elastic analysis will show an earlier onset of cracking than in solid walls of the same aspect (height to length) ratio.

Crack widths may be compared with those in Table 4.4 concerning the classification of severity of cracking, as reproduced from Burland *et al.* (1978).

Table 4.4 Classification of cracking of brickwork

Degree of damage	Description of typical damage	Approximate crack width (mm)
1. Very slight	Hairline cracks of less than about 0.1 mm are classed as negligible. Fine cracks which can easily be treated during normal decoration.	1
2. Slight	Cracks easily filled. Redecoration probably required.	5
3. Moderate	The cracks require some opening up and can be patched by a mason.	5–15
4. Severe	Extensive repair work involving breaking-out and replacing sections of walls, especially over doors and windows.	15–25
5. Very severe	This requires a major repair job involving partial or complete rebuilding.	usually > 25 but depends on no. of cracks

4.8 Concrete walls on sand

A common form of retaining wall is the reinforced concrete cantilever, which has the general profile shown in Figure 4.2.

Example 4.11 Reinforced concrete retaining walls on sand

Walls 7 m, 4 m and 1 m in height are analysed for sensitivity to ground movements caused by a 2 m diameter tunnel driven beneath them. The 7 m and 1 m walls are linearly scaled from the 4 m wall, but all are 6 m long.

The results for all three walls are shown in Figure 4.18. Displacements for the three walls are almost identical, and are plotted in Figure 4.18a. The maximum displacements due to tunnel line B, of 38 mm at one end (with 8 mm heave at the other) give a slightly undesirable degree of tilt of 1:130, but the uniform 14 mm settlement due to tunnel line A is tolerable.

Tensile and shear stresses in the 4 m and 7 m walls are closely grouped in Figure 4.18b, and are less than 0.8 N/mm^2. Stresses of this magnitude are unlikely to initiate cracking. However, tensile stresses of up to 3.5 N/mm^2 in the 1 m wall are very likely to cause cracking.

There is the possibility in walls of all heights that cracking exists prior to tunnelling, caused by differential shrinkage or thermal contraction after the elevated temperatures during early-age hydration. This should be investigated during pretunnelling surveys.

The consequences of cracking are a function of the quantity of reinforcing steel crossing the crack. In the situation of hog bending due to a tunnel on line B, the reinforcement crossing a crack running vertically down from the top surface of the wall is likely to be no more than the 0.25 per cent high yield or 0.3 per cent mild steel required as minimum secondary reinforcement in a wall by CP 110 or BS 5400. A simple calculation shows that a moment sufficient to initiate cracking will cause extensive yield of the reinforcement, so that little resistance is offered to crack widening, with the possibility even of tensile failure.

In this situation an estimate of crack width might be based upon the approach to cracking in brick walls, outlined in section 4.10, although absolute crack width is of secondary importance if structural failure is predicted. In the case of a heavily reinforced wall, then calculations in accordance with the crack width prediction of CP 110 are appropriate. Such an approach is relevant to the case of a wall in sag bending, since the base will be well reinforced.

The problem of cracking and extensive yield appears to be acute in a low wall which clearly has less in-plane bending strength than a high wall. Figure 4.19 shows that walls of less than about 1.5 m are at risk from the example tunnels.

Concrete walls are shown to have considerable strength against cracking. However, if a crack is initiated by hog bending, then a typical wall containing

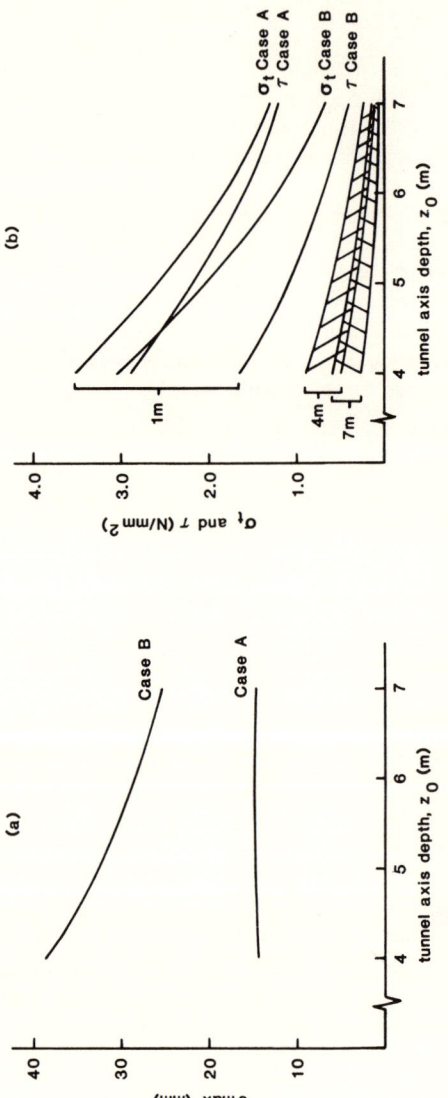

Figure 4.18 Responses of 7 m, 4 m and 1 m high reinforced concrete walls to tunnelling in sand. (a) Maximum settlements of concrete walls on sand. (b) Maximum tensile and shear stresses in concrete walls on sand.

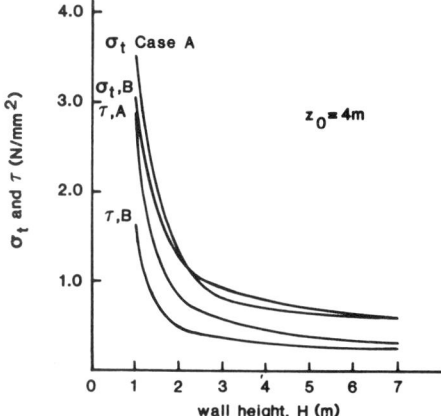

Figure 4.19 Variations in maximum tensile and shear stresses with height, in 6 m long reinforced concrete walls.

minimum secondary reinforcement is likely to suffer severe cracking and damage.

Crack repairs to concrete or brickwork would not normally be undertaken, unless of an emergency nature, until the owners of the structure and of the tunnel are assured that movements of ground and structure have ceased. Such repairing is now a well-established technology undertaken by specialist firms (see Appendix C).

4.9 Modification of the results of wall examples for application to different soils, tunnels and walls

A set of examples of the effects of tunnelling-induced ground movements can cover only a restricted range of parameters. However, by careful reference to the equation $(\mathbf{K}_g + \mathbf{K}_s)\mathbf{d}_s = \mathbf{K}_g \mathbf{w}$, the results of the examples so far presented can be extended.

First, it should be recognized that for walls of about 1m height and above, the $\mathbf{K}_s$ matrix terms are 100 or more times greater than the terms of the $\mathbf{K}_g$ matrix. In consequence, a small change in soil stiffness or in effective soil foundation width will have negligible effect upon $(\mathbf{K}_g + \mathbf{K}_s)$ but will have a linear effect upon the $\mathbf{K}_g \mathbf{w}$ vector. This implies that such a change will also result in an effectively linear change in structural stresses. For example, a 4m high reinforced concrete wall of the proportions of the example, founded on sand of elastic modulus $E = 4 \times 10^6$ MN/m^2 (compared to 10×10^6 MN/m^2 in Example 4.11), will show a reduction in stresses which can be evaluated by factoring by 0.4, so that a tunnel driven on line A at 5 m depth would show a peak tensile stress of some $0.45 \times 0.4 = 0.18$ N/mm^2. In consequence, perhaps a reasonable design approach would be to adopt conservative upper-value laboratory soil moduli.

264   SOIL MOVEMENTS

By similar reasoning, if a larger-diameter tunnel is driven, then the **w** vector is increased by a single factor, and both displacements and stresses can be factored accordingly. This can be verified by reference to equations (4.1) and (4.2), which propose that the vertical displacement, w, is a linear function of surface settlement volume V_s, and the transverse displacements are linear with w. In addition, if a tunnel is driven by means which change the percentage ground loss, such as pregrouting, ground freezing or shield modification (see section 2.2), then again a change in calculated V_s can be transmitted directly as a similar proportional change in wall displacements and stresses. For example, if a 3 m diameter tunnel is to be driven at 7 m depth through clay below a 2 m high brick wall, the surface settlement volume V_s becomes $0.015 \times \pi \times 3^2/4$, or $0.106 \, m^3/m$, compared with the previous value of $0.047 \, m^3/m$. Consequently, for a 3 m tunnel driven on line A, at 7 m below a 2 m high wall, displacement will increase from 3.8 mm to 8.6 mm, and peak shear stress will increase from $0.017 \, N/mm^2$ to $0.038 \, N/mm^2$.

Next, consider changes in wall geometry, such as a single-thickness brick wall of 0.11 m, but on the same foundation as for the 0.22 m solid brick wall analysed previously. Again by reference to $(\mathbf{K_g} + \mathbf{K_s}) \mathbf{d_s} = \mathbf{K_g w}$, there is now a reduction in all terms of the $\mathbf{K_s}$ matrix of the order of, say, 40 per cent. The factoring of stresses is related to the changes in section moduli.

Example 4.12 Incorporation of a changed wall cross-section

Consider a 2m high brick wall on a 1m × 0.3m reinforced concrete strip footing, with modular ratio

$$E_{conc}/E_{brick} = 1.5.$$

For the thicker wall, in brickwork units, the centroidal height $\bar{z}$ is given by

$$\bar{z} = \frac{2 \times 0.22 \times 1.3 + 1.5 \times 0.3 \times 1 \times 0.15}{2 \times 0.22 + 1.5 \times 1 \times 0.3} = 0.72 \, m.$$

Hence the total second moment of area,

$$I = 0.22 \times 2^3/12 + 0.22 \times 2(1.3-0.72)^2$$
$$+ 1.5[0.3^3 \times 1/12 + 0.3 \times 1(0.72-0.15)^2]$$
$$= 0.444 \, m^4.$$

For stresses in the brickwork only, the section moduli for the top (ζ_t) and bottom (ζ_b) courses are

$$\zeta_t = \frac{0.444}{(2.3-0.72)} = 0.281 \, m^3$$

and

$$\zeta_b = \frac{0.444}{(0.72-0.3)} = 1.057 \, m^3.$$

For the thinner wall, $\bar{z} = 0.53$ m,

$$I = 0.272 \text{ m}^4,$$

$$\zeta_t = \frac{0.272}{(2.3 - 0.53)} = 0.154 \text{ m}^3.$$

and

$$\zeta_b = \frac{0.272}{(0.53 - 0.3)} = 1.183 \text{ m}^3.$$

Now the ratio of second moments of area of the thick and thin walls is

$$\frac{I_{\text{thick}}}{I_{\text{thin}}} = \frac{0.444}{0.272} = 1.63.$$

Therefore an increase in d_s by a factor of 1.63 is induced by the thinner wall, since K_s decreases by the factor 1.63. By analogy with simple bending theory,

$$\frac{M}{I} = \frac{E}{R}.$$

and if I decreases by $1/1.63$ while the curvature $(1/R)$ increases by 1×1.63, then the moment on the section is unchanged. Consequently, stresses are increased by the factor of section moduli ratios, that is

$$\frac{\zeta_{t,\text{thick}}}{\zeta_{t,\text{thin}}} \quad \text{for compressive stress, Case A}$$

and

$$\frac{\zeta_{b,\text{thick}}}{\zeta_{b,\text{thin}}} \quad \text{for tensile stress, Case A.}$$

This is, of course, an approximation only, because the walls are similar to short deep beams, for which simple beam bending theory is not applicable with accuracy. Also, if the structure has a wider base than the examples used here, then the structural stiffness K_s will be increased as discussed above, but in addition K_g is increased because of the thicker soil strip, and consequently the $K_g w$ vector must be factored up. This has a linear effect upon displacements and stresses, which must be superimposed on to the modifications due to structural changes.

Another important parameter is the length of the wall, L. Burland and Wroth (1975) and Alexander and Lawson (1981) suggest that the length-to-height ratio has a linear effect upon the permissible bending/settlement ratio Δ/L, above a limit that is a function of shear deformation. Thus, the approach discussed previously in Example 4.12, which proposes increases in stresses in inverse proportion to the modulus ratio, is reasonable in its prediction of stresses in low walls from the results of the examples, since their response is

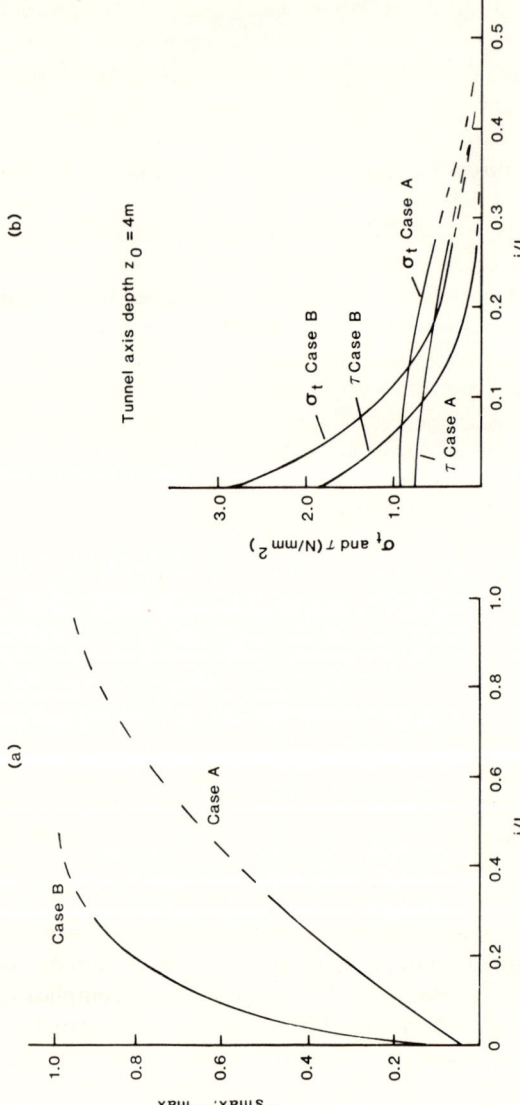

Figure 4.20 The effects of i/L ratio upon the response of a 2 m high brick wall on sand. (*a*) Ratios of maximum wall settlement to free ground settlement for various trough parameter (i) to length ratios. (*b*) Variations of maximum tensile and shear stress with parameter i to length ratios.

primarily in bending not shear. The converse, of poorer estimates of stresses, will apply to high walls with results extrapolated from the worked examples.

Also significant is the ratio of wall length to trough width, which can be represented by the ratio of wall length to i-parameter, L/i. This is demonstrated by the examples, which show that the narrow trough resulting from tunnelling in granular soils is proportionately more severe than the broad trough which occurs when tunnelling in clay soils. Clearly, the effects are most severe when a structure is substantially longer than the trough width parameter, in that end restraint is then applied to the wall segment which is directly affected by the soil deformation, as discussed qualitatively in section 4.13. Investigations on the example 2 m high brick wall on sand with a tunnel at 4 m axis depth show that lengthening of the wall reduces settlement, but increases wall stresses (see Figure 4.20). In the absence of more detailed survey, a simple factoring procedure is recommended for structures on granular soils. If cracking results, then rigid-body deformations may again be applied in an assessment of crack widths. Strictly, the factors from the limited parameter study apply only to the particular case studied, but they can be applied with reasonable confidence to walls on granular soils. They are less satisfactory for clay soils, where the transverse settlement trough is much wider.

4.10 Building frame structures

The infinite range of steel and concrete frame structures makes it unrealistic to attempt to detail a range of examples sufficient to cover all practical cases. Such an attempt is rendered even less viable because in the equation

$$(\mathbf{K}_s + \mathbf{K}_g)\mathbf{d}_s = \mathbf{K}_g\mathbf{w}.$$

the terms in the two stiffness matrices are likely to be of the same order, and in consequence a change to either ground or structure stiffness cannot be treated manually by linear modification as has been the case with walls.

Instead, the details of the interactive stiffness method are discussed with reference to a four-bay frame on clay soil, on a granular soil, on clay soil with a stiffer section frame, and a two-storey frame on clay soil.

In the analysis of frame structures, it is necessary to consider vertical and transverse displacements and also rotation at each column base. This additional, rotational, degree of freedom adds minor complication in the formulation both of $\mathbf{K}_g$ and in the full finite-element soil–structure model.

The finite elements used to model the ground are three-dimensional prismatic or two-dimensional plane-strain, which have nodal degrees of freedom of displacement only. Two possible techniques could be used to compute soil effects with respect to rotation of a footing. One method is to use a mesh with two nodes per footing on to which are imposed a positive and a negative displacement. A preferable method is to use a mesh with three nodes per footing connected by two short very stiff beam elements, on to which a unit rotation can be imposed directly. The second method gives results superior to those of the first method, but the refinement of the mesh precludes a three-dimensional model, and plane-strain elements are used.

Analysis of a four-bay frame requires three interface degrees of freedom at each of five footings. The off-diagonal terms of the resulting $\mathbf{K}_g$ matrix are small, showing low cross-coupling between footings, as is to be expected.

Example 4.13 Steel frame on clay soil

The first frame example is a four-bay single-storey structure, with each bay 8 m wide and 4 m high. Beams are 533 × 210 @ 82 universal beam section, and columns are 305 × 305 @ 198 universal column section. The soil foundation is the clay soil of Table 4.2. Bending moment diagrams induced by tunnels of Cases A and B are shown in Figures 4.21 and 4.22. (Elements of poor aspect ratios are tolerated, but only when remote from the footings.) Maximum induced settlements, and bending stresses in beams and in columns, are plotted in Figure 4.23. These stresses, even when superimposed upon dead and live load effects, are unlikely to cause yield. The differential displacements are small, and damage to the partitioning or glazing units is unlikely.

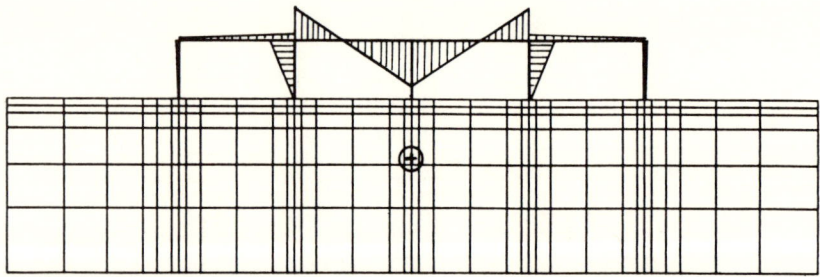

Figure 4.21 Bending moments in a steel frame on clay soil due to tunnel Case A.

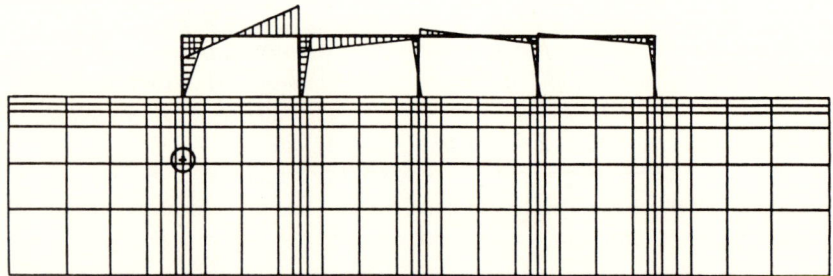

Figure 4.22 Bending moments in steel frame on clay soil due to tunnel Case B.

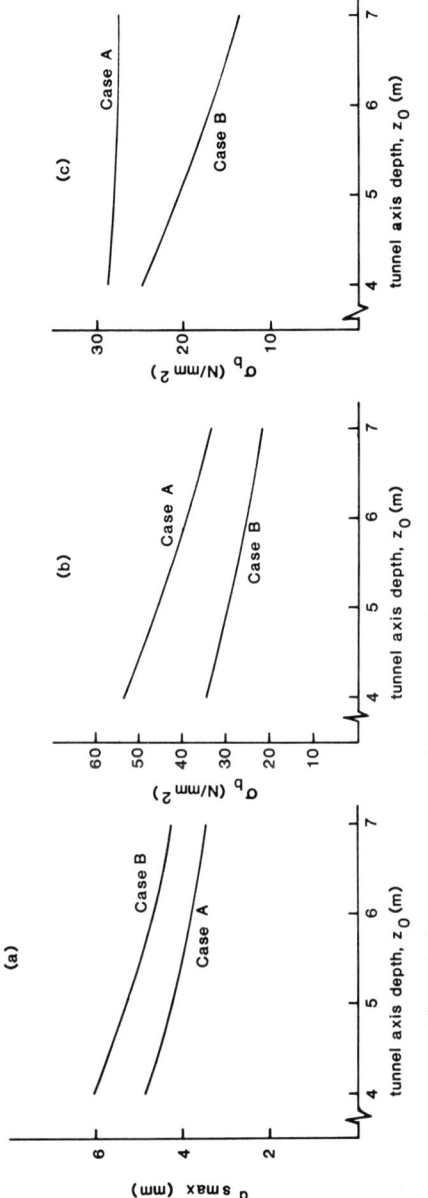

Figure 4.23 Response of a steel frame to tunnels in a clay soil. (a) Maximum settlements of frame. (b) Maximum bending stresses in beam members. (c) Maximum bending stresses in column members.

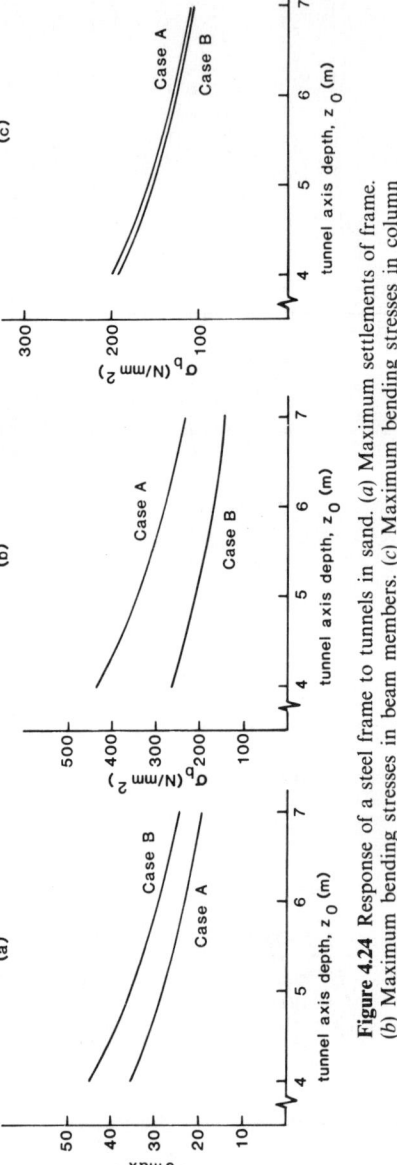

Figure 4.24 Response of a steel frame to tunnels in sand. (*a*) Maximum settlements of frame. (*b*) Maximum bending stresses in beam members. (*c*) Maximum bending stresses in column members.

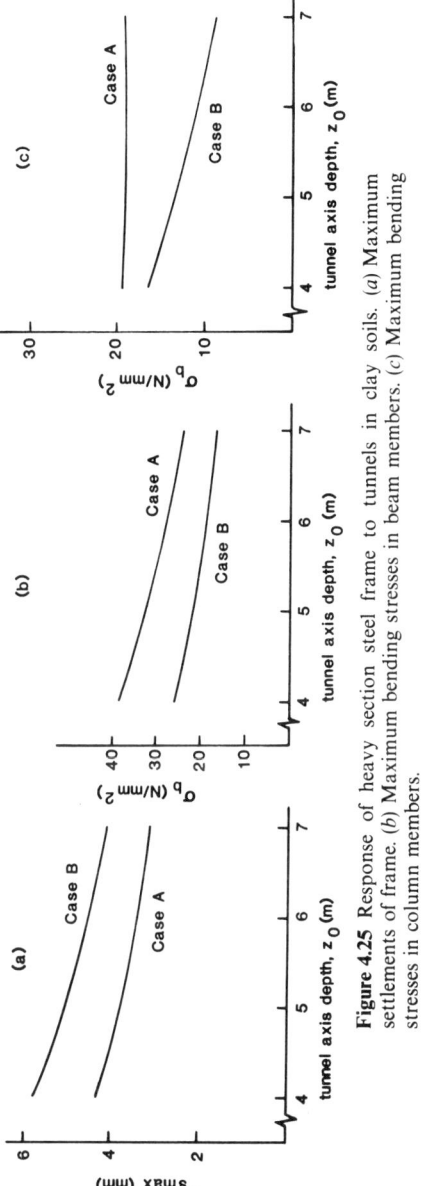

Figure 4.25 Response of heavy section steel frame to tunnels in clay soils. (a) Maximum settlements of frame. (b) Maximum bending stresses in beam members. (c) Maximum bending stresses in column members.

272   SOIL MOVEMENTS

Example 4.14 Steel frame on sand

When the same frame is subjected to movements due to tunnelling in sand, displacements increase to as much as 45 mm, and apparent elastic bending stresses in beams and columns are severe (see Figure 4.24). (Axial and shear stresses, however, are much less significant.) Tunnels of Case A cause inevitable yield in the centre spans, but not in the outer bays. Yield will occur close to beam–column junctions, either in the beams if joints are strong, or in the joints if they are weak. Resulting large beam-to-column rotations, and deflection ratios, Δ/L, of 1/170 will cause damage to partitions and to glazing panels (unless these are removed prior to tunnelling.) Tunnels of Case B will cause similar effects, but in the end bay only.

Example 4.15 Heavy steel frame on clay soil

The first frame example, on clay soil, is now repeated with a heavier steel frame, of 610×229 @ 113 universal beam section, and 356×406 @ 287 universal column section which is close to a doubling of bending stiffness. The results, in Figure 4.25, show little change in displacements, although differential settlements decrease by about 10 per cent. Stresses in beams and columns show reductions of about 30 per cent. As expected, this situation lies between the two extremes of imposed curvature (causing higher stresses in deeper sections), and of imposed force (stresses inversely proportional to section modulus, that is, a reduction of about 60 per cent in the heavier frame). This confirms the earlier statement that $\mathbf{K}_g$ and $\mathbf{K}_s$ are of the same order of magnitude, and consequently no simple relation is apparent between frame stiffness and stress. This prevents manipulation of example results for a specific prototype.

Examples 4.16 Two-storey frame on clay soil

Finally, a two-storey four-bay frame is analysed. The bending moment diagram for Case A is shown in Figure 4.26. Values of displacements and bending stresses

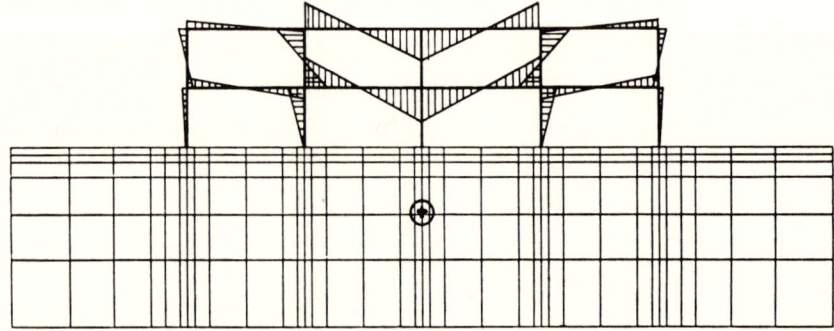

Figure 4.26 Bending moments in two-storey steel frame, Case A.

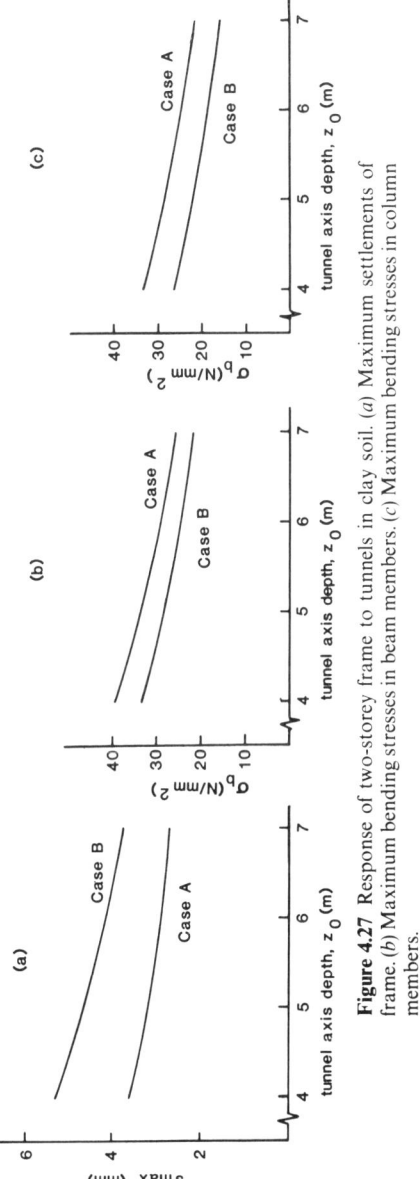

Figure 4.27 Response of two-storey frame to tunnels in clay soil. (*a*) Maximum settlements of frame. (*b*) Maximum bending stresses in beam members. (*c*) Maximum bending stresses in column members.

are plotted in Figure 4.27, and generally show slight reductions from the values for the single-storey frame, except that column moments are slightly increased for shallow tunnels.

In conclusion, steel frames are unlikely to be damaged by settlements caused by tunnelling in clay soils. However, tunnelling in granular soils tends to cause yielding and considerable distortion; the latter may be crudely assessed from imposition of the free-ground settlement profile on to the frame structure. Damage to cladding can then be assessed with respect to Δ/L values (see section 4.13 and also Davies, 1981). The yielding is unlikely to cause collapse, but will cause public concern and will require extensive remedial work.

4.11 Infilled frames

A common form of construction is the infilled frame, which comprises a structural steel or reinforced concrete single or multibay frame with each bay infilled with brickwork or blockwork. Much has been written on the strength and stiffness of infilled frames (see, for example, Stafford-Smith and Carter, 1969; Riddington and Stafford-Smith, 1977; King and Pandey, 1978), and of considerable importance is the paper by Riddington (1984) concerning reduction in stiffness due to small gaps between frame and infill.

Because of the diversity of such structures, a parametric study is not attempted. Instead a single example of a published case study (Attewell *et al.*, 1978; Attewell, 1978*b*) is analysed. It examines the end wall of a factory at Willington Quay in north-east England.

Example 4.17 Infilled frame—Willington Quay

A tunnel of 4.25 m diameter was driven at a depth of 13 m below a steel frame of 300 mm × 125 mm section, infilled with brickwork on a concrete footing, and lying at an angle of 75° to the tunnel centre line. The ground is modelled as soft clay with $E = 2\,\text{MN/m}^2$ (in reality, silty alluvial clay), and a wide shallow free-ground settlement trough is predicted, with maximum settlement of only 8.5 mm, and $i = 6.7$ m (because of the depth of drive). Analysis of the infilled frame under equivalent force vector $\mathbf{K}_g\mathbf{w}$, shows very small stresses in the brickwork, of about $0.01\,\text{N/mm}^2$, and trivial axial stresses in the steelwork of less than $1\,\text{N/mm}^2$. The wall settlements are calculated to be around 7.6 mm.

These results are a serious underestimate of the observed response of the structure which showed severe cracking (see Figure 4.28), and settlements of 25 mm at 23 days increasing to 150 mm to 500 days after tunnelling. The reasons for the major discrepancies include:

(a) Underestimate of volume ground loss. The analysis uses a value of 1.5 per cent, which is low for the type of ground material. Attewell *et al.* (1978) reported the presence of some non-cohesive material encountered during the drive.

(b) Overestimate of trough width parameter, i. Analysis uses a value of 6.7 m, but observed settlements yield a value of i of about 2.7 m.

STRUCTURAL RESPONSE TO TUNNELLING SETTLEMENT 275

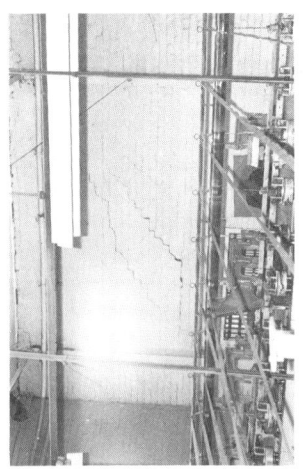

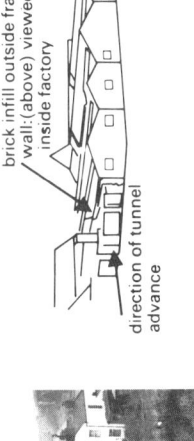

tunnel driven below and between the two wall areas photographed above

brick infill outside frame wall: (above) viewed from inside factory

direction of tunnel advance

Figure 4.28 Cracking of infill wall, Willington Quay, NE England.

(c) Consolidation settlement, which played a major part in continuing settlement increases.

(d) Modelling of the frame as a homogeneous linear elastic structure. Riddington (1984) demonstrates the substantial loss of stiffness on an infilled frame due to very small gaps between frame and infill, which commonly exist in such construction.

Conclusions from this first analysis are:

(1) Consolidation settlement of silty alluvial soils will increase substantially any tunnelling-induced ground movements. In addition, a settlement narrower than predicted may occur, which is disadvantageous to the structure. Reference should be made to section 2.8 and particularly section 2.8.3.

(2) Because of the probability of gaps between a frame and its brick infill (highly likely in this particular case because of the possibility of some pretunnelling differential movement having occurred as a result of the ground conditions and the use of isolated pad footing to the columns—see Attewell *et al.*, 1978: Attewell, 1978*b*), such a frame should be modelled, conservatively, as a continuous brick wall but with the steel frame ignored entirely. This is defensible because as a first proposition the frame and the brickwork should be separated in a stiffness analysis, since the presence of gaps is likely. Hence, if the wall and frame stand side-by-side but not connected, then response to ground

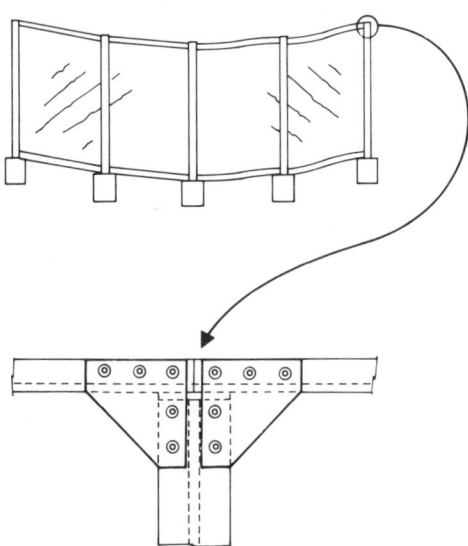

Figure 4.29 Large deformations of an infilled frame. The columns are braced at the base, shear deformation being induced in the brittle panel infilling as a result of differential settlement and the cracking being caused by diagonal tensile strain. Full bending moments are shown for the two right-hand side top and floor cross-beams, with obvious points of contraflexure. The left-hand side cross-beams have rotated and do not support bending moments. There will also be horizontal in plane and out-of-plane distortion superimposed on the system.

movements is modelled adequately. Because the brick wall is substantially stiffer than the frame, then removal of the frame has a negligible effect upon total structural stiffness prior to cracking, and stresses in the wall due to ground movements are predicted adequately. Effects imposed on to the steel frame can be disregarded with safety, since the infill brickwork, even if cracked, will limit frame distortion and resulting stresses to very small values. Clearly, this proposal is inaccurate at large deformations, when gaps will close, and the frame then has the function of controlling crack development towards a shear type failure, rather than a bending response (Figure 4.29).

Re-analysis of the infilled frame at Willington Quay based upon observed rather than predicted free-ground movements, and ignoring the steel frame, shows a substantial increase in tensile stresses in the wall, up to 4.8 N/mm². This implies substantial cracking of the old brickwork, as was observed in reality (Figure 4.28), although crack widths are not easily predicted in this situation.

4.12 Piled foundations

Bearing piles are a common foundation to structures in situations of high loads, weak soils, or both. Tunnelling-induced ground movements may affect pre-existing adjacent bearing piles in two major ways. Generally, soil above a tunnel and in the zone of influence, defined by a wedge of width $\pm\sqrt{2\pi}\,i$ at the surface and with its source at the tunnel lining, will show some downward movement. This may result in negative skin friction on the upper part of an adjacent pile, causing an effective increase in pile load plus reduction in capacity, which will result in additional pile settlement. The second effect is lateral deformation of the pile caused by variation with depth of transverse ground movements. This latter effect is in some ways analogous to the response of buried pipes, as discussed in Chapter 3, in that the stiffening effect of the pile reduces ground movements.

Considering the first consequence, that of negative skin friction, which might, for example, exist over the upper 3 m of a 20 m long friction pile, then the loss of capacity on a 600 kN pile would be of the order of 5 per cent or 30 kN, with a corresponding increase in load. Typically, this might induce pile settlement of around 2 mm, which is unlikely to cause structural distress. A deeper tunnel might cause a substantially higher loss of capacity through the reversal of positive skin friction to negative friction. A simple design approach would be to align tunnelling so that an inferred zone of influence does not intersect existing piles. Such a procedure was adopted by the Northumbrian Water Authority to the foundations of Westgate House in Newcastle upon Tyne, England (see Norgrove *et al.*, 1979, and Figure 4.30). An interesting case of pile design to tolerate ground movements was incorporated in Swan House, Pilgrim Street Roundabout, Newcastle. The piles were designed to accommodate negative skin friction over their upper part, anticipating tunnelling, which followed later, between piled areas of the one building. Where negative skin friction is anticipated, bituminous slip coatings may be of value where soil conditions allow.

The second effect, that of bending induced by transverse soil movements, is less easy to quantify. If the soil wedge model is considered, transverse movement *increases* with depth, but only within the wedge. If an elastic model

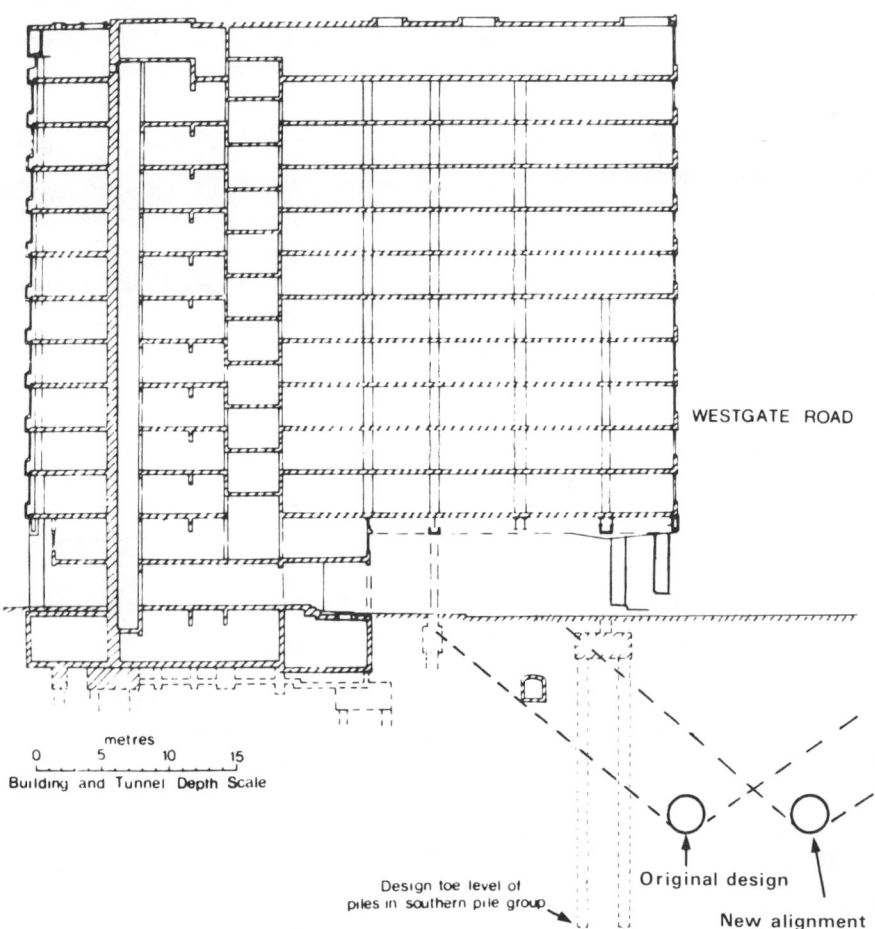

Figure 4.30 Tunnelling close to existing piled foundations—Westgate House, Newcastle upon Tyne, England.

is analysed, then transverse movement *decreases* with depth. The soil wedge is a useful model for granular soils, but clay soils will show a more complex response. Whatever the details of the free-ground movements, the presence of piling will create an interaction, in the same way as do buried pipes (Chapter 3), and structure foundations. Because of these uncertainties, it seems prudent to recommend that, whenever possible, a tunnel be aligned so that concrete piles remain outside its zone of influence. The situation is different with respect to steel piles, which are sufficiently ductile to tolerate substantial deformation without significant loss of capacity. A relevant example of this capability is the piling to the bank-seat abutments of the M180 motorway River Trent bridge in England. The embankment, on poor ground, was expected to move toward the river. Tubular steel piles were driven through the embankment and down to load-bearing stratum. It was demonstrated that the piles could tolerate large lateral movements, even involving some yielding, with only a minor reduction of vertical capacity. This philosophy can be applied directly to the situation of tunnelling-induced ground movements around steel bearing piles.

One aspect which should be considered is the possibility of separation of soil and structure foundation. This may well occur when a stiff structure bridges across a settlement trough, as in Case A of the example of section 4.6, or with a lower probability in Case B of the examples. The resultant problem is the fracture of service pipes between ground and structure. Pipes most at risk are cast iron gas and water pipes, and salt-glazed mortar-jointed and cast iron waste pipes. Pipes of plastics such as polyethylene and uPVC, and of lead are more capable of tolerating small deformations. Failure of gas or water pipes is a potentially dangerous situation, and even if calculations suggest the risk to be small, a careful programme of monitoring of ground levels should follow any tunnelling in urban areas, so that potential danger can be controlled. Problems may also arise due to relative movements of paths and access roads with respect to buildings. The problems may be more severe when buildings on deep piled foundations remain in position while surrounding ground settles. An instructive case history of the problems of ground settlement around piled structures is given by Bowring *et al.* (1984)

4.13 Building damage: practical appraisal

Time and financial constraints, and the sheer number of buildings adjacent to a tunnel route, may dictate that rapid assessments of structural vulnerability be performed at an early site investigation stage. Two methods of data presentation for such a rapid assessment have been used by one of the present writers (Attewell, 1981) for an urban tunnelling contract in England. Only permanent (transverse) settlements and derivatives are shown.

In the first of these methods, surface settlements, transverse displacement and transverse strain profiles are computed and plotted on a transverse

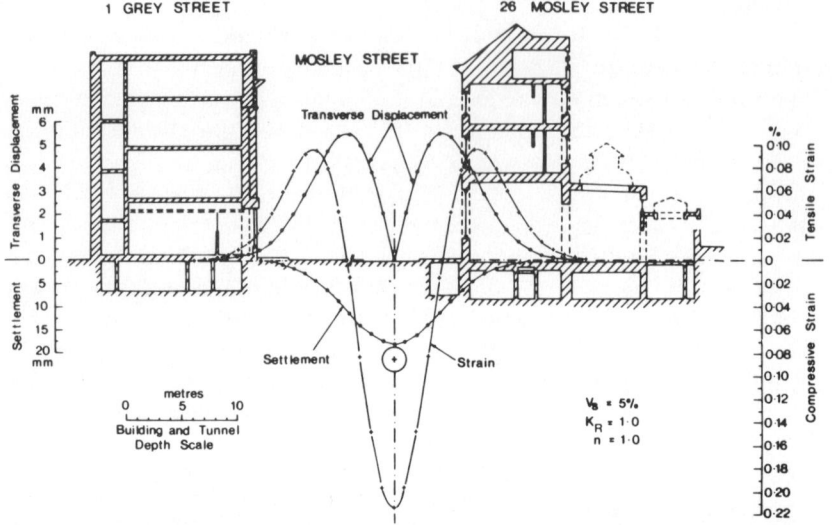

Figure 4.31 Prediction of ground deformation transverse to the centre line of a tunnel driven in soil beneath a major street in Newcastle upon Tyne, England.

vertical cross-section to a scale of 1:200 (see, for example Figure 4.31). The position of the building foundation within the settlement trough is then readily noted, and preliminary inferences with respect to possible structural deformation can be drawn. These inferences are based on one or more damage criteria related to the building as a whole but take no account of individual structural detail. During the property surveys both before and after tunnelling, attention should be directed to such points as zones of stress concentration and potential weakness at the corners of the windows and doors (see Figure 4.32); note, for example, that sub-vertical crack propagation will follow the joints in weak lime mortar or possibly pass through the bricks where the mortar is a strong Portland cement mix; the position of floor joists with respect to the tunnel advance direction (Figure 4.33), the detailed construction of frame buildings, of the panel infilling, stiffening of joints at heads of columns, restriction on column rotation at footings and so on should also be considered. These matters are described in more detail in Attewell (1978a), to which reference should be made.

In the second (often additional) method, an example of which is shown in Figure 4.34, the most important ground-deformation parameters are quantified along key lines drawn on a 1:500 scale plan of an area. Since the predicted normal probability transverse settlement profile is assumed to be symmetrical about the tunnel centre line, the settlement, horizontal displacement and ground strain, and slope predictions located on Figure 4.34 are also reflected on the other side of the centre line.

Building response to settlement depends upon the length of the structure.

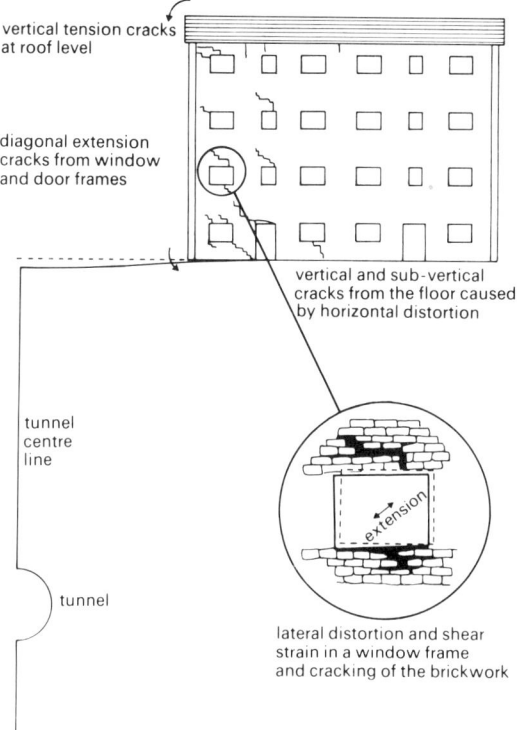

Figure 4.32 Fractures that could develop in a load-bearing brick wall affected by hogging settlement.

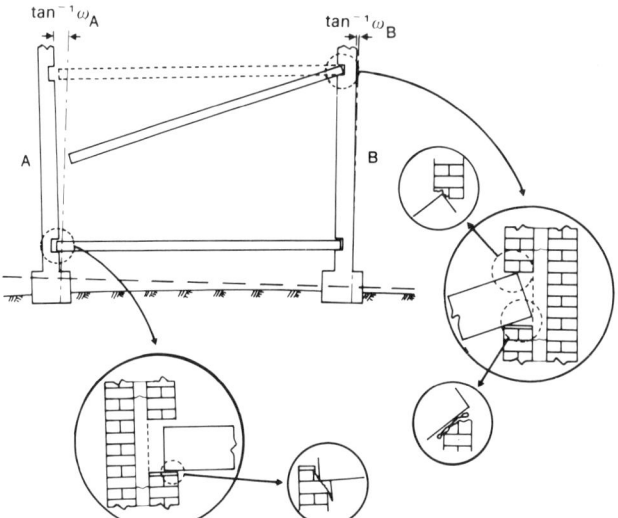

Figure 4.33 Rotation of bearing walls and consequent instability of joists.

282 SOIL MOVEMENTS

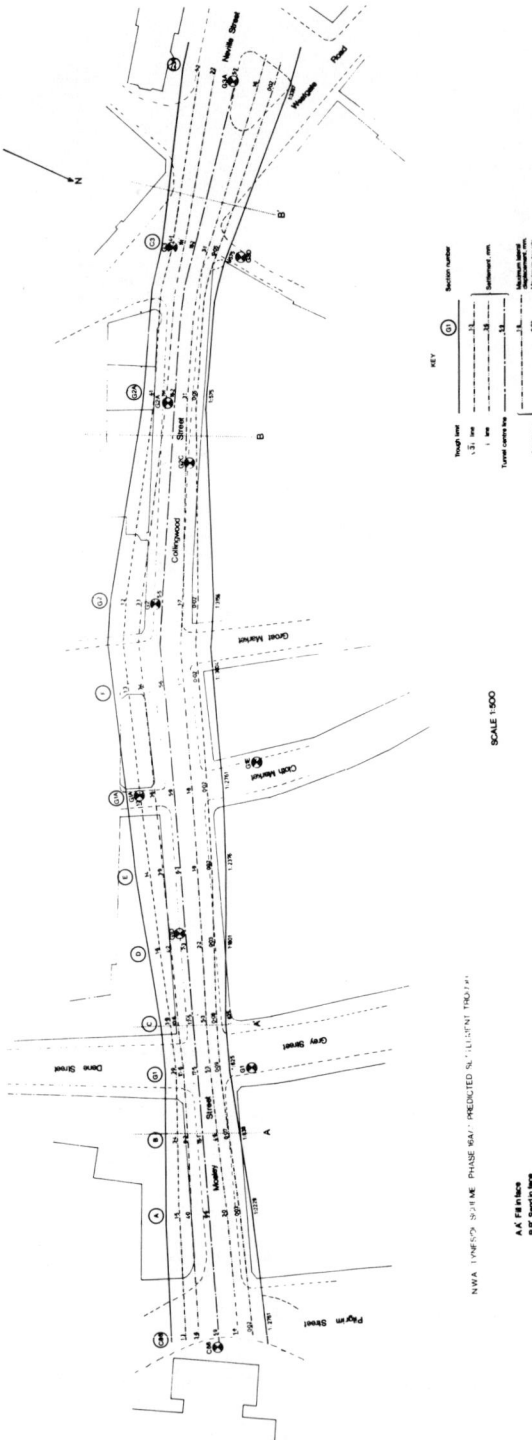

Figure 4.34 Example extract from a site investigation report: method of presenting estimated ground deformation information. Circles, transverse section reference; filled circles, exploratory borehole.

STRUCTURAL RESPONSE TO TUNNELLING SETTLEMENT 283

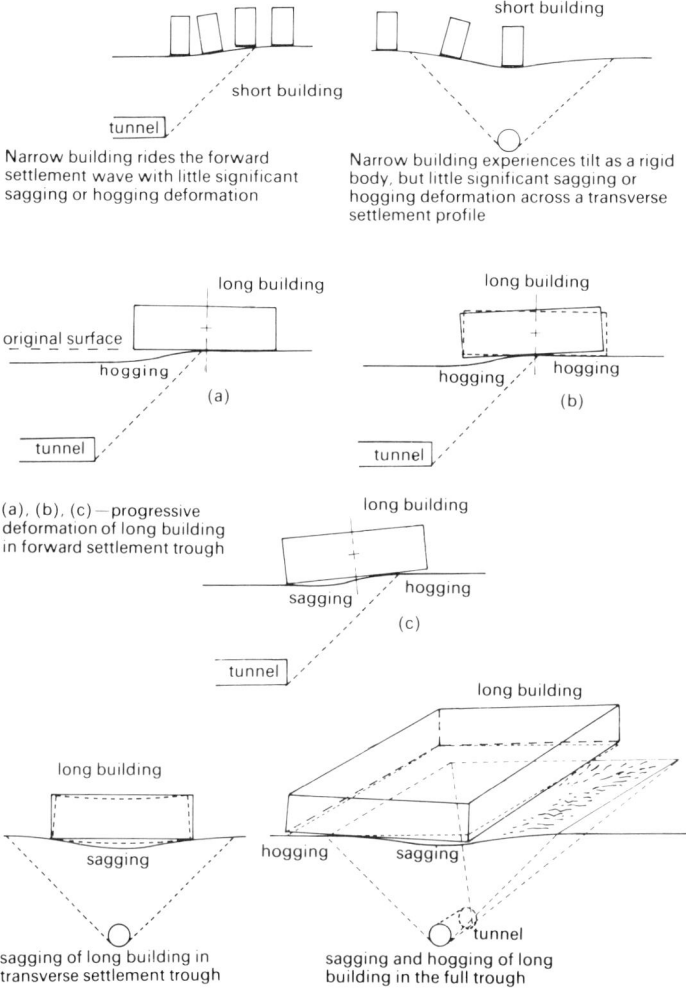

Figure 4.35 Some idealized modes of behaviour—narrow buildings and long buildings.

Figure 4.35 indicates that a short building may 'ride' a settlement trough, rigid-body tilt releasing the structure from much internal distortion (see section 4.8). For an overall assessment, notwithstanding the earlier analyses in this chapter, it is often assumed, pessimistically, at the site-investigation stage that the building foundation will deform conformably with the ground. Three structural damage criteria, based on deflection ratio, angular distortion and horizontal distortion (see Figures 4.36 and 4.37) have been discussed in Attewell (1978a) in which some detailed consideration is given to actual styles of failure. Additional references for further reading are Skempton and MacDonald (1956), Meyerhof (1956), Polshin and Tokar (1957), Bjerrum

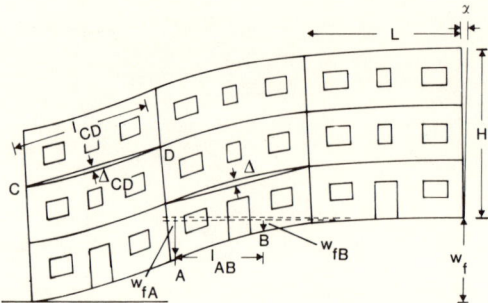

Figure 4.36 Definition of angular distortion and deflection ratio. The building response to settlement is, of course, exaggerated. Δ_{CD} is a relative sagging deflection. Δ (unsubscripted) is a relative hogging deflection. α is the rigid body rotation (tilt). w_{fA}, w_{fB} are the foundation settlements at points A and B respectively. Angular distortion $\omega = (w_{fA} - w_{fB})/l_{AB}$. Δ_{CD} is the relative deflection between points C and D. Deflection ratio $= \Delta_{CD}/l_{CD}$.

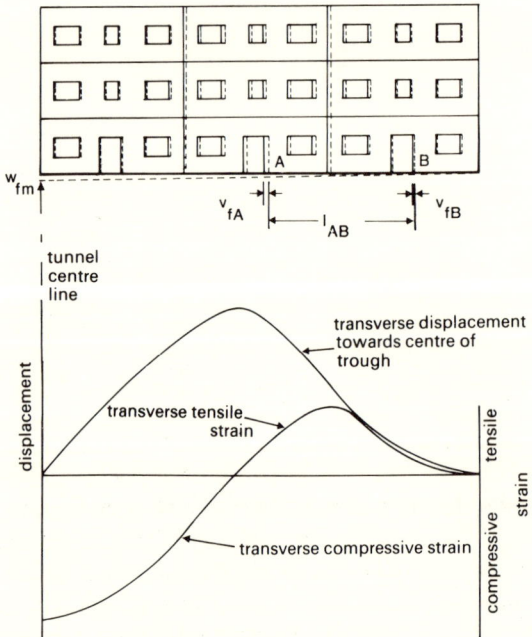

Figure 4.37 Definition of horizontal distortion. v_{fA} and v_{fB} are the transverse (horizontal) displacements of points A and B, respectively (related to the foundation), towards the centre of the settlement trough; l is the distance between A and B. Horizontal distortion ε_{yft} (foundation tensile strain) is $\varepsilon_{yft} = (v_{fA} - v_{fB})/l_{AB}$.

Table 4.5 Criteria for building damage (self-weight settlement)
(a) Angular distortion

Structure	Limit		Notes	Reference
Load bearing walls or panel walls in frame structures	1/300		Cracking likely	Skempton and McDonald (1956)
"	1/150		Structural damage probable	"
"	1/150		Design criterion against cracking	"
Frames with diagonals	1/600		Danger	Bjerrum (1963)
Buildings generally	1/500		Safe limit for no cracking	"
Panel walls	1/300		First cracking	"
"	1/150		Considerable cracking of panel and brick walls	"
Buildings generally	1/150		Danger of structural damage	"
Flexible brick walls L/H > 4	1/150		Safe limit	"
	Sand and hard clay	Plastic clay		
Column foundations, steel and reinforced concrete structures	0.002 (1/500)	0.002 (1/500)		Polshin and Tokar (1957)
Column foundations for end rows of columns with brick cladding	0.007 (1/150)	0.001 (1/1000)		"
Column foundations for structures where auxiliary strain does not arise during non-uniform settlement of foundations	0.005 (1/200)	0.005 (1/200)		

Contd.

Table 4.5 (*Contd.*)
(b) Deflection ratio

Structure	Limit		Notes	Reference
	Sand and hard clay	Plastic clay		
Plane load-bearing brick walls	0.003 (1/3333)	0.004 (1/250)	For multi-storey buildings at $L/H \leqslant 3$	Polshin and Tokar (1957)
"	0.005 (1/2000)	0.0007 (1/1500)	For multi-storey buildings at $L/H \geqslant 5$	"
"	0.001	(1/1000)		Grant *et al.* (1974)

(c) Horizontal distortion

Structure	Limit	Notes	Reference
Load-bearing walls/continuous brick cladding	1/2000 (0.0005)	Onset of cracking	Burland and Wroth (1975)

(1963), Grant et al. (1974), Leonards (1975), Burland and Wroth (1975), O'Rourke et al. (1976) and Wahls (1981). It is recommended that the latter reference be studied first and the values for the damage criteria appropriate to framed structures and load-bearing walls compared with those discussed in Attewell (1978a).

The damage criteria are listed in Table 4.5. Wahls (1981) has pointed out that allowable differential settlements in a structure tend to be significantly smaller when the settlement pattern is concave downwards (hogging mode), in contrast to the concave upwards sagging mode. Leonards (1975) has also made an important point relevant to general framed structures. He notes that whereas it is the usual practice in damage prediction and assessment to remove the rigid-body tilt element of the rotation when adopting an angular distortion criterion, tilting does in fact contribute to the stress and strain in a framed structure.

Obviously, different criteria are required for different types of building (see Table 4.5). As an overall comment, it can be said that for open-frame structures a limiting angular distortion of 1/250 could be adopted. For infill frame structures and panel walls, an angular distortion of about 1/500 should be taken as an allowable limit, 1/300 would be expected to induce some cracking in architectural elements, and 1/150 could cause structural damage. In the case of load-bearing walls and continuous brick cladding, the tolerable and allowable differential settlement would tend to be smaller than those values adopted for frame structures and would depend upon the length/height ratio of the wall. A general limiting angular distortion would be about 1/1000. In the case of a deflection ratio criterion, a limiting value of about 1/3300 has been suggested for load-bearing walls, irrespective of foundation soil.

If compensation claims for attributable damage are made to the tunnel owner then skilled appraisals are necessary to relate the nature of the particular damage to the predicted style of ground movements. Building surveyors responsible for attributing, or otherwise, visible structural damage to nearby tunnel construction operations must consider the following matters in the compensation assessments.

(1) They see the end-product of ground deformation and it may be difficult to infer the origin of deformations which caused the damage.

(2) Some damage may not be directly related but may be consequential upon factors which are attributable to tunnelling. An example is settlement due to groundwater lowering.

(3) Other environmental sources may have contributed to, or have induced damage similar in appearance to, damage caused by tunnelling. For example, buildings on either side of a road along the centre of which a tunnel is driven may have experienced tilt and hogging deformation as a result of compaction from heavy trafficking. The driving of the tunnel causes additional deformations. The trafficking may have caused buildings to reach a strain state just below the onset of cracking, and the tunnel driving would then trigger

cracking and cause severe deformations which might have occurred anyway, over a longer period of time, without the tunnelling.

(4) Virtually all the proposed structural damage criteria relate to self-weight differential settlements. Such deformations are imposed slowly, and also potential deformations will be accommodated, in part, during construction. However, tunnelling-induced deformations are imposed rapidly on to the completed structure. Consequently a more rigorous set of criteria should be adopted for this situation than would be applied to self-weight settlements. One suggestion is given in Table 4.6.

(5) If a building cracks as a result of tunnelling, then it does tend to crack rather badly, especially in the hogging mode.

An important feature of claims related to building damage caused by tunnelling is the time-barring element. The most recent relevant statute in English law is the Limitation Act (1980) which is a consolidation of previous legislation and particularly the Law Reform Committee's 21st Report (1977). The statute states that action, either in contract or in tort, must be brought within six years of the date when the cause of action occurred, otherwise it will be time-barred. Interpretation of the 'date cause of action occurred' was

Table 4.6 Criteria for building damage (brick bearing-wall structures) caused by rapidly-applied ground movements (after O'Rourke et al., 1976)

Type of damage	Distortion*	Comment
Architectural damage threshold	1/1000	As per earlier comment in text related to self-weight settlement (NB. Threshold of lateral strain in mining subsidence is also 1/1000)
Architectural damage a factor: noticeable concentration of cracks having separations of as much as 3–6 mm, also sticking of doors	1/1000 to 1/300	1/350: tunnelling in soft silty clay (Cording et al., 1978)
Functional damage: inconvenience to occupants. Jammed doors and windows, broken window glass, building services disrupted; cracks and separations could be up to 13–25 mm wide	1/300 to 1/150	1/167 to 1/220: tunnelling in soft silty clay (Cording et al., 1978)
Spalling of stone cladding and possible collapse of cornices along facade wall (differential movements parallel to brick bearing-walls)	1/150 to 1/125	

* Angular and lateral distortions are assumed to be equal and are expressed at ground surface.

clarified in the House of Lords, when the defendant's appeal was upheld in the case *Pirelli General Cable Works Ltd v. Oscar Faber* (1983). The opinion of Lord Fraser, that 'time begins to run from the date damage occurred' was authoritative, despite the plaintiff's plea that the term should begin on the date of discovery of damage. This distinction is vital, in that damage is defined to occur at the time of the cause (in our case, perhaps the date of tunnel driving), so that the six-year time-bar acts from this date, even if structural distress is not apparent for some time after driving. However, there is the possibility of a change in the situation resulting from the Law Reform Committee, 24th Report, 'Latent Damage' (1984), which recommends no change in substantive law (i.e. the six years to run from date damage occurred), but that in cases of negligence there should be an extension of three years from the time of 'discoverability', although a long-stop of 15 years should apply. This is not yet law.

4.14 Property schedules

Detailed statements concerning the structural condition of all buildings likely to be affected by tunnelling works should be prepared by, or under the direct supervision of, qualified building surveyors before tunnelling takes place. A copy should be delivered to the owner of the building (or his agent). Because of the detail required, these statements can be quite expensive to prepare. The surveyor must therefore calculate the practical limits of the transverse settlement trough, these limits depending, of course, on the tunnel depth and the type of ground. Reference may be made to Chapter 2 of this book but, for a first estimate, a distance of 1.5 tunnel depths from the tunnel centre line may be taken as defining a reasonable practical limit.

The degree of detail required in a property schedule will obviously depend upon the structural state of that property and to some extent upon the property value. The written schedule should be liberally supplemented by colour photographs, the prints ideally being taken by Polaroid or equivalent 'instant' camera so that they can be marked for reference purposes on the spot. In addition to a camera, tape-measure and notebook, the survey team usually carry a moisture meter, dictaphone or small tape-recorder, and binoculars. They also need to have a builder's level and a pair of steps available, and access to ladders.

The following fictitious schedule on a building in rather poor condition is included to give some idea as to the style which such survey reports should take. After tunnelling operations the property would be resurveyed, each comment in the original schedule re-examined, and conclusions expressly drawn as to differences in property conditions that could be directly attributed to tunnelling operations. There are references to photographs in the fictitious schedule, but photographs are not included here.

It is also recommended that, before tunnelling and with the agreement of the property owner, any significant cracks have 'Demec' points (Figure 2.45) installed on either side of them and initial readings taken for comparison with post-construction readings. If available, tiltmeters (Figure 2.44) may be installed at locations as advised by the Engineer's staff.

DRIFLOW TRANSPORT AUTHORITY
ESTATES DEPARTMENT
EAST–WEST RAPID TRANSIT TUNNEL
PHASE 8B

Schedule of Condition of: 123 Fountain Street, Waterford.

Owner: Mr I.C. Klame

Occupier: Mr and Mrs B.A. Way

Schedule of Condition taken on the following dates:

> Internal: 21 February 1981
> External: 20 February 1981

Notes:

1. The schedule describes the general condition of the property only. It has not been possible to examine those parts of the property which are inaccessible or covered by furniture, carpets, etc.
2. The schedule of condition covers in detail only that part of the building which it is estimated by the Engineer comes within the sphere of influence of the tunnel.
3. Measurements mentioned within this report are taken as accurately as possible but in some instances they have been estimated due to the height and style of the building.

External condition

Front elevation

Roof: Most of the slates are loose and distorted from their true position. The mortar beading to the ridge is generally cracked with small pieces missing. For a general view see photograph number 1. The chimney pots are leaning to the east and the west pot is holed.

Walls: This whole building has moved out of true vertical alignment and facing this elevation it can be seen that the west elevation has bowed outwards to the west approximately one half brick in width. On this elevation there is evidence of previous cracking to the brickwork at the west side of the sill of the second floor window. This has been patched in the past but is now re-opening, dropping through three courses of brick.

Windows: The lintels and sills are generally weathering.

West elevation

Roof: The ridge tiles are out of horizontal alignment, and they sag at the centre. The slates generally are distorted and loose most especially to the south half where there is a slate missing. The chimney stack at the north edge is out of original alignment having twisted towards the northwest/southeast. For a general view see photograph number 2.

Walls: As previously noted, this elevation has bowed outwards. Most of the brickwork has been repointed in the past.

Rear elevation

Walls: This elevation bows out severely at the basement and ground floor levels to the extent of approximately one brick in width. A steel plate has been placed on the exterior of this bulge. The brickwork to the centre of the bulging portion is more recent than the remainder of the brickwork. This recent brickwork is showing signs of stress with a number of bricks showing full-height pencil-line cracks. There is a more predominant crack to the west side of the steel plate. This crack steps up through the mortar joints and brick, approximately 20 courses at heavy pencil-line thickness. The majority of the brickwork is in need of repointing. The brickwork has washed out and there are damaged bricks to the east side of the dilapidated door. There is heavy cracking/open jointing generally around the wood beam supports. There is one half missing brick and brickwork has washed out of the bottom west corner. The small window has distorted and the pane is broken. For a general view of this elevation see photographs numbers 3, 4, and 5.

East elevation

Roof: The whole roof sags into the centre.

Walls: There is evidence that there has been a line of split bricks at eaves level. Most of those bricks are now missing. This elevation is generally in poor condition and most of the bricks are in need of repointing. At the south end there has been some movement in the building and there is a serious structural crack rising from the roof line of the adjacent lock-up garage. This crack steps vertically up to approximately mid-height on the first floor level. There is evidence that the brickwork has been repointed around this crack which covers approximately one-third of this elevation. There is also evidence that an attempt has been made in the past to fill this crack. It has, however, re-opened and at its widest is approximately one half-brick width. For a general view see photographs numbers 6 and 7.

Windows: The lintels and sills have generally weathered and the second floor window sill has a chipped portion missing from the south end. The second-floor window has three broken panes.

Internal condition

Second-floor room

This room has been divided by a wood panel partition at approximately room centre. The panels have warped with the dampness. The ceilings are unlined boards with wood battens. These generally have warped and twisted out of original level. The walls have been wood-panel clad to a height of 1.25 m. The remaining portion of wall above this height up to the ceiling line is generally showing evidence of dampness with flaking decoration on all sides. There is water penetrating at time of inspection on the west side in the front portion. In the front portion there are cracks evident to the top portion of wall. On the west there are three vertical pencil-line cracks, and on the south there is one vertical pencil-line crack. At the southeast corner there is a heavy pencil-line crack to full height and a portion of the plasterwork has chipped away. On the east there are two full-height diagonal ascending pencil-line cracks and one vertical hairline crack. In the rear portion there are five full height pencil-line cracks to the west portion of the wall above the panelling. This section of the wall is bulging out along its entire length, with evidence of dampness and flaking decoration. The floor tends to fall towards the northeast corner. On the east window there are three broken panes. The hearth to the fireplace has three full-width cracks through it, one heavy pencil-line thickness, one of chalkline thickness and one of maximum width 5 mm. For a general view of this room see photograph number 8.

Stairs down to first floor

At the head of the stairs at the northwest corner there is evidence of stress cracking/crazing to full height, the cracking ranging from hairline to chalkline thickness. At approximately 1.2 m from the floor there is bulging of the plasterwork associated with the aforementioned cracking and covering a height of approximately 0.3 m—see photograph number 9.

There are two cracks descending from the ceiling at this corner, and associated with these is hollow-backed plasterwork. One extends down to the doorhead at heavy pencil-line/chalkline thickness. The other, at chalkline thickness tailing down to heavy pencil-line thickness, extends down 0.6 m at which point it then runs into the north wall, ascending diagonally up to the ceiling line and then returning down the ceiling line to the northeast corner. There is some minor flaking decoration to full height at the northwest corner. On the north wall there are two further diagonal cracks situated at approximately centre. One is associated with cracking at the northwest corner and extends from the northwest corner, rising 0.65 m towards the northeast. The other is associated with a crack in the northeast corner and this descends from the northeast corner 0.6 m towards the west. There is a pencil line crack rising 1.35 m from the base at the northeast corner at which point it then ascends diagonally into the east wall, rising up to the ceiling line, associating with crazing and hollow-backed plaster, and covering an area 0.3 m × 0.15 m in height. On the east wall at the north end there is a hairline diagonal ascending crack extending up from the northeast corner 0.54 m and this is situated approximately 0.8 m in height at the northeast corner. On the east wall above the window there is a diagonal ascending crack extending from the stairs up to the ceiling line. Its approximate length is 1.36 m and its maximum width is 1.5 mm. Associated with this there is a hairline diagonal ascending crack rising from approximately mid-distance and extending 150 mm towards the north. At approximately mid-height on the north reveal of the window there is a full-width pencil-line crack which then extends through the east wall diagonally down to the stairs, 0.47 m in length, maximum width 0.9 mm. At approximately centre on the head reveal of the window there is a full-width crack ranging from hairline through to chalkline. This returns above the window into the east wall to the ceiling line 2.1 mm width and 50 mm length. There is a further hairline crack full width of this reveal at approximately centre and again this extends up to the ceiling line on the east wall. There is open jointing generally between the reveals and the window frames. The east wall is suffering from dampness with blistering and flaking decoration overall. On the landing the walls and ceilings are wood panelled.

Floor: The stairs lean from east to west.

First-floor room

Ceiling: The ceilings are unlined plasterboard panels which again have bevelled due to dampness. One of the boards at the northwest corner has been removed. See photograph number 10.

Walls: The walls have all been panel clad. In the cupboard under the stairs a small portion of one of the panels has been broken away revealing a crack in the east wall. This is associated with a major crack in the structure which will be dealt with at the next floor level. At this level the crack has a maximum width of 2.1 mm.

Window: There is one broken pane.

Floor: The floor appears to slope from west down to east.

Stairs down to ground floor

Ceiling: There is evidence of dampness to the ceiling with staining evident and minor flaking decoration.

Walls: The walls are very damp with flaking/peeling decoration and efflorescence overall. See photographs numbers 11 and 12. On the east wall at the top south end there is a severe diagonal ascending crack, 0.67 m in length and 5.4 mm maximum width. This crack then runs along the ceiling line to the north at heavy pencil-line thickness and then descends down the northeast corner on the north wall to the doorhead at heavy pencil-line/chalkline thickness. At the southeast corner there is a heavy crack to two-thirds the height associated with the aforementioned severe crack. Associated with this there is a full-width diagonal ascending hairline/pencil-line crack at approximately half-height on the east wall. At the base of the southeast corner situated on the south wall there is a severe diagonal ascending crack rising from the skirting level and extending to the southeast corner. It is 0.55 m in length, maximum width 8.2 mm and maximum depth 30 mm. At the base of the south wall there are two diagonal severe cracks through the plasterwork and the plasterwork is hollow-backed. The most easterly of the cracks has a maximum width of 20.6 mm and a maximum depth of 51 mm. The bottom crack has a maximum width of 11.1 mm and a maximum depth of 63 mm. See photograph number 13.

On the north wall there is a full-width horizontal heavy pencil-line crack running level with the top step. At the ground-floor entrance door the floor slopes down from west to east. There is a hole in the floor approximately half the width of the passage and one board wide.

Ground-floor room

Ceiling: The ceiling has been panel clad with unlined plasterboard. It is generally sagging towards the centre and one board has been damaged. For a general view see photograph number 14.

Walls: Generally the decoration is peeling due to dampness. The west wall is out of vertical alignment and leans towards its top. There is a horizontal pencil-line crack on the west wall extending from the southwest corner at approximately a height of 1.65 m, and extending to about 1.5 m in length. On the west wall there is a further horizontal pencil-line/hairline crack extending approximately 1.6 m in length. This is situated in the centre of the wall at a height of approximately 1.8 m from the floor. At the northwest corner there is full-height heavy pencil-line cracking. Associated with this on the west wall at approximately mid-height there is a descending diagonal crack of heavy pencil-line thickness and approximately 0.6 m in length.

A severe crack rises on the west face of the chimney breast to approximately half height. Its maximum width is 9.5 mm and its depth is in excess of 155 mm. On the east wall at the north end there is a severe structural crack occupying the full height of this wall. Daylight is visible over the entire length of the crack. There are numerous heavy pencil-line/chalkline vertically orientated cracks associated with this particular crack. These cracks pass from the ceiling line to approximately half depth. The plasterwork surrounding this major crack has been damaged along the full length. Generally the maximum width of the crack between the brickwork is approximately 30 mm; however, at the base this extends in width to approximately one half-brick. For a general view of this crack see photographs numbers 15 and 16. On the east face of the chimney breast there is a severe structural crack extending over approximately two-thirds the height at the corner. Associated with this there is a descending diagonal heavy chalkline crack situated at approximately mid-height and extending over a length of approximately 0.5 m. For a general view see photograph number 17.

At the south end of the east wall there is severe dampness with flaking decoration and efflorescence overall. At the base at the south end of this wall there is a damaged portion of plasterwork with a missing section covering an area of approximately 0.7 m in height × 0.9 m in length.

Floor: The whole of the ground floor slopes west to east. There is a portion of a board missing at the northeast corner. The floor at the southeast corner vibrates. There is a steel tie bar running across the floor from the southeast corner through to the north wall at its northeast corner.

Stairs down to basement

Walls: The south wall is very damp, with disintegrating plasterwork at the east side. See photographs numbers 18 and 19. The top two steps are loose due to the fact that the supports have rotted. The bottom two steps are loose.

Basement room

Once again the ceiling comprises unlined plasterboard. These boards are damp and have generally bevelled, three of the panels having been damaged. The walls generally are of unlined brickwork. For a general view of this area see photograph number 20.

At the northeast corner there is a severe crack associated with the aforementioned severe structural crack on the ground floor level. This crack rises full height, floor to ceiling, and is of average maximum width 16 mm. See photograph number 21.

In the northeast alcove there is a full-height severe chalkline crack to the northwest corner.

For practical appraisals of damage, and notes on other possible contributory factors which must also be assessed in conjunction with tunnelling-induced movements, reference may be made to the Building Research Establishment Digests (1979*a,b*; 1981*b*).

4.15 Bridge structures

The typical short-span highway bridge is a highly articulated structure, which behaves rather differently from building structures (see Sims and Bridle, 1966; Sims, 1982; Selby, 1985).

296 SOIL MOVEMENTS

Figure 4.38 Typical short-span highway bridges in the UK.

In the UK, a substantial number of bridges exists in the 12–16 m span range (see Figure 4.38). Codes and memoranda relevant to their design include BS 153, CP 117, BE 1/73, BE 5/73 (DoE, 1973a; 1973b), now superseded by BS 5400. While many of these bridges are individual because of special requirements of skew, ground conditions, or aesthetics, a substantial proportion of this stock conforms in the main to a simple configuration. Substructure end supports are typically a vertical abutment wall on a spread footing base. The abutment wall serves to carry the loads from the deck bearings down to ground, and also retains the embankment. To each side of the abutment are wing walls which are free-standing retaining walls sloping down from the abutment wall, to control the embankment slopes. Alternative forms (such as bank seat, spill-through, cellular, counterfort) exist, but are not considered here. Intermediate supports (piers, walls or columns) are unlikely to introduce additional problems, and are not considered specifically. A shelf on the abutment carries bearings which in turn support the deck. The bearings must allow rotation due to flexure of the deck, and may also be required to allow longitudinal movements. The deck, which sits on the bearings, will probably consist of precast prestressed concrete beams with infill or with slab decking, or steel universal beams with composite concrete deck (see Figure 4.39 and Somerville and Tiller, 1970; Manton and Wilson, 1971; and Ward et al., 1970). This is topped with a waterproofing layer and wearing surface. At each end of the deck a movement joint will be provided which should prevent moisture penetration into the gap between deck and abutment. Along the deck edges

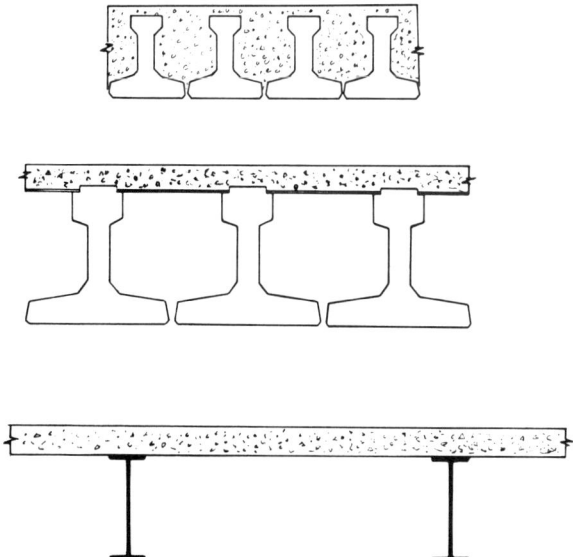

Figure 4.39 'T' beam, 'M' beam, and steel/concrete composite bridge decks.

runs a parapet, and beneath the footpath or suspended from the deck will be service ducts and pipes.

Example 4.18 Responses of bridge components to tunnelling

The response of a typical highway bridge, with its alternative deck constructions, will be considered with respect to settlement caused by two example tunnels. The first is a 1.5 m diameter sewer at an axis depth of 4 m below ground, in a clay soil. The second is an underground transport or metro tunnel of 4.8 m diameter, driven at an axis depth of 12 m through a medium dense, non-cohesive soil. In assessment of potential damage to the several bridge components, the line of the tunnel will be chosen to cause maximum distress.

For the chosen example of a 1.5 m tunnel at 4 m depth in a clay soil with an undrained cohesion of 0.15 N/mm^2, a realistic value of surface settlement volume V_s is 15 per cent of the face area. A value for i, the trough parameter, is chosen as 2.8 m. Hence the maximum settlement can be calculated from equation (4.1) as 3.8 mm. Transverse displacement at $y = i$ is given by equation (4.2) as 1.6 mm. The maximum transverse slope is 0.0008 radians, from equation (4.3).

For the case of a metro tunnel, 4.8 m diameter, driven at 12 m depth through a medium sand, the surface settlement volume V_s is taken as 5 per cent of the face area, and the trough parameter, i, is taken to be 3.2 m. These give a maximum settlement of 113 mm, a transverse displacement at $y = i$ of 18 mm and a maximum transverse slope of 0.021 radians.

It should be noted that these settlement profiles are calculated on the assumption of a level ground surface with no restraint from a structure.

The interactive stiffness method is used to analyse walls on soils.

Walls

A wing wall element is chosen for assessment of possible damage to retaining wall members of the typical bridge, due to the two example tunnels. This is 12 m long with a base 3 m wide and 600 mm thick, and a wall 5.6 m high at one end sloping to 1.6 m and 400 mm thick. Two lines for each tunnel are investigated, running at right angles to the wall and either beneath the mid-length of the wall or beneath the low end of the wall: Cases A and B. These conditions cause maximum bending of the wall as a deep beam, either as simply supported or cantilevered over the trough. To avoid the expense of a full three-dimensional analysis of the ground, it is assumed that the ground can be modelled by a two-dimensional strip, 4 m thick, 22 m deep and 48 m long. Elastic moduli, E, (constrained *in-situ*) of 5 N/mm^2, for the clay and 8 N/mm^2 for the sand are chosen.

Calculations of load of $\mathbf{K_g}\mathbf{w}$ are made, and used as input to a plane-stress finite-element analysis of wall and ground. A summary of maximum wall stresses is given in Table 4.7.

Table 4.7 shows that the settlement stresses induced by the 1.5 m sewer tunnelled in clay soil are trivial. Even the very severe case of the metro tunnel in the sand showed surprisingly small wall stresses, which might be tolerated depending on the reinforcement in the wall.

STRUCTURAL RESPONSE TO TUNNELLING SETTLEMENT 299

Table 4.7 Response of wing walls to tunnelling

	Settlement max. (mm)	Wall rotation (deg)	Concrete stresses (N/mm²)			Ground stresses (N/mm²)		
			σ_c	σ_t	τ	σ_c	σ_t	τ
Sewer in clay								
Case A	1.0	0.0	0.053	0.030	0.027	0.004	0.0033	0.0017
Case B	2.4	0.015	0.018	0.033	0.016	0.0018	0.0018	0.0007
Metro in sand								
Case A	30.8	0.02	2.1	1.4	1.4	0.132	0.14	0.049
Case B	90.0	0.52	0.82	1.3	0.68	0.059	0.072	0.027

Certain tunnel lines would cause relative settlement and tilt of abutment and wing walls. A detail common in bridge design is a short overhang of a wing wall over the abutment base. A preferable, but slightly less economical, arrangement is to form a single vertical joint between abutment and base to wing wall and base. This alternative arrangement is more tolerant of differential settlement, although no better in absorbing relative in-plane tilt (see Figure 4.40).

Embankment behind the end supports

Because of a cheap and plentiful source of lightweight pulverized fuel ash, embankments behind bridge abutments have often been constructed of this material. Pulverized fuel ash (pfa) shows self-hardening, pozzolanic behaviour through hydration of the free lime content (Sutherland *et al.*, 1968). Thus, a strong, stiff, lightweight embankment may be constructed. A note of caution is voiced here, however, concerning susceptibility of compacted pfa to cracking during imposed deformation.

Compacted and aged pfa might be expected to crack at a shear strain of about 0.03. The example tunnels would generate, respectively, tensile strains of 0.001

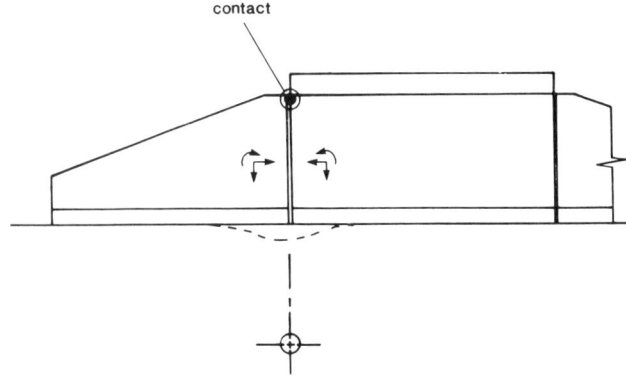

Figure 4.40 Potential damage by movement joint closure.

and 0.01, which might thus cause severe cracking with potential damage to the embankment.

Bearings

Purpose-designed bearings are incorporated into the typical bridge design, as components which rest on a shelf on the abutments and upon which rest the deck beams. One bearing is used at each end of every beam. The purpose of the bearings is to provide a compliant support to the beams which is tolerant of rotation due to beam flexure, and of longitudinal movement due to thermal changes. The longitudinal movement is accommodated at the expansion end, and the deck is restrained in position at the fixed end either by restraint bearings or by vertical dowel bars. A bearing in the typical bridge would probably be an elastomeric prism (perhaps reinforced with steel plates), although proprietary alternatives exist. Current practice is to specify bearings by performance, rather than by material, so bearings will now be examined within this context.

Tunnelling-induced settlements can affect bridge bearings in three ways. First, load may be redistributed due to in-plane tilting of the abutment wall. Second, the bearing may be required to absorb additional relative rotation due to forward (or backward) tilting of the abutment and relative settlement between abutments. Third, additional longitudinal movement may be enforced both by the forward tilt of the abutment and by the transverse horizontal movement of soil toward the trough centre. The load redistribution is a function both of bearing stiffness and of the deck warping stiffness, and will therefore be treated together with the deck assessment.

As a first step, the bearing expansion and rotation requirements will be calculated for normal highway conditions. Consider a 14 m span deck constructed of precast prestressed inverted T beams, type T7, giving an overall deck thickness of 780 mm. End slope due to weight of *in-situ* concrete plus surfacing will be 0.013 radians (using an elastic modulus of concrete of 0.5 × short-term value). End slope due to HA loading will be approximately 0.001 radians. The longitudinal movement required of the bearing from thermal strains only is ±4 mm.

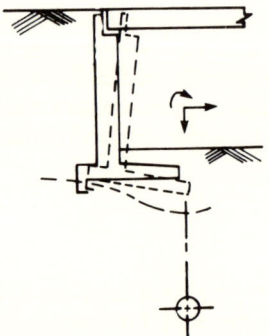

Figure 4.41 Forward tilting of abutment, causing failure of bearings and of abutment/deck joints.

Now consider movements induced by forward tilting and movement of an abutment towards a tunnelling trough (see Figure 4.41). The maximum side slope of the trough is defined by equation (4.3) at $y = i$, and for the tunnel in clay soil its value is 0.0008 radians. To this must be added the change in deck slope, due to settlement of the abutment, of 0.0002 radians, giving a total of 0.001 radians. This is comparable with the HA rotation and can be tolerated easily. The longitudinal movement required of the bearing comprises the effect of tilt and of forward movement, which are 5 mm and 1.6 mm, respectively, and totalling 6.6 mm. This exceeds the entire design requirement, and would probably result in distress to the bearings, which might require to be reset. This emphasizes the desirability of designing so as to allow access to bearings.

Similar calculations for movement induced by a metro tunnel in a granular soil show an increase in rotation of 0.028 radians, which would distress the bearings, and a massive longitudinal movement of 120 mm, which could not be tolerated, either by bearings or by the deck/abutment joint.

Consider, however, alternative lines for tunnelling transversely to the bridge. If the tunnel centre line were at midspan of even a 12 m bridge, then rotations would be halved to tolerable values. The longitudinal movements, however, would be reduced scarcely at all, since both abutments would now tilt forward and move towards each other. Tunnel lines either directly beneath an abutment, or at some distance ($> 4i$) behind an abutment, are potential solutions.

At the fixed end of the deck, stainless steel dowel bars are commonly included to locate and restrain the deck. These are firmly cast into the abutment top, and are cast into the deck end diaphragm with a deformable surround so as to allow rotation. This deformable surround should be sufficient to absorb additional rotation, but even if it lacks capacity, the dowel bars are sufficiently ductile to accept plastic deformation, and remain serviceable thereafter.

Expansion joints

Seals are provided above the gaps between abutment and deck end, in an attempt to prevent water penetration. At the fixed end, a buried joint will be used which would accept the additional rotation. At the expansion end, however, either a buried joint with rubber seal, or an epoxy-nosing joint with mastic seal would fail under the additional longitudinal movements discussed in the section on bearings. A resealing could be implemented cheaply and easily for the sewer tunnel example, but obviously not for the metro tunnel. Indeed, if the metro tunnel were driven transverse to the bridge span, the gaps between the deck end and abutment would close and the abutment upstand walls would be severely damaged.

Deck

The three most common types of deck structure in the 12 m to 16 m span are probably precast prestressed inverted T beams with infill *in-situ* concrete, precast prestressed M beams with a thin *in-situ* deck slab, and universal steel beams with composite concrete slab. The T-beam deck of T7 beams is effectively a solid slab for these calculations. The other two decks are significantly stiffer longitudinally than they are transversely. Live-load bending moment is of the order of 300 kN m/m width, and live-load shear will be up to 90 kN/m width.

Potential damage to the deck slabs is through warping caused by in-plane tilting of one abutment only, as a result of tunnelling beneath one end. The rare situation of a tunnel passing diagonally beneath a bridge causing opposite abutment corners to settle will not be considered here, although that particular case would cause double the following distress.

To assess the damage due to tilting of one abutment, it is assumed that the sewer in clay causes a slope of 0.0003 radians, while the metro in sand causes 0.01 radians, these figures being taken from the wing wall calculations. The three deck slabs, of the types mentioned above, are assessed by torsion and warping theory (Timoshenko and Goodier, 1970) and checked (with close agreement) using finite-element analyses of plate plus asymmetric beam elements.

The infilled T-beam deck, being effectively a closed-section homogeneous slab, is relatively sensitive to imposed warping deflection. The warp induced by the sewer in clay is sufficient to cause surface shear stresses of 0.21 N/mm^2 (cf. 0.37 N/mm^2 permissible, 4.10 N/mm^2 ultimate for torsional shear and direct shear). This is tolerable, especially since these stresses might be alleviated by up to 50 per cent if bearings of low vertical stiffness were used. The analysis, assuming rigid bearings, shows bearing loads to be increased, but only by 9 kN (cf. design capacities of 125 kN). The effect of the metro tunnel is severe, resulting in shear stresses up to 6.7 N/mm^2, which would cause cracking and damage to the deck and also to the bearings.

The M-beam deck shows much greater tolerance to warping, with trivial surface shear stresses of up to 0.045 N/mm^2 due to the sewer in clay soil, and negligible direct stress. Even the metro tunnel causes only 1.5 N/mm^2 of shear stress which could be absorbed by many designs, especially if the bearings were compliant. Even stiff bearings would attract only 3 kN additional load.

The steel universal beam and composite concrete slab would react in a similar way to the M-beam deck, with concrete shear stresses of only 0.045 N/mm^2 in the deck slab, and 0.04 N/mm^2 in the steel. Again, the metro tunnel settlement would cause shear stress of 1.5 N/mm^2 in the deck, and trivial shear in the steel.

Parapets

Parapets are constructed along the deck edges to limit the possibility of vehicles or pedestrians falling off the deck. The design is influenced by the vehicle design speed of the road over the bridge, but typically they consist of three longitudinal box-section members, carried at 2–3 m intervals by posts, and with mesh infill. The members are of steel or aluminium alloy. Less commonly, concrete walls are constructed. Movement joints are provided in the longitudinal members, above the deck joints, to accommodate thermal movements. The additional movements in the deck joints, due to tunnelling, could be absorbed for the sewer in clay soil, but not for the metro in granular soil. However, repairs to damaged parapets are relatively cheap and simple, since all components are accessible.

Services

Commonly, bridges carry electricity power supply cables, telephone lines, and perhaps water and gas pipes. Electricity lines are usually loosely laid inside PVC ducts, and are fully tolerant of the possible bending or stretching due to tunnelling-induced movements. Gas and water pipes are usually of spun iron or steel, welded or with rubber-seal bolted joints, and are carried over the bridge beneath the footpath in loose sand, or on hanger brackets beneath the deck, or in a large duct. The hanger brackets are sufficiently flexible to allow relative rotation to be spread over some distance. Pipes of less than about 100 mm diameter are unlikely to suffer damage even during extreme movements, but a 300 mm pipe, following a deck/abutment rotation of 0.028 radians, might suffer an elastic stress of 300 N/mm^2. Fortunately the ductility of steel, and even of spun iron, is ample for absorption of imposed strains. The details of joints are non-standard, but pipe joints would best be remote from the deck joints to avoid potential damage and leaking. The axial deformation which a bridge could tolerate, of say 50 mm, with reset bearings and joints, might cause tensile strain of 0.004 in a pipe, which would cause yield, but not failure. Again, however, any joints might be damaged causing leakage.

In conclusion, the 1.5 m prototype sewer tunnel in stiff clay could be driven beneath the bridge structure along any line, without causing damage to the main structural elements. However, this tunnel could cause forward tilt and movement of the abutments, which would necessitate maintenance to the bearings and the deck movement joints and seals. The line of such a tunnel, transverse to the bridge span, might cause least damage if it were chosen to run beneath one abutment, so as to cause settlement of this one abutment without tilt or horizontal movement. A tunnel parallel to the bridge span could be aligned almost at will, the only place to avoid being beneath the abutment-to-wing-wall joint, which would then be forcibly closed.

Location of the line of a 4.8 m prototype metro tunnel beneath the bridge would require considerable planning. If this tunnel were running parallel to the bridge span, then a line either under the abutment centre, or under the outer third of a wing wall, would cause no damage. Such a tunnel driven transversely to the span would cause irremediable damage to bearings, joints and to the upper abutment walls, unless it were situated 10 m or more behind an abutment wall. A diagonal drive would require careful alignment, particularly if the deck were of concrete beam and infill construction, and, in addition, monitoring and analysis of settlement during driving would be necessary.

Appendix A

Infinite beam on elastic foundation—Winkler foundation model
Formulae for deflection, angular deflections, bending moment and shear force
(after Hetenyi, 1946)

The following are positive: downward deflection (w_b), clockwise angular deflection (θ), shear force acting upwards on the left of the elemental section (Q), bending moment on the left of the element in the direction of positive shear force (M), downward acting load (P, q).

The functions A_x, B_x, C_x and D_x are defined as follows:

$$A_x = e^{-x}(\cos x + \sin x), \quad B_x = e^{-x} \sin x,$$
$$C_x = e^{-x}(\cos x - \sin x), \quad D_x = e^{-x} \cos x,$$

where the value of the parameter x is always taken as positive and in the trigonometric functions the parameter x is in radians.

Semi-infinite uniformly distributed loading

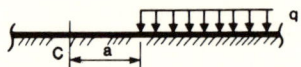
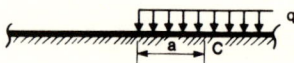

At point C to left of loading:

$$w_{bc} = \frac{q}{2K}(D_{\lambda a}), \quad \theta_c = \frac{q\lambda}{2K}(A_{\lambda a})$$

$$M_c = -\frac{q}{4\lambda^2}(B_{\lambda a}), \quad Q_c = \frac{q}{4\lambda}(C_{\lambda a})$$

At point C under loading:

$$w_{bc} = \frac{q}{2K}(2 - D_{\lambda a}), \quad \theta_c = \frac{q\lambda}{2K}(A_{\lambda a})$$

$$M_c = \frac{q}{4\lambda^2}(B_{\lambda a}), \quad Q_c = \frac{q}{4\lambda}(C_{\lambda a})$$

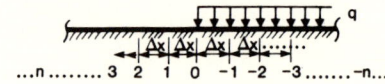

At point n the bending moment is given by:

$$M_n = \frac{q}{4\lambda^2} I_n,$$

where I_n is an influence factor given in the following table.

↓Δx \ n→	0	1	2	3	4	5	6	7	8	9	10	11	12
3/λ	0	−.0071	+.0007										
2/λ	0	−.1230	+.0139	+.0007	−.0003								
1/λ	0	−.3096	−.1230	−.0071	+.0139	+.0065	+.0007	−.0006	−.0003				
1/1.5λ	0	−.3175	−.2562	−.1230	−.0318	+.0068	+.0139	+.0094	+.0039	+.0007			
1/2λ	0	−.2908	−.3096	−.2226	−.1230	−.0492	−.0071	−.0106	+.0139	+.0108	+.0065	+.0029	+.0007
1/2.5λ	0	−.2610	−.3223	−.2807	−.2018	−.1230	−.0613	−.0204	+.0024	+.0121	+.0139	+.0117	+.0082
1/3λ	0	−.2344	−.3175	−.3096	−.2562	−.1880	−.1230	−.0701	−.0318	−.0071	+.0068	+.0128	+.0139

$|I_n| < .0001$ for entries not shown.

↓Δx \ n→	13	14	15	16	17	18	19	20	21	22	23	24	25
1/2λ	−.0003	−.0006	−.0005	−.0003	−.0002								
1/2.5λ	+.0049	+.0023	+.0007	−.0002	−.0006	−.0006	−.0005	−.0003	−.0002	−.0001			
1/3λ	+.0122	+.0094	+.0065	+.0039	+.0020	+.0007	−.0001	−.0005	−.0006	−.0006	−.0005	−.0003	−.0002

$|I_n| < .0001$ for entries not shown.

Note: $I_{-n} = -I_n$

Triangular loading

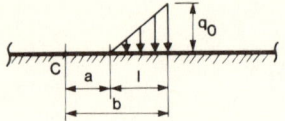 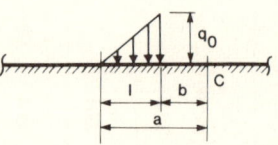

At point C to the left of the loading: At point C to the right of the loading:

$$w_{bc} = \frac{q_0}{4\lambda K}\frac{1}{l}(C_{\lambda a} - C_{\lambda b} - 2\lambda l D_{\lambda b})$$
$$w_{bc} = \frac{q_0}{4\lambda K}\frac{1}{l}(C_{\lambda a} - C_{\lambda b} + 2\lambda l D_{\lambda b})$$

$$\theta_c = \frac{q_0}{2K}\frac{1}{l}(D_{\lambda a} - D_{\lambda b} - \lambda l A_{\lambda b})$$
$$\theta_c = \frac{-q_0}{2K}\frac{1}{l}(D_{\lambda a} - D_{\lambda b} + \lambda l A_{\lambda b})$$

$$M_c = \frac{-q_0}{8\lambda^3}\frac{1}{l}(A_{\lambda a} - A_{\lambda b} - 2\lambda l B_{\lambda b})$$
$$M_c = \frac{-q_0}{8\lambda^3}\frac{1}{l}(A_{\lambda a} - A_{\lambda b} + 2\lambda l B_{\lambda b})$$

$$Q_c = \frac{-q_0}{4\lambda^2}\frac{1}{l}(B_{\lambda a} - B_{\lambda b} + \lambda l C_{\lambda b})$$
$$Q_c = \frac{q_0}{4\lambda^2}\frac{1}{l}(B_{\lambda a} - B_{\lambda b} - \lambda l C_{\lambda b})$$

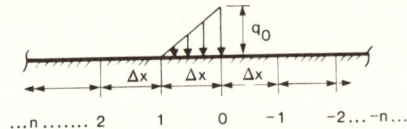

At point n the bending moment is given by:

$$M_n = \frac{q_0}{4\lambda^2} I_n$$

where I_n is an influence factor given in the following table.

APPENDIX

↓Δx \ n→	8	7	6	5	4	3	2	1	0	-1	-2	-3	-4	-5	-6	-7	-8
3/λ				$\lvert I_n \rvert < .001$			+.007	-.167	+.174	-.014							
2/λ						+.006	-.037	-.110	+.233	-.100	+.007	+.001		$\lvert I_n \rvert < .001$			
1/λ		$\lvert I_n \rvert < .001$	+.002	+.004	-.006	-.047	-.098	+.064	+.246	-.089	-.069	-.015	+.003	+.001			
1/1.5λ	-.001	+.014	-.002	+.015	-.041	-.070	-.042	+.107	+.211	-.019	-.067	-.050	-.023	-.005	-.004	+.002	+.001
1/2λ	-.001	-.007	-.018	+.034	-.049	+.047	-.005	+.114	+.177	+.024	-.040	-.051	-.040	-.024	-.015	+.002	+.002
1/2.5λ	-.008	-.015	-.023	+.031	-.032	-.021	+.015	+.087	+.122	+.034	+.012	-.031	-.032	-.026	-.018	-.010	-.005
1/3λ	-.018	-.024	-.036	-.032	-.026	+.006	+.027	+.102	+.130	+.058	+.002	-.027	-.036	-.031	-.028	-.020	-.014

↓Δx \ n→	16	15	14	13	12	11	10	9	-9	-10	-11	-12	-13	-14	-15	-16
1/2λ	$\lvert I_n \rvert < .001$				+.001	+.002	+.002	+.002	+.002	+.002	+.001	+.001	$\lvert I_n \rvert < .001$			
1/2.5λ	+.001	+.001	+.001	+.001	+.001	+.001	.000	-.003	-.001	+.001	+.001	+.001	+.001	+.001	+.001	+.001
1/3λ	+.001	+.001	+.001	+.001	.000	-.002	-.006	-.011	-.008	-.004	-.001	+.001	+.001	+.002	+.001	+.001

Note: $\sum_{-n}^{n} I_n \to 0$, as $n \to \dfrac{6}{\lambda \Delta x}$ or greater.

Appendix B

Beam on elastic foundation—Winkler foundation model, concentrated loading
Formulae for deflection, angular deflection, bending moment and shear force
(after Hetenyi, 1946)

Notation and sign convention as Appendix A.
Infinite beam   *Semi-infinite beam*

Concentrated force

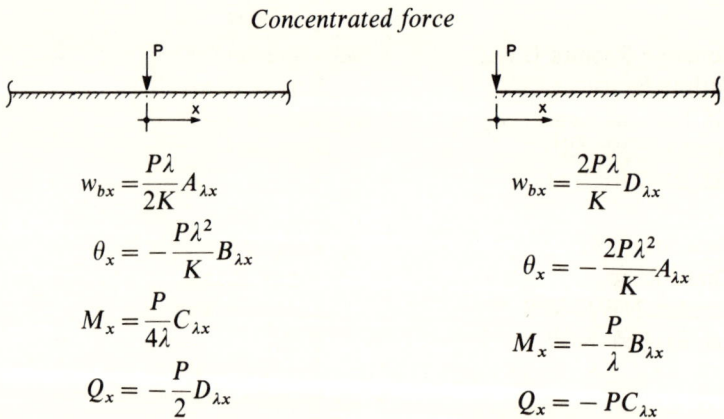

$$w_{bx} = \frac{P\lambda}{2K} A_{\lambda x}$$

$$\theta_x = -\frac{P\lambda^2}{K} B_{\lambda x}$$

$$M_x = \frac{P}{4\lambda} C_{\lambda x}$$

$$Q_x = -\frac{P}{2} D_{\lambda x}$$

$$w_{bx} = \frac{2P\lambda}{K} D_{\lambda x}$$

$$\theta_x = -\frac{2P\lambda^2}{K} A_{\lambda x}$$

$$M_x = -\frac{P}{\lambda} B_{\lambda x}$$

$$Q_x = -P C_{\lambda x}$$

Concentrated moment

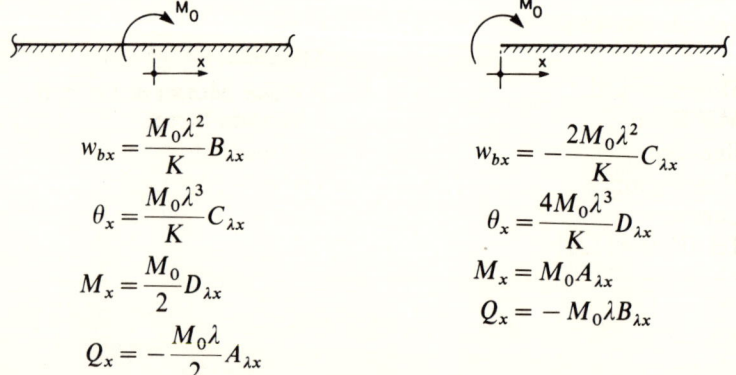

$$w_{bx} = \frac{M_0 \lambda^2}{K} B_{\lambda x}$$

$$\theta_x = \frac{M_0 \lambda^3}{K} C_{\lambda x}$$

$$M_x = \frac{M_0}{2} D_{\lambda x}$$

$$Q_x = -\frac{M_0 \lambda}{2} A_{\lambda x}$$

$$w_{bx} = -\frac{2M_0 \lambda^2}{K} C_{\lambda x}$$

$$\theta_x = \frac{4M_0 \lambda^3}{K} D_{\lambda x}$$

$$M_x = M_0 A_{\lambda x}$$

$$Q_x = -M_0 \lambda B_{\lambda x}$$

Appendix C Some specialist crack-repair firms in the UK

1. Almondeck Ltd.,
   48 Newlands Drive,
   Sheffield.
   Tel. 0742-651977

2. Cementation Ltd.,
   Cementation House,
   Denham Way,
   Maple Cross,
   Rickmansworth,
   Herts.
   Tel. 0923-76666

3. Concrete Repairs U.K. Ltd.,
   Cathite House,
   748 Fulham Road,
   London SW6 5SN.
   Tel. 01-731-2141

4. Colcrete Ltd.,
   Bryant House,
   Strood,
   Rochester,
   Kent.
   Tel. 0634-722571

5. Expandite Ltd.,
   Chase Road,
   London NW10 6PS
   Tel. 01-965-8877

6. Fosroc Ltd.,
   Vinny Road,
   Leighton Buzzard,
   Beds.
   Tel. 0525-37646

7. Sika Ltd.,
   Abbeydale Road,
   Wembley,
   Middlesex.
   Tel. 01-998-0961

8. SDB Products Ltd.,
   Denham Way,
   Maple Cross,
   Rickmansworth,
   Herts.
   Tel. 0923-77311

9. Stonecare Ltd.,
   Mill Road,
   Buckden,
   Huntingdon,
   Cambs.
   Tel. 0480-811291

10. Whitley Moran & Co. Ltd.,
    7 Victoria Street,
    Liverpool.
    Tel. 051-227-1702

References

Abu-el-Magd, S.A. and MacLeod, I.A. (1980) Experimental tests on brick beams under in-plane bending conditions. *Structural Engineer*, **58B** (3), 62–66.

Alexander, S.J. and Lawson, R.M. (1981) Design for movement in buildings, Tech. Note 107, CIRIA, London.

The All England Law Reports (1983) Case—Pirelli General Cable Works Ltd. v. Oscar Faber and Ptrs (Firm), **1**, p. 65, Butterworths, London.

American Concrete Institute (1983) ACI 318–83, Building code requirements for reinforced concrete, ACI, Detroit.

American Institute of Steel Construction (1978) Specifications for the design, fabrication and erection of structural steel for buildings. AISC, New York.

American Society for Testing and Materials (1974) ASTM C34–74, Structural clay load-bearing wall tile, ASTM, Philadelphia.

Anderson, T. and Misund, A. (1983) Pipeline reliability: an investigation of pipeline failure characteristics and analysis of pipeline failure rates for submarine and cross-country pipelines. *J. Petroleum Technology, 'AIME*, 709–717.

Andrews, E.N. (1972) *Cast Iron Pipelines—Their Manufacture and Installation*. Stanton and Staveley, British Steel Corporation, Ilkeston, Publication Ref. No. PJF 149/7.

Angus, H.T. (1976) *Cast Iron: Physical and Engineering Properties*, Butterworths, London.

Atkinson, J.H. and Mair, R.J. (1981) Soil mechanics aspects of soft ground tunnelling. *Ground Engineering*, **14** (5), pp. 20–24, 26, 38.

Atkinson, J.H. and Potts, D.M. (1976) Subsidence above shallow circular tunnels in soft ground, Dept. of Engineering, University of Cambridge Report CUED/C-Soils/T.R.27.

Attewell, P.B. (1978a) 'Ground movements caused by tunnelling in soil', in *Proc. Conf. on large Ground Movements and Structures*, Cardiff, July 1977, ed. J.D. Geddes, Pentech Press, London, pp. 812–948.

Attewell, P.B. (1978b) 'Large ground movements and structural damage caused by tunnelling below the water table in a silty alluvial clay,' in *Proc. Conf. on Large Ground Movements and Structures*, Cardiff, July 1977, ed. J.D. Geddes, Pentech Press, London, pp. 307–356.

Attewell, P.B. (1981) 'Engineering contract, site investigation and surface movements in tunneling works,' in *Soft Ground Tunneling*, eds. D. Reséndiz and M.P. Romo, A.A. Balkema, Rotterdam, pp. 5–12.

Attewell, P.B. and Boden, J.B. (1971) Development of stability ratios for tunnels driven in clay. *Tunnels and Tunnelling*, **3**, 195–198.

Attewell, P.B. and Farmer, I.W. (1973) Measurement and interpretation of ground movements during construction of a tunnel in laminated clay at Hebburn, Country Durham, Rept to Transport and Road Research Laboratory, Contract No. ES/GW/842/68.

Attewell, P.B. and Farmer, I.W. (1974a) Ground disturbance caused by shield tunnelling in a stiff, overconsolidated clay. *Engineering Geology* **8**, 361–381.

Attewell, P.B. and Farmer, I.W. (1974b) Ground deformations resulting from shield tunnelling in London Clay. *Canadian Geotech. J.* **11**, 380–395.

Attewell, P.B. and Farmer, I.W. (1976) *Principles of Engineering Geology*, Chapman and Hall, London.

Attewell, P.B., Farmer, I.W. and Glossop, N.H., (1978) Ground deformation caused by tunnelling in a silty alluvial clay. *Ground Engineering*, **11** (8), 32–41.

Attewell, P.B., Farmer, I.W., Glossop, N.H. and Kusznir, N.J. (1975) 'A case history of ground deformation caused by tunnelling in laminated clay,' in *Proc. Conf. on Subway Construction*, Budapest—Balatonfüred, pp. 165–178.

Attewell, P.B., Farmer, I.W., and Wickson, J.L. (1976) 'Measurements of ground-lining interaction

REFERENCES

pressure in an underwater tunnel in Coal Measures rock,' in *Proc. Tunnelling*' 76, ed. M.J. Jones, IMM, London, pp. 255–263; 273–4.

Attewell, P.B. and Norgrove, W.B. (1984a) Survey of United Kingdom tunnel contract and site investigation costs, Construction Industry Research and Information Association Report RP324, London.

Attewell, P.B. and Norgrove, W.B. (1984b) A survey of some site investigation and contractual matters based primarily but not entirely on Northumbrian Water tunnelling experience, Construction Industry Research and Information Association Report RP324, London.

Attewell, P.B. and Norgrove, W.B. (1984c) A flow-chart guide to site investigation for tunnelling. *Municipal Engineer*, **1**, 297–302.

Attewell, P.B. and Woodman, J.P. (1982) Predicting the dynamics of ground settlement and its derivatives caused by tunnelling in soil. *Ground Engineering*, **15**, (8), 13–22, 36.

Audibert, J.M.E. and Nyman, K.J. (1977) Soil restraint against horizontal motion of pipes. *J. Geotech. Eng. Div, ASCE* **103**, (GT 10) 1119–1142.

Barratt, D.A. and Tyler, R.G. (1975) Measurements of ground movements and lining behaviour on the London Underground at Regent's Park. Dept. of the Environment, TRRL Report LR 684, Crowthorne, Berks.

Beeby. A.W. (1983) Cracking, cover and corrosion of reinforcement. *Concrete Int.* **5** (2), 35–40.

Biot, M.A. (1937) Bending of an infinite beam on an elastic foundation, *J. App. Mech., Trans. ASME* **59**, A1–A7.

Bjerrum, L. (1963) Allowable settlement of structures. In *Proc. Eur. Conf. on Soil Mechanics and Foundation Engineering, Wiesbaden*, vol. III, p. 135–137.

Boden, J.B. (1969) 'Site investigation and subsequent analysis for shallow tunnels.' Dissertation, M.Sc. Advanced Course in Engineering Geology, University of Durham.

Bowring, R.W.J., May J., and Wilford, M.J.C. (1984) 'A case history of structures on deep alluvium', in *Symposium on Soil-Structure Interaction*, I. Struct.E/ICE London, pp. 33–42.

British Standards Institution (1972) Steel girder bridges, BS 153.

British Standards Institution (1969) The use of structural steel in building, Part 2, Metric Units, BS 449.

British Standards Institution (1979) Specification for weldable structural steels, BS 4360.

British Standards Institution (1976) Code of practice for the structural use of concrete for retaining aqueous liquids, BS 5337.

British Standards Institution (1978 on) Steel, concrete and composite bridges, Parts 1 to 10, BS 5400.

British Standards Institution (1978) Structural use of masonry, BS5628 Part 1.

British Standards Institution (1981) Code of Practice for Site Investigations, BS 5930.

British Standards Institution (1970) Structural recommendations for load-bearing walls, Code of Practice CP111.

British Standards Institution (1967) Composite construction in structural steel and concrete, Part 2, Beams for bridges, Code of Practice CP 117.

Broms, B.B. and Bennermark, H. (1967) Stability of clay in vertical openings. *J. Soil Mech Found. Div. ASCE* **93**, 71–94.

Building Research Establishment (1979a) Estimation of thermal and moisture movements and stresses, Part 2. *BRE Digest* **228**.

Building Research Establishment (1979b) Estimation of thermal and moisture movements and stresses, Part 3. *BRE Digest* **228**.

Building Research Establishment (1981a) Concrete in sulphate-bearing soils and groundwaters. *BRE Digest* **250**.

Building Research Establishment (1981b) Assessment of damage in low-rise buildings with particular reference to progressive foundation movement, *BRE Digest* **251**.

Burland, J.B., Broms, B.B. and de Mello, V.F.B. (1978) Behaviour of foundations and structures, *BRE Current Paper* CP51/78, 495–546.

Burland, J.B. and Wroth, C.P. (1975) Settlement of buildings and associated damage, *BRE Current Paper* CP33/75.

Butler, R.A. and Hampton, D. (1975) Subsidence over soft ground tunnel, *J.Geotech. Eng. Div., ASCE*, **101**, 35–49.

Casagrande, A., (1948) Classification and identification of soils. *Trans. Amer. Soc. Civ. Engrs.* **113**, 901–992.

Casson, D.R. (1985) 'Settlement loading on buried pipelines', in *Ground Movements and Structures: Proc. 3rd Internat. Conf., Cardiff, 1984*, ed. J.D. Geddes, Pentech Press, London, pp. 80–93.

Collins, H.H., Fuller, A.G. and Harrison, J.J. (1973) 'Corrosion characteristics and protection of buried ductile iron pipe', in *12th World Gas Conference, Nice, 1973*, London International Gas Union.

Compston, D.G., Cray, P., Schofield, A.N., and Shann, C.D. (1978) Design and construction of buried thin-wall pipes, CIRIA Report **78**.

Cooke, R.W. (1975) The settlement of friction pile foundations. *BRE Current Paper* CP12/75.

Cording, E.J. and Hansmire, W.H. (1975) 'Displacements around soft ground tunnels', in *General Report: Session IV, Tunnels in Soil, 5th Panamerican Congress on Soil Mechanics and Foundation Engng.*, Buenos Aires.

Cording, E.J., Hansmire, W.H., MacPherson, H.H., Lenzini, P.A. and Vonderohe, A.D. (1976) Displacements around tunnels in soil. Final report by the University of Illinois on Contract No. DOT FR 30022 to the Office of the Secretary and Federal Railroad Administrator, Dept. of Transportation, Washington DC 20590, USA.

Cording, E.J., O'Rourke, T.D. and Boscardin, M. (1978) 'Ground movements and damage to structures', in *Internat. Conf. Evaluation and Prediction of Subsidence*, Pensacola Beach, Florida, 1978, ASCE, New York, pp. 516–537.

Cottington, J. and Akenhead, R. (1984) *Site Investigation and the Law*, Thomas Telford Ltd., London.

Couldery, P.A.J. (1983) 'The effects of tunnelling-induced settlement upon frame structures with shallow foundations.' Dissertation, M.Sc. Advanced Course in Engineering Geology, University of Durham.

Craig, R.N. (1983) Pipe jacking: a state of the art review. CIRIA Technical Note **112**, London.

Davies, B.L. (1981) 'The design of rigid frame steel structures in mining areas', in *Proc. Conf. on Ground Movements and Structures*, Cardiff, ed. J.W. Geddes, Pentech Press, London, pp. 241–250.

Deere, D.U., Peck, R.B., Monsees, J.E. and Schmidt, B., (1969) Design of tunnel liners and support systems. Final report by the Dept. of Civil Engineering, University of Illinois at Urbana, USA for the Office of High Speed Ground Transportation, US Dept. of Transportation, Washington DC 20591, Contract No. 3–0152.

Department of Energy (1977) Report of the Inquiry into Serious Gas Explosions, HMSO, London.

Department of the Environment (1973*a*) BE1/73, Reinforced concrete for highway structures. Technical Memo (Bridges).

Department of the Environment (1973*b*) BE/5/73, Standard highway loadings. Technical Memo (Bridges).

Department of the Environment (1981) Assessment of damage in low-rise buildings. *BRE Digest* **251**.

Dobson, C., Cooper, I, Attewell, P.B. and Spencer I.M. (1979) 'Settlement caused by driving a tunnel through fill', in *Proc. Midland Geotech. Soc. Symp. Engineering Behaviour of Industrial and Urban Fill*, Birmingham, pp. E41–E50.

Dorling, C. (1984) Pipe strains and vibrations caused by percussive moling: site experiment number 1. Water Research Centre, Engineering Centre External Report **137E**, Swindon.

Dumbleton, M.J. and West, G. (1976*a*) Preliminary sources of information for site investigation in Britain. Transport and Road Research Laboratory Report **403** (revised edition), Crowthorne. Berks.

Dumbleton, M.J. and West, G. (1976*b*) A guide to site investigation procedures for tunnels. Transport and Road Research Laboratory Report **740**, Crowthorne, Berks.

Edgell, G.J. and Yarwood, D.J. (1981) Underground pipes: a bibliography. British Ceramic Research Association Special Publication Number **102**, Stoke-on-Trent.

Einstein, H.H. and Baecher, G.B. (1983) Probabilistic and statistical methods in engineering geology. *Rock Mechanics* **16**, 39–72.

Elson, W.K. (1984) Design of laterally-loaded piles. CIRIA Report **103**. London.

Freeman, L. (1983) New mains for old—will a gas system work for water? *Water Bulletin* **65** (July) National Water Council, London, pp. 6–7.

Fry, R. and Rumsey, P.B. (1983) Ground movements caused by trench excavation and the effect on adjacent buried pipelines. Water Research Centre, Engineering Centre External Resport **100E**, Swindon.

Geddes, J.D. and Kennedy, D. (1985) 'Structural implications of horizontal ground strains', in *Ground Movements and Structures: Proc. 3rd Internat. Conf. Cardiff*, 1984, ed. J.D. Geddes, Pentech Press, London, pp. 610–629.

Ghaboussi, J. and Ranken, R. (1975) Tunnel design considerations: analysis of stress and deformations around advancing tunnels. Final Report prepared by the University of Illinois for the Dept. of Transportation, Washington D.C., No. FRA-OR&D 75–84.

Gibson, R.E. (1967) Some results concerning displacements and stresses in a non homogeneous elastic halfspace. *Géotechnique* **17**, 58–67.

Glossop, N.H. (1978) 'Soil deformations caused by soft-ground tunnelling', Ph.D. Thesis, University of Durham, England.

Glossop, N.H., (1980) Ground deformation caused by tunnelling in soft ground at Grimsby, Report to Transport and Road Research Laboratory and the Dept of the Environment, University of Newcastle upon Tyne.

Glossop, N.H. and Farmer, I.W. (1977) Ground deformation during construction of a tunnel in Belfast, Report to the Dept of the Environment for Northern Ireland and the Transport and Road Research Laboratory, No. R6/77, October 1977, by the University of Newcastle upon Tyne.

Glossop, N.H. and Farmer, I.W. (1979) Settlement associated with removal of compressed air pressure during tunnelling in alluvial clay. *Géotechnique* **29**, 67–72.

Glossop, N.H. and O'Reilly, M.P. (1982) Settlement caused by tunnelling through soft marine silty clay. *Tunnels and Tunnelling* **14**, (9), 13–16.

Glossop, N.H., Saville, D.E., Moore, J.S. Benson, A.P., and Farmer, I.W. (1979) 'Geotechnical aspects of shallow tunnel construction in Belfast Estuarine deposits,' in *Proc. Tunnelling 79 Conf.*, London, Ed. M.J. Jones, IMM, London, pp. 45–50.

Grant, R., Christian, J.T., and Vanmarke, E.H. (1974) Differential settlement of buildings. *J. Geotech. Eng. Div. ASCE* **100** (GT9), 973–991.

Gregory. M.S. (1966) *Linear Framed Structures*. Longmans, London.

Gumbel, J.E. and Wilson, J. (1981) Interaction design of buried flexible pipes—a fresh approach from basic principles. *Ground Engineering* **14** (4), 36–40.

Hansen, B.J. (1961) The ultimate resistance of rigid piles against transversal forces. Geoteknisk Institut. Bull No **12**, Copenhagen.

Hansmire, W.H. (1975) 'Field measurements of ground displacements about a tunnel in soil', Ph.D. thesis, University of Illinois at Urbana–Champaign.

Hansmire, W.H. and Cording, E.J. (1972) 'Performance of a soft ground tunnel on the Washington Metro', in *Proc. 1st North American Rapid Excavation and Tunneling Conf.*, AIME, **1**, pp. 371–389.

Harris, C.W. and O'Rourke, T.D. (1983) Response of jointed cast iron pipelines to parallel trench construction. Report to New York Gas Group, Geotechnical Engineering Report 83–5, School of Civil and Environmental Engineering, Cornell University, Ithaca, N.Y.

Hetenyi, M. (1946) *Beams on Elastic Foundation—Theory with Applications in the Fields of Civil and Mechanical Engineering*. University of Michigan Press, Ann Arbor.

Hough, B.K. (1957) *Basic Soils Engineering*. The Ronald Press Co., New York.

Howe, M. (1985a) 'The protection of buried plant from adjacent deep works', in *Ground Movements and Structures: Proc. 3rd Internat. Conf. Cardiff, 1984*, ed. J.D. Geddes, Pentech Press, London, pp. 62–79.

Howe, M. (1958b) 'Discussion of statistical studies of failure rates in grey iron mains close to recent excavations', in *Ground Movements and Structures: Proc. 3rd Internat. Conf. Cardiff, 1984*. ed. J.D. Geddes, Pentech Press, London, pp. 814–817.

Howe, M., Hunter, P. and Owen, R.C. (1980) 'Ground movements caused by deep excavations and tunnels and their effects on adjacent mains', in *Proc. 2nd Conf. on Ground Movements and Structures, Cardiff, 1980*, ed. J.D. Geddes, Pentech Press, London, pp. 812–840.

Howe, M. and Hunter P. (1983) Discussion on the effect of tunnelling operations on buried plant, in *Tunnelling '82 Symp., Trans. IMM* **92**, A35–A41.

Hurrell, M.R. (1983) Monitoring of ground and structural response to shield tunnelling in soft ground beneath Collingwood Street in Newcastle upon Tyne. Internal Report to Northumbrian Water Authority and WRC Engineering Centre, England, Engineering Geology Laboratories, University of Durham, June 1983.

Hurrell, M.R. (1984a) Results of a programme of monitoring and pipe response to shield-driven tunnelling in soft ground at Norton, Stockton-on-Tees, Cleveland. Report to Northumbrian

Water Authority and the Water Research Centre, Swindon, by the Department of Engineering, University of Durham.

Hurrell, M.R. (1984b) 'The empirical prediction of long-term surface settlements above shield-driven tunnels in soils', in *Ground Movements and Structures: Proc. 3rd Internat. Conf., Cardiff*, 1984, ed. J.D. Geddes, Pentech Press, London, pp. 161–170.

Hurrell, M.R. (1985) Results of a programme of monitoring ground and pipe response to shield-driven tunnelling in soft ground at Norton, Stockton-on-Tees, Cleveland. Internal Report to WRC Engineering Centre, Northumbrian Water Authority and Stockton Borough Council, England, Department of Engineering, University of Durham, April 1985.

Institution of Civil Engineers (1983) *ICE Conditions of Contract for Ground Investigation*. Thomas Telford Ltd., London.

Joyce, M.D. (1982) *Site Investigation Practice*. E and F.N. Spon, London.

King, G.J.W. and Pandey, P.C. (1978) The analysis of infilled frames using finite elements, *Proc. Inst. Civ. Engrs.*, (2) **65**, 749–760.

Kirsch, G. (1898) Die Theorie der Elastizität und die Bedurfnisse der Festigkeitslehre. *Zeit. Ver, Deut. Ing.* **42** (29), 797–807.

Lambe, T.W. and Whitman, R.V. (1969) *Soil Mechanics*. John Wiley and Sons, New York.

Law Reform Committee (1977) Final report on limitations of actions. L.R.C. 21st Report, HMSO, London.

Law Reform Committee (1984) Latent damage. L.R.C. 24th Report, HMSO, London.

Leach, G. (1983) 'Discussion of theoretical studies on the effects of tunnel excavations on buried mains', in *Tunnelling '82 Symp. Trans. IMM* **92**, A41–44.

Leach, G. (1984) Pipeline response to tunnelling. British Gas Corporation Engineering Research Station, Report E463, Newcastle upon Tyne.

Leach, G. (1985) Pipeline response to tunnelling. Unpublished paper presented to the North of England Gas Association, January 1985.

Lee, I.K., White, W. and Ingles, O.G. (1983) *Geotechnical Engineering*. Pitman, Melbourne.

Leonards, G.A. (1975) Discussion on paper by R. Grant, J.T. Christian and E.H. Vanmarke 'Differential settlement of buildings', *J. Geotech. Eng. Div.*, ASCE **101**, 700–702.

MacLeod, I.A. and Abu-El-Magd, S.A. (1980) The behaviour of brick walls under conditions of settlement, *Structural Engineer* **88A** (9), 279–286.

MacLeod, I.A. and Abu-El-Magd, S.A. (1982) Discussion on MacLeod and Abu-El-Magd (1980), *Structural Engineer* **60A** (5).

Mair, R.J. (1980) 'Centrifugal modelling of tunnel construction in soft clay', Ph.D. thesis, University of Cambridge.

Manton, B.H. and Wilson, C.B. (1971) M.O.T./C&CA Standard bridge beams, Publ. 46.012, C&CA, London.

Marsland. A. (1980) Discussion on design parameters in geotechnical engineering, in *Proc. VII Europ. Conf. S.M.F.E., Brighton*, **4**, pp. 159–163.

Matlock, H. and Reese, L.C. (1961) 'Foundation analysis of offshore pile supported structures', in *Proc. 5th Inernat. Conf. Soil Mechanics and Foundation Engineering*, Paris, **2**, pp. 91–97.

Matyas, E.L. and Davis, J.B. (1983) Experimental study of earth loads on rigid pipes. *J. Geotech. Eng. Div., ASCE* **109**, 202–209.

Meyerhof, G.G. (1956) Discussion on paper by A.W. Skempton and D.H. MacDonald: 'The allowable settlements of buildings', *Proc. Inst. Civ. Engrs* **5**, 774–775.

Mikoaka, T. (1978) 'Pipelines laid in difficult conditions including earthquake zones', in *Proc. 12th Congress International Water Supply Association, Kyoto*.

Mindlin, R.D. (1936) Forces at a point in the interior of a semi-infinite solid. *Physics* **7**, 195.

Mitchell, R.J. (1983) *Earth Structures Engineering*. Allen and Unwin Inc., Boston.

Monie, W.D. and Clarke, C.M. (1974) Loads on underground pipe due to frost penetration. *J. Amer. Water Works Association* **66** 353–358.

Morton, J.D. and Dodds, R.B. (1979) Ground subsidence associated with machine tunnelling in fluvio deltaic sediments, Part 2. *Tunnels and Tunnelling*, 23–28.

Nath, P. (1977) Finite element analysis of a large diameter buried steel pipeline. Transport and Road Research Laboratory Report **778**, Crowthorne, Berks.

Nath, P. (1983) Trench excavation effects on adjacent buried pipes: finite element study. *J. Geotech. Eng. Div., ASCE* **109**, 1399–1415.

National Water Council (1977) Sewers and Water mains—a national assessment. Department of the Environment Standing Technical Committee, Report No. 4, National Water Council.

REFERENCES

Needham, D. and Howe, M. (1979) Why pipes fail. Communication **1103**, Institution of Gas Engineers, London.
Neville, A.M. (1981) *Properties of Concrete*. 3rd edn., Pitman, London.
Norgrove, W.B. and Attewell, P.B. (1984) Assessing the benefits of site investigation for tunnelling. *Municipal Engineer* **1**, (2), 99–106.
Norgrove, W.B., Cooper, I. and Attewell, P.B. (1979) 'Site investigation procedures adopted for the Northumbrian Water Authority's Tyneside Sewerage Scheme, with special reference to settlement prediction when tunnelling through urban areas, in *Proc. Tunnelling '79*, ed. M.J. Jones, IMM, London, pp. 79–104.
Nyman, K.J. (1984) Soil response against oblique motion of pipes, *J. Transportation Eng. ASCE* **110**, 190–202.
O'Reilly, M.P. and New, B.M. (1982) 'Settlements above tunnels in the United Kingdom—their magnitude and prediction', In *Proc. Tunnelling '82*, ed. M.P. Jones, IMM, London, pp. 137–181.
O'Rourke, T.D., Cording, E.J. and Boscardin, M. (1976) Ground movements related to braced excavations and their influence on adjacent buildings. Final Report, U.S. Dept of Transportation Report DOT-TST 76T-23, Prepared for US DOT, Office of the Sec. and Federal Railroad Admin., Washington D.C. by the Univ. of Illinois, Dept. of Civ. Engrg. (UILU-ENG-76-2023), Urbana.
O'Rourke, T.D. and Trautmann, C.H. (1980) Analytical modelling of buried pipeline response to permanent earthquake displacements. Geotechnical Engineering Report 80-4, School of Civil and Environmental Engineering, Cornell University, Ithaca, NY.
O'Rourke, T.D. and Trautmann, C.H. (1982) Buried pipeline response to tunnelling ground movements', in *Europipe '82 Conf.*, Basle, Switzerland, pp. 9–15.
Osborne, G. (1985) The use of vibrating wire strain gauges for long term measurements on buried gas mains. *Strain* **21**, 23–27.
Owen, R.C. (1985) 'Vegetation and seasonal ground movement effects on buried mains', in *Ground Movements and Structures: Proc. 3rd Internat. Conf. Cardiff. 1984*, ed., J.D. Geddes, Pentech Press, London, pp. 145–160.
Padfield, C.J. and Sharrock, M.J. (1983) Settlement of structures on clay soils. CIRIA Sp. Pub. 27/PSA Civ. Eng. Tech. Guide **38**.
Peck, R.B. (1969) 'Deep excavations and tunneling in soft ground.' State of the art report, in *7th Int. Conf. On Soil Mechanics and Foundation Engrg.* Mexico City, pp. 225–290.
Peck, R.B., Hanson, W.E. and Thorburn, T.H. (1953) *Foundation Engineering*. John Wiley and Sons, New York.
Pearson, F.H. (1977) Beam behavior of buried rigid pipelines. *J. Environmental Eng. Div. ASCE* **103**, 767–783.
Pocock, R.G. Lawrence, G.J.L. and Taylor, M.E. (1980) Behaviour of shallow buried pipeline under static and rolling wheel loads. Transport and Road Research Laboratory Report **954**, Crowthorne, Berks.
Polshin, D.E. and Tokar, R.A. (1957) 'Maximum allowable non-uniform settlement of structures', in *Proc. 4th Internat. Conf. Soil Mechanics and Foundation Engineering*, London, **1**, pp. 402–405.
Poulos, H.G. and Davis, E.H. (1980) *Pile Foundation Analysis and Design*. John Wiley and Sons, New York.
Public Utilities Street Works Act (1950) Section 26. HMSO, London.
Randolph, M.F. (1981) The response of flexible piles to lateral loading, *Géotechnique* **31**, 247–259.
Reed, E.C. (1978) 'Renovation of old pipelines', in *Proc. 12th Congress International Water Supply Association*, Kyoto.
Reeves, M.J., Attewell, P.B. and Woodman, J.P. (1983) Graphical methods of estimating ground movements caused by tunnelling in soil. Internal Report, Engineering Geology Laboratories, University of Durham.
Richardson, M. and Scruby, J. (1981) Earthworm systems will threaten conventional tunnel jacking. *Tunnels and Tunnelling*, April, 29–32.
Riddington, J.R. (1984) The influence of initial gaps on infilled frame behaviour. *Proc. Inst. Civ. Engrs.* (2) **77**, 295–310.
Riddington, J.R. and Stafford-Smith, B. (1977) Analysis of infilled frames subject to racking with design recommendations. *Structural Engineer* **55**, 263–268.
Roark, R.J. (1965) *Formulas for Stress and Strain*, 4th edn. McGraw-Hill, New York.

Roark, R.J. and Young, W.C. (1975) *Formulas for Stress and Strain*, 5th edn. McGraw-Hill, New York.
Roberts, N.P. and Regan, T. (1974) Causes of fractures in grey cast iron water mains. Dept of Civil Engineering, City University, and Metropolitan Water Board, London.
Roberts, N.P. and Regan, T. (1977) Fractures in water mains. Internal Report to Thames Water Authority, Dept of Civil Engineering, City University, London.
Rumsey, P.B., Cooper, I. and Kyrou, K. (1981) 'Ground movement and pipe strain associated with trench excavation', in *Proc. Internat. Conf. on Maintenance, Repair, Renovation and Renewal of Sewerage Systems*, ICE, London, pp. 91–101.
Rumsey, P.B. and Dorling, C. (1985) 'The prediction of ground movement and induced pipe strain caused by trench excavation', in *Ground Movements and Structures: Proc. 3rd Internat. Conf. Cardiff, 1984*, ed. J.D. Geddes, Pentech Press, London, pp. 24–49.
Samarasinghe, W., Page, A.W. and Hendry, A.W. (1981) Behaviour of brick masonry shear walls. *Structural Engineer* **59B** (3), 42–48.
Savin, G.N. (1961) *Stress Concentration Around Holes*. Pergamon Press, Oxford.
Sawko, F. and Rouf, A. (1984) On the stiffness properties of masonry, TN381, *Proc. Inst. Civ. Engrs.*, (2) **77**, 1–12.
Schmidt, B. (1969) 'Settlements and ground movements associated with tunneling in soil', Ph.D. thesis, University of Illinois.
Schmidt, B. (1974) 'Predictions of settlements due to tunnelling in soil: three case histories', in *Rapid Excavation and Tunnelling Conf.*, San Francisco. ASCE., New York.
Scott, Wilson, Kirkpatrick and Ptrs. (1982) Report on soil modulus to the Water Research Centre, External Report 78E, Water Research Council Engineering Centre, Swindon.
Sears, E.C. (1968) Comparison of the soil corrosion resistance of ductile iron pipe and grey cast iron pipe. *Materials Protection* **7** (October), 33–36.
Selby, A.R. (1985) 'Tolerance of highway bridges to ground movements induced by tunnelling in soil', in *Ground Movements and Structures: Proc. 3rd Internat Conf. Cardiff, 1984*, ed. J.D. Geddes, Pentech Press, London, pp. 630–642.
Shiraishi, S. (1968) Recent major shield-driven tunnels through soft ground in Japan. *Soils and Foundations* **9**, 16–34.
Sims, F.A. (1982) Treatment for subsidence in bridge design. Seminar on Movements in Concrete Bridges L.S. 2083, C&CA, Slough.
Sims, F.A. and Bridle, R.J. (1966) Bridge design in areas of mining subsidence. *Proc. Inst. Highway Engrs.* (November), 19–34.
Singhal, A.C. and Benavides, J.C. (1983) Axial and bending behavior of pipeline joints. *J. Amer. Water Works Association* **75**, 572–578.
Singhal, A.C. and Meng, C.L. (1983) Junction stresses in buried jointed pipelines. *J. Transportation Eng. ASCE.* **109**, 450–461.
Sizer, K.E. (1976) 'The determination and interpretation of ground movements caused by shield tunnelling in silty alluvium at Willington Quay. North East England', M.Sc. thesis, University of Durham.
Skempton, A.W. (1951) 'The bearing capacity of clays', in *Building Research Congress*, London, Div. 1, pp. 180–189.
Skempton, A.W. and Macdonald, D.H. (1956) Allowable settlement of buildings. *Proc. Inst. Civ. Engrs.* **5**, 727–768.
Smith, W.H. (1976) Frost loadings on underground pipe. *J. Amer. Water Works Association* **68**, 673–674.
Smyth-Osbourne, K.R. (1971) Discussion on Muir Wood and Gibb (1971). *Proc. Inst. Civ. Engrs.* **50**, 187–203.
Somerville, G. and Tiller, R.M. (1970) Standard bridge beams for spans 7 m to 36 m. Publn, 46,005, C&CA, London.
Spencer, I.M. (1978) 'Soft ground tunnelling on contracts 16A and 276 of the Tyneside Sewerage Scheme'. Dissertation, M.Sc. Advanced Course in Engineering Geology, University of Durham.
Stafford-Smith, B. and Carter, C. (1969) A method of analysis for infilled frames. *Proc. Inst. Civ. Engrs.* **44** (September), 31–48.
Stanton and Staveley (1979) Ductile Iron Pipelines—Embedment Design. Stanton and Staveley, British Steel Corporation, Ilkeston, Publication Ref., No. PJF 268, 1979 Section 5.
The Statutes (1980) The Limitations Act, c58, *Index to the statutes*, HMSO, London.

Sutherland, H.B., Finlay, T.N. and Cram, I.A. (1968) Engineering and related properties of pulverised fuel ash. *J. Inst. Highway Engineers* **15** (6), 19–27, 29–35.
Széchy, K. (1970) 'Surface settlements due to the shield tunnelling method in cohesionless soils', in *Proc. Conf. on Subway Construction*, Budapest-Balatonfüred, pp. 615–624.
Széchy, K. (1973) *The Art of Tunnelling*, 2nd edn., Akademiai Kiado, Budapest.
Takagi, N., Shimamura, K. and Nishio, N. (1985) 'Buried pipe response to adjacent ground movements associated with tunnelling and excavations', in *Ground Movements and Structures: Proc. 3rd Internat. Conf.*, Cardiff, 1984, ed. J.D. Geddes, Pentech Press, London, pp. 97–112.
Terzaghi, K. (1943) *Theoretical Soil Mechanics*, Wiley, New York.
Terzaghi, K. (1955) Evaluation of coefficients of subgrade reaction. *Géotechnique* **5**, 297–326.
Terzaghi, K. and Peck, R.B. (1967) *Soil Mechanics in Engineering Practice*, 2nd edn., John Wiley and Sons, New York.
Thorburn, S. (1984) 'Soil structure interaction,' in *Symposium on Soil-Structure Interaction*, I.Struct.E./ICE, London, 5–8.
Timoshenko, S.P. and Goodier, J.N. (1970) *Theory of Elasticity*, 3rd edn., McGraw-Hill, Japan.
Todd, J.D. (1974). *Structural Theory and Analysis*. Macmillan, London.
Tsutsumi, M. (1983) 'Tunnelling in Soil—Movements and structures', Ph.D. Thesis, University of Durham.
Valliappan, S. and Raja-Sekar, H.L. (1985) 'Stability of buried pipes due to adjacent excavations', in *Ground Movements and Structures: Proc. 3rd Internat. Conf. Cardiff, 1984*, ed. J.D. Geddes, Pentech Press, London, pp. 50–61.
Vesic, A.B. (1961) Bending of beams resting on isotropic elastic solid. *J. Eng. Mech. Div., ASCE* **87**, 35–53.
Vesic, A.B. (1971) Breakout resistance of objects embedded in ocean bottom. *J. Soil Mech. Foundation Eng. Div., ASCE*, **97**, 1183–1205.
Wahls, H.E. (1981) Tolerable settlement of buildings. *J. Geotech. Eng. Div., ASCE*, **107**, 1489–1504.
Ward, F.G., Bryant, E.G. and Pound, R.P. (1970) Simply supported bridges in composite construction. Publication BD2, British Construction Steelwork Association, London.
WAA/BGC (1984) Model Consultative Procedure for Pipeline Construction Involving Deep Excavation. 1st Revision, June 1984, Water Research Centre, Swindon: British Gas Corporation Engineering Research Station, Newcastle upon Tyne.
Weltman, A.J. and Head, J.M. (1983) Site Investigation Manual. CIRIA Special Publication **25**.
West, G., Carter, P.G., Dumbleton, M.J. and Lake, L.M. (1981). Site investigation for tunnels *Int. J. Rock Mech. Mining Sci. and Geomech. Abs.* **18**, 345–367.
Whipp, S.H. and Glennie, E.B. (1982) The renovation of sewers by sliplining with polyethylene, a discussion document. External Report Water Research Centre, Swindon.
Whitman, R.V. (1984) Evaluating calculated risk in geotechnical engineering. *J. Geotech. Eng. Div., ASCE*, **110**, 145–188.
Wilson, J. (1984) 'Road of Light' brings more data, *Consulting Engineer*, Fed. 1984.
Winkler, E. (1867) *Die Lehre von Elastizität und Festigkeit* (On Elasticity and Flexibility.) Prague.
Wood, A.M., Muir (1970) 'Soft ground tunnelling', in *Proc. TUNCON 70*, Johannesburg, **1**, pp. 167–174.
Wood, A.M. Muir, and Gibb, T.R. (1971) 'Design and construction of the cargo tunnel at Heathrow Airport, London', *Proc. Inst. Civ. Engrs.* **48**, 11–35.
Yeates, J. (1985a) Discussion of ground movement due to parallel trench construction and effects on buried pipeline, in *Ground Movements and Structures: Proc. 3rd Internat. Conf. Cardiff, 1984* ed. J.D. Geddes, Pentech Press, London, pp. 798–804.
Yeates, J. (1985b) Discussion of factors affecting risk to gas mains from parallel trench construction, in *Ground Movements and Structures: Proc. 3rd Internat. Conf. Cardiff, 1984*, ed. J.D. Geddes, Pentech Press, London, pp. 814–816.

Another professional book from Blackie

Ground Movements and their Effects on Structures

Edited by Professor P B Attewell and Dr R K Taylor

This book reviews the most common sources of ground movements. The authors concentrate on movements which are reasonably predictable, and which can therefore be accommodated by design.

Contents:
Settlement of natural ground under static loadings. Settlement of fill. Slopes and embankments. Deep trenches and excavations in soil. Trenches in soil. Tunnelling in soil. Mining subsidence. Structural design and ground movements. Mineralogical controls on volume change. The effects of clay soil volume changes on low-rise buildings. Settlement and stability of embankments on soft subsoils. Seismic movements. Dynamic ground movement — man-made vibrations. References. Index.

1984　　　　　460 pages　　　　　0-903384-36-1

Published by Blackie and Son Ltd, Bishopbriggs, Glasgow G64 2NZ, UK.

Technology in Construction! Tunnel design has to be based on reliable data. It's as well to remember that one company has been doing an excellent job for 40 years.

Wimpey Laboratories.

Thorough professionalism is the way to sum up work, with no half measures or cut corners. Thoroughness means that every stage of the investigation is planned and carried out with immense care.

It involves:

- geotechnical advice
- desk studies and site inspections
- designing and specifying the work needed
- carrying out the site work using experienced supervisors and personnel and our own plant equipment
- monitoring and interpreting data as field work proceeds

Too much ground investigating today barely scratches the surface

Wimpey digs deeper

- testing samples in the largest and best equipped commercial soil and rock laboratory in the UK

The result of all that thoroughness and care is a comprehensive report of the site ground conditions on which you can rely. Those half measures and cut corners so often taken today may make a tender look more attractive on the surface. But they can prove very expensive indeed in the long run.

It'll pay you to let Wimpey Laboratories dig deeper.

Wimpey Laboratories Limited,
at Beaconsfield Road,
Hayes, Middlesex UB4 0LS.
Tel: 01-573 7744.

382 Newport Road, Cardiff.
Tel: 0222-498172.

Uphall Depot, Broxburn,
West Lothian EH52 5NT.
Tel: (0506) 38883.

Author Index

Abu-el-Magd, S. A. 248, 249, 258
Akenhead, R. 99
Alexander, S. J. 248, 265
Anderson, T. 122
Andrews, E. N. 203
Angus, H. T. 196
Atkinson, J. H. 40, 65
Attewell, P. B. 6, 8, 9, 23, 25, 28, 29, 30, 32, 33, 37, 41, 42, 50, 51, 52, 53, 54, 57, 58, 59, 60, 61, 62, 63, 64, 65, 66, 73, 74, 78, 82, 83, 87, 89, 91, 92, 93, 96, 99, 102, 104, 108, 232, 242, 274, 275, 277, 279, 280, 283
Audibert, J. M. E. 165, 166
Baecher, G. B. 99
Barratt, D. A. 92
Beeby, A. W. 249
Benavides, J. C. 200, 203
Bennermark, H. 63
Benson, A. P. 83
Biot, M. A. 129
Bjerrum, L. 283, 285
Boden, J. B. 23, 25, 30
Boscardin, M. 288
Bowring, R. W. J. 279
Bridle, R. J. 295
Broms, B. B. 62, 260
Bryant, E. G. 297
Building Research Establishment 105, 295
Burland, J. B. 74, 248, 260, 265, 286, 287
Butler, R. A 31
Carter, C. 274
Carter, P. G. 99
Casson, D. R. 165
Christian, J. T. 287
Clarke, C. M. 224
Collins, H. H. 122
Compston, D. G. 189
Cooke, R. W. 126
Cooper, I. 64, 74, 78, 83, 96, 102, 224, 232, 277
Cording, E. J. 28, 29, 31, 47, 52, 288
Cottington, J. 99
Couldery, P. A. J. 242
Craig, R. N. 49
Cray, P. 189
Davies, B. L. 274
Davis, E. H. 128, 169, 170, 171, 175
Davis, J. B. 165
Deere, D. U. 23
Department of Energy 198

Department of the Environment 122, 224, 225, 226
Dobson, C. 64, 83
Dodds, R. B. 83
Dorling, C. 123, 128, 151, 221, 227, 235
Dumbleton, M. J. 99
Edgell, G. J. 194
Einstein, H. H. 99
Elson, W. K. 128, 187, 188
Farmer, I. W. 6, 23, 29, 30, 32, 33, 41, 52, 65, 73, 83, 89, 92, 102, 108
Finlay, T. N. 299
Freeman, L. 235
Fry, R. H. 187, 222
Fuller, A. G. 122
Geddes, J. D. 167
Ghaboussi, J. 28
Gibb, T. R. 83
Gibson, R. E. 247
Glennie, E. B. 235
Glossop, N. H. 29, 41, 61, 81, 83, 89, 90, 91, 92
Goodier, J. N. 302
Grant, R. 286, 287
Gregory, M. S. 156
Gumbel, J. E. 128
Hampton, D. 31
Hansen, B. J. 165
Hansmire, W. H. 28, 31, 47, 51, 52, 83
Hanson, W. E. 33
Harris, C. W. 200, 201, 202
Harrison, J. J. 122
Head, J. M. 99
Hendry, A. W. 248
Hetenyi, M. 133, 239, 242, 304, 308
Hough, B. K. 33
Howe, M. 122, 123, 221, 224, 226, 227
Hunter, D. 221
Hurrell, M. R. 52, 83, 88, 95, 221, 233
Ingles, O. G. 94
Joyce, M. D. 99
Kennedy, D. 167
King, G. J. W. 274
Kirsch, G. 40
Kusznir, N. J. 29
Kyrou, K. 224
Lake, L. M. 99
Lambe, T. W. 33
Lawrence, G. J. L. 200
Lawson, R. M. 248, 265
Leach, G. 62, 65, 169, 221

Lee, I. K. 94
Lenzini, P. A. 28, 47
Leonards, G. A. 287
MacDonald, D. H. 74, 283, 285
MacLeod, I. A. 248, 249, 258
MacPherson, H. H. 28, 47
Mair, R. J. 40, 62
Manton, B. H. 297
Marsland, A. 187
Matyas, E. L. 165
May, J. 279
de Mello, V. F. B. 260
Meng, C. L. 167
Meyerhof, G. G. 283
Mikaoka, T. 234
Mindlin, R. D. 130, 169, 170
Misund, A. 122
Mitchell, R. J. 63, 93
Monie, W. D. 224
Monsees, J. E. 23
Moore, J. S. 83
Morton, J. D. 83
Nath, P. 128, 189
National Water Council 122
Needham, D. 122
Neville, A. M. 249
New, B. M. 64, 65, 76, 246
Nishio, N. 200, 221
Norgrove, W. B. 74, 78, 96, 99, 102, 232, 277
Nyman, K. J. 165, 166
O'Reilly, M. P. 64, 65, 76, 89, 90, 246
O'Rourke, T. D. 200, 201, 202, 203, 221, 287, 288
Osborne, G. 235
Owen, R. C. 224
Page, A. W. 248
Pandey, P. C. 274
Pearson, F. H. 223
Peck, R. B. 9, 23, 33, 54, 63, 64, 83, 94, 104, 164
Pocock, R. G. 200
Polshin, D. E. 248, 283, 285, 286
Potts, D. M. 65
Poulos, H. G. 128, 169, 170, 171, 175
Pound, R. D. 297
Raja-Sekar, H. L. 127
Randolph, M. F. 158
Ranken, R. 28
Reed, E. C. 235
Reeves, M. J. 73
Regan, T. 122, 128, 165
Richardson, M. 50
Riddington, J. R. 274
Roark, R. J. 193, 194
Roberts, N. P. 122, 128, 165
Rumsey, P. B. 128, 151, 187, 221, 222, 224, 227

Samarasinghe, W. 248
Saville, D. E. 83
Savin, G. N. 40
Schmidt, B. 23, 54, 58, 61, 63
Schofield, A. N. 189
Scruby, J. 50
Sears, E. C. 122
Selby, A. R. 295
Shann, C. D. 189
Shimamura, K. 200, 221
Shiraishi, S. 31
Sims, F. A. 295
Singhal, A. C. 167, 200, 203
Sizer, K. E. 57, 83
Skempton, A. W. 74, 283, 285
Smith, W. H. 224
Smythe-Osborne, K. R. 83
Somerville, G. 64, 83
Spencer, I. M. 64, 83
Stafford-Smith, B. 274
Stanton and Staveley 189
Sutherland, H. B. 299
Széchy, K 33, 64
Takagi, N. 200, 221
Taylor, M. E. 200
Terzaghi, K. 94, 104, 131, 164
Thorburn, T. H. 33, 258
Tiller, R. M. 297
Timoshenko, S. P. 302
Tokar, R. A. 248, 283, 285, 286
Trautmann, C. H. 201, 203, 221
Tsutsumi, M. 8, 9
Tyler, R. G. 92
Valliappan, S. 127
Vanmarke, E. H. 287
Vesic, A. B. 128, 130, 165
Vonderohe, A. D. 28, 47
Wahls, H. E. 287
Ward, F. G. 297
Weltman, A. J. 99
West, G. 99
Whipp, S. H. 235
White, W. 94
Whitman, R. V. 33, 99
Wickson, J. L. 6
Wilford, M. J. C. 279
Wilson, C. B. 297
Wilson, J. 128
Winkler, E. 129, 130, 133, 238, 242, 304, 308
Wood, A. M. M. 31, 83
Woodman, J. P. 58, 59, 64, 65, 66, 73, 82, 87, 96
Wroth, C. P. 74, 248, 265, 287
Yarwood, D. J. 194
Yeates, J. 60, 61, 221, 230, 242
Young, W. C. 193

Subject Index

Principal references are indicated in **bold face.**

alluvial clay *see* clay, alluvial
anchors 112
angular deflection 304, 308
angular distortion 78, 265, 284, 287
artesian (and sub-artesian) pressure 103
asbestos cement (pipes)
    see pipe, asbestos cement
axial (ground) loss *see* loss, axial

backfilling *see* trench backfilling
basements 101
Bayesian statistics 99
bead 23, 28, 29
beam 238, 265, 297, 300, 302, 304
    bending moment in 239, 304, 306, 308
    shear in 240, 304
beam elements 267
bearing capacity factors 40, 164, 165
bearings
    bridge 300, 301, 302
    elastomeric 300
benchmark (surveying) 102, 108
bending moment diagrams 268, 272
bending moment formulae 304, 306, 308
bending moment/settlement ratio
    see angular distortion
bleed wells 118
    grout in 32
bolted gland mechanical joint 199
borehole cores 103
borehole logs 102
borehole packers 101
boreholes 99
    closed circuit television in 102
    cone penetration tests in 102
    man-entry 102
    permeability of 102
    photography in 102
    pressuremeter tests in 102
    pumping-in/out tests in 101
    rising head tests in 102
    SPT in 102
        (*see also* standard penetration test)
    vane tests in 102
boulder clay 91, 232 (*see also* tills)
boulders 103
box-heading 7
brick wall **250-60**, 264, 266, 276, **285, 286**, 287

length of 265, 267
tensile strength of 248
    (*see also* brickwork)
brickwork 237, 247, 248, 250, 280
bridges 279, **295-303**
    joints in *see* joints, bridge
    walls in *see* wall, bridge
brine (freezing) 44
building damage 288
building surveyors 105, 287
    (*see also* property surveys)
buildings 102, 115, 121
bulk density 102, **104**

Caisson method *see* shafts
carbonation, of concrete 249
cement 248
chemical analysis 102, **105**
chlorides 105
clay
    alluvial 93, 256, 274, 276
    laminated 34, 90, 107, 114
    London 23
    organic 91
    shrinkable 222, 224
    silty 63, 64, 91, 93, 142
    soft 169
    stiff 64, 89, 90, 142, 169, 208, 225, 245
    stiff fissured 63
    stony 34, 89, 90, 107, 114
clay pipes *see* pipes, clay
clay soil 23, **50**, 60, 247, 250, 267, 268, 269, 272, 273, 274, 279, 302, 303
clay spades 2
coefficient of subgrade reaction
    see subgrade reaction
coffer dam, sheet piled 118
cohesive soils 60, 65, 118, 146, 174, 246
collapse (soil) 104, 222
common law *see* consultation
compressed air **44**, 63, 104
compression index 93, 94
concrete 250
    prestressed 249
    reinforced 249
    stresses in 299
concrete cantilever retaining wall 247, 261, 262, 263

SUBJECT INDEX

concrete foundation 247, 250, 255
concrete roads 249
concrete strip footing 264
conditions of contract 98, 99, 106
cone penetration 102
consolidation 41, 45
consolidation settlement
    see time-dependent settlement
consultation **226-34**
contract documents 102, 103
corrosion
    of concrete reinforcement 249
    of pipes 204, 207, 223, 224
crack 238, 248, 255, 256, 259, **260**, 261, 267, 280, 287, 288
    repairs to 263
crack width 249, **258-60**, 261
creep, soil *see* soil creep
    structural 248
critical-state soil mechanics
    see soil mechanics
critical-state boundary 104
cumulative probability curve **54**, 81-7
curvature **70, 71**, 95, 259, 265
curve, of tunnel 31

damage criteria 248
damping factor
    soil-pipe 176
    for wall on elastic foundation 241
deck (bridge) 297, 298, 300, 301, 302, 303
deflection ratio 284, 287
deformation
    modulus of 263
    modulus of, undrained 63, 102, **105**
    rate of 23, **28**, 114
    soil (intrusion) 28
deformation time 29
Demec points 116
design curves
    for centre-line (tunnel) displacement **67**
    for centre-line (tunnel) strain **70**
    for settlements and derivatives **66**, 74
    for transverse displacement of **68**
    for transverse strain in **71**
    for vertical strain in **72**
design stress (pipe-bending tension) 197
dial gauge 114
diaphragm wall 121 (*see also* shafts)
dilation 51, 52
displacement (ground)
    axial (tunnel centre line) **59**, 67, 73
    transverse **60**, 68, 73, 171, 279, 280
    vertical (settlement) **59**, 67, 73, 267, 269, 270, 271, 273, 274, 279, 280

doors, of houses 280, 281
drawdown 88, 93
drilling rate 102
ductile iron (pipes) *see* pipe, ductile iron

earth load 222
effective stress 93
elastic modulus
    of brickwork 247
    of clays 247
    of concrete 247
    of pipe/beam 129, 140
        (*see also* pipe, elastic modulus)
    of sands 247
    of soil 134, 171, 187, 188, 189
        (*see also* deformation modulus)
    of steel 247
English law 288, 289
error function 73
expansion joints (bridge) 301, 302
extrusion test **26, 27**, 114
face loss *see* loss, face
failure criteria 80
failure rate 225, 226
filled ground 63, 89, 90, 91
fills
    household refuse 64
    industrial waste 64
finite elements *see* numerical methods
flow net 118
footing 242, 243, 244, 259, 267
    strip 264
foundation
    pipe 150
foundation (pipe) pressure 129
frame
    brick infill steel 247, 274, 276, 287
    concrete 267
    multibay multistorey 243, 272
    steel 238, 242, 247, 250, 267, 272
frame structure 242, 244, 245, **267-77**, 287
freezing **44**, 264
friction piles 126
frost 224
frost heave 222

gas mains 165, 223, 224, 226, 227, 232, 233
geology 98
geomorphology 98
gradient, ground **70**
granular soil 33, **51**, 63, 65, 118, 146, 246, 247, 256, 267, 302
grey iron (pipes) *see* pipe, grey iron
ground investigation *see* site investigation
ground loss, *see* loss, ground
ground treatment 234 (*see also* grout)

SUBJECT INDEX

groundwater table *see* water table
groundwater lowering 287
grout
  cement 43
  compaction of 43
  injection of 43, 264
  organic polymer 43
  particle size of 43
  silicate 43

heave (ground) 44
hinge 248, 259, 260
horizontal distortion 284
house 251, 256
hydrology 98

i-parameter **54**, **64**, 65, 81-7, 136, 228, 267, 274
impact (mechanical) 222, 226
in situ tests *see* boreholes
inclinometer 35, 108, 109
  cable drum of 109
  torpedo of 109
inclinometer tubing 108, 109
index properties (soil) 102, **104**
interaction
  soil-pipe 221, 222
  soil-structure 237

jet grouting method *see* shafts
joints, bridge 301
joists, floor 280, 281

laminated clay *see* clay, laminated
latent damage *see* English law
lateral distortion 76
law *see* English law
lead-yarn joint 198, 200, 201, 202, 203
length
  of structure 265, 267, 280, 286
  of pipeline 80
levelling station 106, 233
liner plates 6
linings 4
  bending of 4
  bolted wedge block 6
  expanded 5
  flexible 4
  'One-Pass' 4, 5
  rigid 4
  secondary, precast concrete 4, 5
  secondary, cast iron 6
liquid limit 94
liquid nitrogen 44
London Clay *see* clay, London
loss
  axial 23, 112, 113
  face 23, **28**, 33

  ground 23, **60**, 230, 264
  post-grout 25
  pre-grout 25, 33
  radial 23, **29**, 33
  shield 23
  volume 8, **34**, 228
Lytag 32, 62

mapping, tunnel face 99
marine warp 89, 90
mini-tunnelling 61
model, physical, of ground 102
modulus of subgrade reaction
  see subgrade reaction
Mohr diagram 104
moisture content 102, **104**
mortar 248, 260, 280
  lime 248, 299

negative skin friction 277
nodes
  foundation 243, 244
  ground 243
  structural 243, 244
normal probability curve **54**, **55**, 65, 149, 280
normally consolidated soils 237
numerical methods 96-8, 151, 169, 170, 212, 242, 244, 245, 246, 250, 251, 267

organic clay *see* clay, organic
overconsolidated soils 104, 131, 237
overlays 75-80

parapets 302
peat 90
percussive rig 101
permeability 90, 93, 94
permeability tests 102
pH 105
piezometer 91, 100
piled foundation 277
piles 277
  tubular steel 279
pipe
  asbestos cement 225
  cast iron 79, 120, 303
    (*see also* pipe, grey iron)
  clay 79
  clayware 225
  ductile iron 190, 192, 194, 225, 235, 236
  grey iron 120, 186, 190, 192, 194, **195**, 196, 204, 207, 208, 223, 224, 227, 228, 232, 234, 235
  mild steel 190, 192, 225, 234, 235, 236, 303

SUBJECT INDEX

polyethylene 120, 122, 191, 207, 208, 225, 235, 236
polyethylene sleeving 185, 208, 235
pipe area ratio 171
pipe diameter 130, 144, 149, 152, 153, 159, 173, 178, 179, 180, 185, 188, 192, 223, 226, 230, 235
  uPVC 225
pipe elastic modulus 171
pipe failure 193
pipe jacking **49**, 50
pipe joints 61, 62, 78, 122, 124, 125, 134, 151, 153, **154**, 159, 160, 161, 181-5, **198-204**, 236
pipe joint leakage 235
pipe joint rotation 153, 155
pipe strain, direct and bending 78, 80, 114, 115, 233
  (see also pipeline bending strain)
pipe tensile strain 228, 229
pipeline 78, 97, 102, 114, 120, 236
  plastic 185
    (see also pipe, polyethylene, uPVC)
pipeline backfilling see trench backfilling
pipeline bedding 235
pipeline bending curvature 136, 138, 142, 143, 163
pipeline bending moment 136, 211, 222, 223
pipeline bending strain 136, 151, 153, 160, 211, 212, 215, 218, 220, 222, 223
pipeline branches 126, 147, **209-21**, 236
pipeline creep 140
pipeline deflection 133, 148, 149
pipeline flexural rigidity 127, 128
pipeline joints see pipe joints
pipeline length 157, 158, 160, 163
pipeline longitudinal bending 124, 128, **129**, 156, 204
pipeline networks **209**, 222
pipeline renewal 229
pipeline repair 224
pipeline shear 133
pipeline strain (parallel) 144, 145, 177, 179, 185 (see also pipe strain)
pipeline strain (transverse) 137, 140, 141, 151, 172, 173, 174, 185 (see also pipe strain)
pipeline stress 225
pipeline tensile strain (branch) 219
  (see also pipeline branches)
pitch, excess 31
plan see survey plan
plasticity chart 104
plasticity index 104
Poisson's ratio
  for foundation/ground 129
  for soil 187
polyethylene pipes see pipe, polyethylene

polyethylene sleeving
  see pipe, polyethylene sleeving
post-grout loss see loss, post-grout
pre-grout loss see loss, pre-grout
probing 99
property schedules 105, **289**
property surveys 248, 280
pulverized fuel ash (pfa) 299
PUSWA 105, 226

radial loss see loss, radial
raft 238
risk assessment 228, 235
rock head 102
rubber gasket push-in joint 199, 200, 202, 203
rupture factor (pipe) 194

sample, soil 99
sand 33, 51, 90, 237, 240, 247, 255, 261, 262, 263, 267, 272
seasonal (thermal) stress 224
second moment of area 129, 135, 240, 264, 265
section modulus, wall 264, 265
service connection, pipe 197
service holes 204, 208, 223
settlement see displacement, vertical
  self-weight 288
settlement probes 23
settlement rings 108, 233
settlement volume 50, 54, **60**, 73, 246
shafts 99, **117-20**
  base slab to 119
  caisson method of construction of **118**
  diaphragm wall method of construction of **119**
  jet grouting method of construction of **119**
  lining segments in 117
  underpinning method of construction of **117, 118**
shear force formulae 304
shear strength 104
shear stress (soil-pipe) 180, 181
sheet piling 121
shield 3, 114, 264
  bentonite 47, 48
  enclosed **46**
  slurry 47, 48
shield dive (off-grade) 31
shield extrados 41
shield length 28
shield loss see loss, shield
shield radius 28
shield squat 31, 33
shield tail 31
shrinkable clay see clay, shrinkable

SUBJECT INDEX

silt 89, 104 (*see also* clay, silt)
silty clay *see* clay, silty
simple overload factor *see* stability ratio
site investigation **98-105**, 228, 232, 234, 283
sleech 89
slope 280, 282
smearing, in clay soils 100
soft clay *see* clay, soft
soil creep 140
soil mechanics, critical-state 40
stability numbers 40
stability ratio 61, 62, 63, 88, 89, 90, 91, 96
standard deviation *see* i-parameter
standard penetration test 33, 63, 101, 102, 188
steel, mild (reinforcement) 261
  (*see also* concrete, reinforced)
steel band, surveying 108
steel frame *see* frame, steel
steel pipes 190, 192, 225, 234, 235, 236
stiff clay *see* clay, stiff
stiff fissured clay *see* clay, stiff fissured
stiffness, structural 265
stiffness (soil-pipe) factor 171, 174, 179, 180, 184, 185
stiffness matrix **238**
stony clay *see* clay, stony
strain
  axial (tunnel centre line) **60**, 68, **70**, 73
  bending 76, 77
  ground 53, 68-74, 280
  shear **70**, **73**
  tensile 282, 284
  transverse **60**, 68, **71**, **73**, 279
  vertical **60**, 68, **72**, **73**
  volumetric 93
strength, soil 102, **104**
stress
  bending (steel frame) 270, 271, 272, 273
  bending tensile (brickwork) 248
  compressive (walls) 265
  shear (walls) 250-7
  tensile (walls) 250-7, 265-77
stress concentration (pipe) 193, 197
stress path testing 105
strip foundations 249
subgrade reaction 128, **129**, **132**, 134, 238, 239, 243, 304

submerged unit weight 93
sulphates 105
surcharge pressure
  see surface surcharge pressure
surface surcharge pressure 60, 62, 64
survey, building *see* property surveys
survey plan 75, 77, 102, 280

tensile failure 76
thermal stress *see* seasonal (thermal) stress
tills 63 (*see also* boulder clay)
tilt
  of abutment and wing walls (bridges) 299
  of bridge abutment 301, 303
  rigid body 283, 284, 287
tiltmeter 115, 116, 233
time-dependent settlement **87-96**, 222, 276
topography 98
transverse settlement 55, 94, 96
trench backfilling 235, 236
trenching 151, 221, 222, 230
triaxial tests 104, 187

underpinning 121 (*see also* shafts)
undrained shear strength 188
uni-tunnel system 50
uPVC pipes *see* pipe, uPVC

vacuum ejectors 118
vane shear strength 169
vane tests 101, 102
vault, building 101
vehicle wheel loading 222
vibrating wire strain gauges 114, 115
void ratio 93
volume loss *see* loss, volume

wall 238, 240, 244, 245, 247, 281
  bridge 298, 299, 302, 303
  reinforced concrete 240
water mains 223, 224, 227
water strike 99
water table 93, 102
water-retaining structures 249
wellpoint dewatering **42**
windows 280, 281
Winkler model *see* subgrade reaction

yield stress, pipe 193

RAYMOND H. FOGLER LIBRARY
DATE DUE

BOOKS ARE SUBJECT TO RECALL AFTER TWO WEEKS

JAN 06